Atlantis Studies in Mathematical Physics: Theory and Applications

Volume 2

Series editor

Norbert Euler, Lulea, Sweden

More information about this series at http://www.atlantis-press.com

Gianfausto Dell'Antonio

Lectures on the Mathematics of Quantum Mechanics II: Selected Topics

ATLANTIS PRESS

Gianfausto Dell'Antonio
SISSA
Rome
Italy

ISSN 2211-8055 ISSN 2211-8063 (electronic)
Atlantis Studies in Mathematical Physics: Theory and Applications
ISBN 978-94-6239-114-7 ISBN 978-94-6239-115-4 (eBook)
DOI 10.2991/978-94-6239-115-4

Library of Congress Control Number: 2015937751

Printed on acid-free paper

Il ne faut pas toujours tellement epuiser
un sujet q'on ne laisse rien a fair au lecteur.
Il ne s'agit pas de fair lire, mais de faire
penser.

Charles de Secondat, Baron de Montesquieu

A Caterina, Fiammetta, Simonetta

Contents

Introduction to Both Volumes

These books originated in lectures that I have given for many years at the Department of Mathematics of the University of Rome, La Sapienza, and at the Mathematical Physics Sector of the SISSA in Trieste.

I have tried to give a presentation which, while preserving mathematical rigor, insists on the conceptual aspects and on the unity of Quantum Mechanics.

The theory which is presented here is Quantum Mechanics as formulated in its essential parts on one hand by de Broglie and Schrödinger and on the other by Born, Heisenberg, and Jordan with important contributions by Dirac and Pauli.

For editorial reason the book in divided in two parts, with the same main title (to stress the unity of the subject).

The present second volume consists of "Lecture 1: Wigner Functions. Coherent States. Gabor Transform. Semiclassical Correlation Functions" to "Lecture 16: Measure (Gage) Spaces. Clifford Algebra, C.A.R. Relations. Fermi Field". Each lecture is devoted to a specific topic, often still a subject of advanced research, chosen among the ones that I regard as most interesting. Since "interesting" is largely a matter of personal taste other topics may be considered as more significant or more relevant.

I want to express here my thanks to the students that took my courses and to numerous colleagues with whom I have discussed sections of this book for comments, suggestions, and constructive criticism that have much improved the presentation.

In particular I want to thank my friends Sergio Albeverio, Giuseppe Gaeta, Alessandro Michelangeli, Andrea Posilicano for support and very useful comments.

I want to thank here G.G. and A.M. also for the help in editing.

Content of Volume II

Lecture 1: Wigner functions. Husimi distribution. Semiclassical limit. WKB states. Coherent states. Gabor transform. Semiclassical limit of joint distribution functions.
Lecture 2: Pseudo-differential operators. Calderon–Vaillantcourt Theorem. hbar-admissible operators. Berezin, Kohn–Nirenberg, Born–Jordan quantizations.
Lecture 3: Compact, Shatten-class, Carleman operators. Compactness criteria. Radon–Nikodym theorem. Hadamard inequalitily. Bouquet of inequalities.
Lecture 4: Periodic potentials. Theory of Bloch-Floquet-Zak Wigner-Satz cell. Brillouen zone. Bloch waves. Wannier functions.
Lecture 5: Connection with the properties of a crystal. Born–Oppenheimer approximation. Peierls substitution. The role of topology. Chern number. Index theory. Quantum pumps.
Lecture 6: Lie–Trotter–Kato formula. Wiener process. Stochastic processes. Feynman–Kac Formula.
Lecture 7: Elements of probability theory. Sigma algebras. Chebyshev and Kolmogorov inequalities. Borel–Cantelli lemma. Central limit theorems. Construction of brownian motion. Girsanov formula.
Lecture 8: Diffusions. Ornstein–Uhlenbeck process. Covariance. The infinite-dimensional case. Markov structure. Semigroup property. Paths over function spaces. Markov and Euclidean fields.
Lecture 9: Standard form for algebras and spaces. Cyclic and separating vector. K.M.S. conditions. Tomita–Takesaki theory. Modular operator. Non-commutative Radon–Nikodym derivative.
Lecture 10: Scattering theory. Time-Dependent formalism. Wave operators. Chain rule.
Lecture 11: Scattering Theory. Time dependent formalism. Limit absorption principle. Lippmann-Schwinger equations. The method of Enns. Ruelle's theorem. Inverse scattering. Reconstruction
Lecture 12: Enns' propagation estimates. Mourre method. Conjugate operator. Kato smoothness. Double commutator method. Algebraic scattering theory.
Lecture 13: N-Body system. Clusters. Zhislin's theorem. Spectral structure. Thresholds. Mourre compact operator. double commutator estimates.
Lecture 14: Positivity preserving maps. Ergodicity. Positive improving maps. Contractions. Markov semigroups. Contractive Dirichlet forms.
Lecture 15: Hypercontractivity. Logarithmic Sobolev inequalities. Harmonic group.
Lecture 16: Measure (gage) spaces. Perturbation theory. Non-commutative integration theory. Clifford algebra. C.A.R. relations. Free Fermi field.

Content of Volume I: Conceptual Structure and Mathematical Background

This first volume consists of Lectures 1 to 20. It contains the essential part of the conceptual and mathematical foundations of the theory and an outline of some of the mathematical instruments that will be most useful in the applications. This introductory part contains also topics that are at present subject of active research.

Lecture 1: Elements of the History of Quantum Mechanics I.
Lecture 2: Elements of the History of Quantum Mechanics II.
Lecture 3: Axioms, States, Observables, Measurement. Difficulties.
Lecture 4: Entanglement, Decoherence, Bell's inequalities, Alternative Theories.
Lecture 5: Automorphisms. Quantum Dynamics. Theorems of Wigner, Kadison, Segal. Generators.
Lecture 6: Operators on Hilbert spaces I: basic elements.
Lecture 7: Quadratic forms.
Lecture 8: Properties of free motion. Anholonomy. Geometric phases.
Lecture 9: Elements of C^*-algebras. G.N.S representation. Automorphisms and Dynamical Systems.
Lecture 10: Derivations and Generators. K.M.S. condition. Elements of modular structure. Standard form.
Lecture 11: Semigroups and dissipations. Markov approximation. Quantum Dynamical semigroups I.
Lecture 12: Positivity preserving contraction semigroups on C^*-algebras. Complete dissipations.
Lecture 13: Weyl system. Weyl algebra. Lifting symplectic maps. Magnetic Weyl algebra.
Lecture 14: Representations of Bargmann-Segal-Fock. Second quantization. Other quantizations.
Lecture 15: Semiclassical limit. Coherent states. Metaplectic group.
Lecture 16: Semiclassical approximation for fast oscillating Phases. W.K.B. method. Semiclassical quantization rules.
Lecture 17: Kato-Rellich comparison theorems. Rollnick and Stummel classes. Essential spectrum.
Lecture 18: Weyl's criterium. Hydrogen and Helium atoms.
Lecture 19: Estimates of the number of bound states. The Feshbach method.
Lecture 20: Self-adjoint extensions. Relation with quadratic forms. Laplacian on metric graphs. Boundary triples. Point interactions.

References for Volumes I and II

V. Amrein, J. Jauch, K. Sinha, *Scattering theory in Quantum Mechanics* (V.Benjamin, Reading, 1977)

J. Baez, I.E. Segal, Z. Zhou, *Introduction to Algebraic and Constructive Field Theory* (Princeton University Press, Princeton, 1992)

O. Bratteli, Derivations, *Dissipations and Group Actions on C*-algebras*. Lecture Notes in Mathematics, vol. 1229 (Springer, 1986)

O. Bratteli, D.W. Robinson, *Operator Algebras and Quantum Statistical Mechanics I, II* (Springer, New York, 1979/1987)

J. Brezis, *Analisi Funzionale e Applicazioni* (Liguori, Napoli, 1986)

G. Cassinelli, E. De Vito, A. Levrero, P.J. Lahti, *The Theory of Symmetry Actions in Quantum Mechanics* (Springer, Berlin, 2004)

H.L. Cycon, R.H. Froese, W. Kirsch, B. Simon, *Schrödinger operators with application to Quantum Mechanics and global geometry Text and Monographs in Physics* (Springer, Berlin, 1987)

E.B. Davies, *Quantum Theory of Open Systems* (Academic Press, New York, 1976)

J. Diximier, *Les Algebres d'operateurs dan l'espace hilbertien* (Gauthier-Villars, Paris, 1969)

J. Doob, *Stochastic Processes* (Wiley, New York, 1953)

G.B. Folland, *Harmonic analysis in Phase Space* (Princeton University Press, Princeton, 1989)

A. Galindo, P. Pasqual, *Quantum Mechanics II* (Springer, Berlin, 1991)

S. Gustafson, I.M. Segal, *Mathematical Concepts of Quantum Mechanics* (Springer, Berlin, 2006)

E. Hille, R.S. Phillips, *Functional Analysis and Semigroups* (American Mathematical Society, Providence, 1957)

P.D. Hislop, I.M. Sigal, *Introduction to Spectral Theory, with Application to Schrödinger Operators* (Springer, New York, 1996)

R.V. Kadison, J.R. Ringrose, *Fundamentals of the Theory of Operator Algebras. vol. I–IV* (Academic Press, 1983/1986)

T. Kato, *Perturbation Theory for Linear Operators* (Springer, Berlin, 1980)

P. Kuchment, *Floquet Theory for Partial Differential Equations* (Birkhauser, Basel, 1993)

M. Jammer, *The Conceptual Development of Quantum Mechanics* (Mc Graw-Hill, New York, 1966)

G.W. Mackey, *Mathematical Foundations of Quantum Mechanics* (Benjamin, New York, 1963b)

V. Maslov, *The Complex W.K.B. method part 1*. Progress in Physics. vol. 16 (Birkhauser, Verlag, 1984)

E. Nelson, *Topics in Dynamics, I: Flows Notes* (Princeton University Press, 1969)

G.K. Pedersen, *C*-algebras and their Automorphism Groups* (Academic Press, London 1979)

M. Reed, B. Simon, *Methods of Modern Mathematical Physics*. vol. I–IV (Academic Press, New York, 1977, 1978)

S. Sakai, *C*-algebras and W*-algebras*. (Springer, Berlin, 1971)

I.E. Segal, *Mathematical Problems of Relativistic Physics* (American Mathematical Society, Providence R.I., 1963)

B. Simon, *Quantum Mechanics for Hamiltonians defined as Quadratic Forms* Princeton Series in Physics (Princeton University Press, 1971)

B. Simon, *Functional Integration and Quantum Physics* (Academic Press, 1979)

Y.G. Sinai, *Probability Theory Springer Textbooks* (Springer, Berlin, 1992)

M. Takesaki, *Tomita's Theory of Modular Hilbert Algebras*. Lecture Notes in Mathematics, vol. 128 (Springer, Heidelberg, 1970)

M. Takesaki, *Theory of Operator Algebras*, vol. 1 (Springer, New York, 1979)

S. Teufel, *Adiabatic Perturbation Theory in Quantum Dynamics*. Lecture Notes in Mathematics, vol. 1821 (Springer, Verlag, 2003)

J. von Neumann, *Mathematische Grundlage der Quantenmechanik (Mathematical Foundation of Quantum Mechanics)* (Springer, Berlin, 1932) (Princeton University Press, 1935)
B.L. Van der Werden (ed.), *Sources of Quantum Mechanics* (Dover, New York, 1968)
K. Yoshida, *Functional Analysis* (Springer, Berlin, 1971)
H. Weyl, *Quantenmechanik und Gruppentheorie*. Z. Phys. **46**, 1–59 (1927)

Lecture 1: Wigner Functions. Coherent States. Gabor Transform. Semiclassical Correlation Functions

In Classical Mechanics a pure state is described by a Dirac measure supported by a point in phase space.

We have seen that in quantum Mechanics a pure state is represented by complex-valued functions on configuration space, and functions that differ only for a constant phase represent the same pure state.

Alternatively one can describe pure states by complex-valued functions in momentum space.

To study the semiclassical limit it would be convenient to represent pure states by real-valued functions on phase-space, and that this correspondence be one-to-one. These requirements are satisfied by the Wigner function W_ψ associated to the wave function $\psi \in L^2(R^N)$.

The function W_ψ is not positive everywhere (except for coherent states with total dispersion $\geq \hbar$) and therefore *cannot be interpreted as probability density.*

Still it has a natural connection to the Weyl system and good regularity properties.

To a pure state described in configuration space by the wave function $\psi(x)$ one associates the *Wigner function* W_ψ which is a *real function* on R^{2N} defined by

$$W_\psi(x,\xi) = (2\pi)^{-N} \int_{R^N} e^{-i(\xi,y)} \psi\left(x+\frac{y}{2}\right) \bar{\psi}\left(x-\frac{y}{2}\right) d^N y \quad x,\xi \in R^N. \tag{1}$$

We shall say that W_ψ is the *Wigner transform* of ψ and will call *Wigner map* the map $\psi \to W_\psi$.

It is easy to verify that the function W_ψ is real and that $W_\psi = W_{e^{ia}\psi}$ $\forall a \in R$.

Therefore the Wigner map maps rays in Hilbert space (pure states) to real functions on phase space.

Moreover we will see that the integral over momentum space of $W_\psi(x,\xi)$ is a *positive function* that coincides with $|\psi(x)|^2$ and the integral over configuration space coincides with $|\hat{\psi}(p)|^2$ where $\hat{\psi}$ is the Fourier transform of ψ.

G. Dell'Antonio, *Lectures on the Mathematics of Quantum Mechanics II: Selected Topics*, Atlantis Studies in Mathematical Physics: Theory and Applications 2, DOI 10.2991/978-94-6239-115-4_1

The correspondence between W_ψ and the integral kernel $\bar{\psi}(x)\ \psi(x')$ permits to associate by linearity a Wigner function W_ρ to a density matrix ρ,

$$\rho = \sum_k c_k P_k, \qquad c_k \geq 0, \qquad \sum c_k = 1 \tag{2}$$

where P_k is the orthogonal projection on ψ_k one has

$$W_\rho = \sum_k c_k W_{\psi_k} \tag{3}$$

Explicitly

$$W_\rho(x,\xi) = \sum_k c_k (2\pi)^{-N} \int e^{-i(\xi,y)} \psi_k\left(x+\frac{y}{2}\right) \bar{\psi}_k\left(x-\frac{y}{2}\right) dy \tag{4}$$

If ρ is a density matrix (positive trace-class operator of trace one) with integral kernel

$$\rho(x,y) = \sum_n c_n \bar{\phi}_n(x) \phi_n(y) \tag{5}$$

its Wigner function is

$$W_\rho(x,\xi) = (2\pi)^{-N} \int e^{-i(\xi,y)} \rho\left(x+\frac{1}{2}y, x-\frac{1}{2}y\right) dy \tag{6}$$

where the sum converges pointwise in x, ξ if $\rho(x,x')$ is continuous and in the L^1 sense otherwise.

The definition can be generalized to cover Hilbert–Schmidt operators when the convergence of the series is meant in a suitable topology.

From (6) one has

$$W_\rho(x,\xi) \in L^2(R^N \times R^N) \cap C_0(R^N_y, L^1(R^N_x)) \cap C_0(R^N_x, L^1(R^N_y)) \tag{7}$$

Through (6) one can extend by linearity the definition of Wigner function to operators defined by an integral kernel; this can be done in suitable topologies and the resulting kernels are in general distribution-valued.

When ρ is a Hilbert–Schmidt operator and one has

$$\| W_\rho \|_2^2 = (2\pi)^{-N} \| \rho \|^2 \tag{8}$$

If $\psi(x,t)$ is a solution of the free Schroedinger equation

$$i\frac{\partial \psi}{\partial t} = -\frac{1}{2} \Delta \psi \tag{9}$$

the function W_ψ solves the transport (or Liouville) equation

$$\frac{\partial W}{\partial t} + \xi \cdot \nabla_x W = 0 \tag{10}$$

Introducing Planck's constant one rescales Wigner's function as follows

$$W_\psi^\hbar(x, \xi, t) = \left(\frac{i}{\hbar}\right)^N W_\psi\left(x, \frac{\xi}{\hbar}, t\right) \tag{11}$$

and Liouville equation is satisfied if ψ satisfies

$$i\hbar \frac{\partial \psi}{\partial t} = -\frac{1}{2}\hbar^2 \Delta\psi.$$

Consider now the equation that is satisfied by W_ϕ if ϕ satisfies Schroedinger's equation with hamiltonian $H = -\frac{1}{2}\Delta + V$.

We have seen in Volume I that under the condition

$$V \in L^2_{loc}(R^N), \qquad V^- \in St(R^N), \qquad \int_{|x|<R} |V(x)|^2 dx \le c(1+R)^m \tag{12}$$

(St denotes Stummel class) the operator H is self-adjoint with domain

$$D(H) \equiv \{\phi \in L^2, \ \ |V|\phi \in L^1_{loc}, \ \ -\Delta\phi + V\phi \in L^2\} \tag{13}$$

Let ρ_0 be a density matrix and set $\rho(t) \equiv e^{-iHt}\rho_0 e^{iHt}$. Denote by $W_{\rho(t)}(x, \xi; t)$ the Wigner function of $\rho(t)$.

Under these conditions the following theorem holds (the easy proof is left to the reader).

Theorem 1 *If V satisfies (12) then $W_{\rho(t)}$ belongs to the space*

$$C\left(R_t, L^2\left(R_x^N \times R_\xi^N\right)\right) \cap C_b\left(R_t \times R_x^N, \mathcal{F}L^1\left(R_\xi^N\right)\right) \cap C_b\left(R_t \times R_\xi^N, \mathcal{F}L^1\left(R_x^N\right)\right) \tag{14}$$

(we have denoted by $\mathcal{F}L^1$ the space of functions with Fourier transform in L^1) and satisfies

$$\frac{\partial W}{\partial t} + (\xi, \nabla_x W) + K * W = 0 \tag{15}$$

where K is defined by

$$K(x, \xi) \equiv \left(\frac{i}{2\pi}\right)^N \int e^{-i(\xi, y)} \left(V\left(x + \frac{y}{2}\right) - V\left(x - \frac{y}{2}\right)\right) dy \tag{16}$$

and

$$(K * W)(x, \xi) \equiv \int K(x, \eta) W(x, \xi - \eta) d\,\eta \tag{17}$$

◇

Setting $f^{\hbar}(t) = W^{\hbar}_{\psi}(t)$ one derives

$$\frac{\partial f^{\hbar}}{\partial t} + \xi \cdot \nabla_x f^{\hbar} + K_{\hbar} * f^{\hbar} = 0, \qquad \rho^{\hbar}(t=0) = \rho_0(\hbar) \tag{18}$$

where

$$K_{\hbar}(x, \xi) = \left(\frac{i}{2\pi}\right)^N e^{-i\xi \cdot y}\hbar^{-1}\left[V\left(x + \frac{\hbar y}{2}\right) - V\left(x - \frac{\hbar y}{2}\right)\right] dy \tag{19}$$

If the potential V it is sufficiently regular it reasonable to expect that if the initial datum $f_0^{\hbar}$ converges when $\hbar \to 0$ in a suitable topology to a positive measure f_0, then the (weak) limit $f \equiv lim_{\hbar \to 0} f^{\hbar}$ exists, is a positive measure and satisfies (weakly)

$$\frac{\partial f}{\partial t} + \xi \cdot \nabla_x f - \nabla V(x) \cdot \nabla_{\xi} f = 0 \qquad f(0) = f_0 \tag{20}$$

We shall prove indeed that when V satisfies suitable regularity assumptions, then for every $T > 0$ there exists a sequence $\hbar_n \to 0$ such that $f^{\hbar_n}(t)$ converges uniformly for $|t| < T$, in a weak * sense for a suitable topology, to a function $f(t) \in C_b(R^N)$ which satisfies (20) as a distribution.

Under further regularity properties $f(t)$ is the unique solution of (20) and represents the transport of f_0 along the free flow

$$\dot{x} = \xi, \qquad \dot{\xi} = -\nabla V \tag{21}$$

Under these conditions the correspondence $\psi \to W^{\hbar}_{\psi}$ is a valid instrument to study the semiclassical limit.

We shall give a precise formulation and a proof after an analysis of the regularity properties of the Wigner functions.

We have remarked that in general the function $W_{\psi}(x, \xi)$ is not positive. It has however the property that its marginals reproduce the probability distributions in configuration space and in momentum space of the pure state represented by the function ψ. Indeed one has the following lemma (we omit the easy proof).

Lemma 1

$$\int (W_{\psi})(x, \xi) dx = |\hat{f}(\xi)|^2, \qquad \int (W_{\psi})(x, \xi) d\xi = |f(x)|^2 \tag{22}$$

◇

In a strict sense (22) holds if $\phi \in L^1 \cap L^2, \ \hat{\phi} \in L^1 \cap L^2$. In the other cases one must resort to a limiting procedure.

We also notice that

$$W_{e^{i(ax-b\frac{\partial}{\partial x})}\psi} = W_\psi(x-b, \xi - a) \qquad W_f = W_g \Leftrightarrow f(x) = e^{ic} g(x) \qquad c \in R \quad (23)$$

and that

$$(W_\psi, W_\phi) = (\psi, \phi) \tag{24}$$

The essential support of a Wigner function cannot be too small; roughly its volume cannot be less than one in units in which $\hbar = 1$.

In particular for any Lebesgue-measurable subset $E \in R^{2N}$ one has

$$\int_E W_f(x, \xi) dx d\xi \leq |f|_2^2 \, \mu(E) \tag{25}$$

where $\mu(E)$ is the Lebesgue measure of E.

This statement is made precise by the following proposition [1, 2].

Proposition 1 (Hardy) *Let*

$$C_{a,b}(x, \xi) = e^{-\frac{a\xi^2}{2} - b\frac{x^2}{2}} \quad x, \xi \in R^N \quad a, \ b > 0. \tag{26}$$

Then for any $f \in L^2(R^{2N}$ *one has*
(1) If $ab = 1$ *then* $(W_f^*, C_{a,b})(x, \xi) \geq 0$
(2) If $ab > 1$ *then* $(W_f^*, C_{a,b})(x, \xi) > 0$
(3) If $ab < 1$ *there are values of* $\{x, \xi\}$ *for which* $(W_f^*, C_{a,b})(x, \xi) < 0$. ◇

1 Coherent States

If $ab = 1$ the functions $C_{a,b}$ defined above and suitably normalized are called *coherent states*.

Coherent states play a relevant role in geometric optics and also, as we saw in Volume I, in the Bargman–Segal representation of the Weyl system and in the Berezin–Wick quantization.

Introducing Planck's constant the coherent states are represented in configuration space R^n by

$$C_{q,p;\Delta}(x) = c_N e^{-\frac{(x-q)^2}{2\Delta^2}} e^{\frac{ix.p}{\hbar}\Delta}, \quad q, p \in R^N, \quad \Delta > 0 \tag{27}$$

where c_N is a numerical constant.

These states have dispersion Δ in configuration space and $\frac{\hbar}{\Delta}$ in momentum space, and therefore the product of the dispersions in configuration and momentum space is $\hbar$, the minimal value possible value due to Heisenberg inequalities.

The Wigner function of the coherent states is positive

$$W_{q,p;\Delta}(x,\xi) = c_N e^{-\frac{(x-q)^2}{2\Delta^2} - \frac{\Delta^2(\xi-p)^2}{2\hbar^2}} \tag{28}$$

As a consequence of the theorem of Hardy it can be proved [2] that the Wigner function W_ψ associated to a wave function ψ is positive if and only if $\psi(x)$ is a gaussian state of the form (26) with $\Delta_0 \cdot \Delta \geq \hbar$.

The Wigner function W_ψ *is not positive in general* but its *average over each coherent state* is a non-negative number.

Since coherent states are parametrized by the points in phase space, one can associate to the function ϕ the *positive function* on phase space

$$H_\phi(q,p) = \int dx\, d\xi W_{q,p;\Delta}(x,\xi) W_\phi(x,\xi) dx d\xi$$

This is the *Husimi distribution* associated to the function ϕ.

Since the coherent states form an over-complete system, one may want to construct a positive functions associated to the function ϕ by integrating over a smaller set of coherent states, but still sufficient to characterize completely the function ϕ.

This is the aim of *Gabor analysis* [3] a structure that has gained prominence in the field of signal analysis. We shall outline later the main features and results in this field.

Not all phase-space functions are Wigner functions W_ρ for some state ρ.

A simple criterion makes use of the *symplectic Fourier transform*; we shall encounter it again when in the next Lecture we will introduce the *pseudo-differential operators*

If $f \in L^2(R^{2N})$ define its *symplectic Fourier transform* f^J by

$$f^J(z) = \int_{R^{2N}} f(\xi) e^{-iz^T J\xi} d\xi, \quad z \in R^{2N} \tag{29}$$

where J is the standard symplectic matrix.

The symplectic Fourier transform $f^J(z)$ is said to be of β-positive type if the $m \times m$ matrix M with entries

$$M_{i,j} = f^J(a_i - a_j) e^{i\frac{\beta}{2}(a^T Ja)} \qquad a = \{a_1, \dots, a_m\} \tag{30}$$

is hermitian and non negative.

With these notations the necessary and sufficient condition for a phase space function to be a Wigner function is [4, 5]

(i) $f^J(0) = 1$

(ii) $f^J(z)$ is continuous and of $\hbar$-positive type.

2 Husimi Distribution

For a generic density matrix ρ the positive function

$$H_\rho(q,p) \equiv (W_\rho, W_{\phi_{q,p;\hbar}}) \tag{31}$$

is called a *Husimi transform* (or also *Husimi distribution*) of the density matrix ρ.

If the density matrix has trace one, the corresponding Husimi distribution has L^1 norm one.

The correspondence $H_\rho \leftrightarrow \rho$ is one-to-one.

One verifies that ρ is of trace-class if and only if $H_\rho \in L^1(R^{2N})$ and that $Tr\rho = \int H_\rho dx^{2N}$.

Denote by $\mathcal{S}$ and $\mathcal{S}'$ the Schwartz classes of functions.

The Fourier transform acts continuously in these classes and one can derive the following regularity properties

$$\rho(x,y) \in \mathcal{S}\left(R_x^N \times R_\xi^N\right) \Leftrightarrow W_\rho(x,\xi) \in \mathcal{S}\left(R_x^N \times R_\xi^N\right) \tag{32}$$

$$\rho \in \mathcal{S}'\left(R_x^N \times R_\xi^N\right) \Leftrightarrow W_\rho \in \mathcal{S}'\left(R_x^N \times R_\xi^N\right) \tag{33}$$

More generally, for any pair of functions f, g one can consider the quadratic form

$$W_{f,g}(x,\xi) = (2\pi)^{-N} \int e^{-i(\xi,y)} f\left(x + \frac{1}{2}y\right) \bar{g}\left(x - \frac{1}{2}y\right) dy \tag{34}$$

From the properties of Fourier transform one derives Lemma 2.

Lemma 2 *If* $f, g \in \mathcal{S}(R^N) \times \mathcal{S}(R^N)$ *then* $W_{f,g} \in \mathcal{S}(R^{2N})$.
If $f, g \in \mathcal{S}'(R^N) \times \mathcal{S}'(R^N)$ *then* $W_{f,g} \in \mathcal{S}'(R^{2N})$.
If $f, g \in L^2(R^N) \times L^2(R^N)$ *then*$W_{f,g} \in L^2(R^{2N}) \cap C_0(R^{2N})$

Moreover

$$(W_{f_1,g_1}, W_{f_2,g_2}) = (f_1, f_2)(g_2, g_1) \qquad |W_{f,g}|_\infty \le \|f\|_2 \|g\|_2 \tag{35}$$

◇

We study next the limit when $\epsilon \to 0$ of a one-parameter family of functions u_ϵ. Consider the corresponding Wigner functions

$$W_{u_\epsilon}^\epsilon(x,\xi) = \left(\frac{1}{2\pi\epsilon}\right)^N \int_{R^N} e^{-\frac{i}{\epsilon}(\xi,y)} u_\epsilon\left(x + \frac{y}{2}\right) \bar{u}_\epsilon\left(x - \frac{y}{2}\right) dy \tag{36}$$

$$= \left(\frac{1}{2\pi}\right)^N \int_{R^N} e^{-i(\xi,z)} u_\epsilon\left(x + \frac{\epsilon z}{2}\right) \bar{u}_\epsilon\left(x - \frac{\epsilon z}{2}\right) dz \tag{37}$$

Let $H_{u_\epsilon}^\epsilon(q,p)$ be the corresponding Husimi functions.

If the family u_ϵ is bounded in $L^2(R^N)$, the family $H^\epsilon_{u_\epsilon}$ consists of non-negative functions in $L^1(R^N)$ which define, if considered as densities, a family of measures μ_{u_ϵ}.

We shall study limit point of this set of measures, in the sense of the weak* topology of Borel measures.

In order to be able to use compactness results it is convenient to introduce a topological space in which the $W^\epsilon(u_\epsilon)$ are *uniformly bounded.*

To this end, we introduce the following Banach algebra

$$\mathcal{A} \equiv \left\{u \in C_0\left(R^N_x \times R^N_\xi\right), (\mathcal{F}_\xi u)(x,z) \in L^1\left(R^N_z, C_0\left(R^N_x\right)\right)\right\} \tag{38}$$

with norm

$$||\mathcal{F}_\xi(u_x)||_\mathcal{A} = \int_{R^N} sup_x|\mathcal{F}_\xi u|(x,z)dz \tag{39}$$

$\mathcal{A}$ is a separable Banach algebra that contains densely $S(R^N_x \times R^N_\xi)$, $C^\infty_0(R^N_x \times R^N_\xi)$ and every finite linear combination of $u_1(x)u_2(\xi)$, with $u_k \in C^\infty_0$ or $\hat{u} \in C^\infty_0$.

In (38) we have used the notation $\mathcal{F}_\xi u$ to denote Fourier transform of u with respect to ξ.

With these notation one has

Proposition 2 *The family $W^\epsilon_{u_\epsilon}$ is equibounded in $\mathcal{A}$.* ◇

Proof A simple estimate gives

$$\int_{R^{2N}} (W^\epsilon_{u_\epsilon}\phi)(x,\xi)dxd\xi = \frac{1}{(2\pi)^N}\int_{R^N} (\mathcal{F}_\xi\phi)(x,y)u_\epsilon\left(x+\frac{\epsilon z}{2}\right)\bar{u}_\epsilon\left(x-\frac{\epsilon z}{2}\right)dx\,dy\,dz \tag{40}$$

It follows

$$\begin{aligned}
&|\int_{R^{2N}} (W^\epsilon_{u_\epsilon}\phi)(x,\xi)dxd\xi| \\
&\leq \left(\frac{1}{2\pi}\right)^N \int_{R^N} (sup_x|(\mathcal{F}_\xi\phi)(x,y)|dy)|\ |sup_z u_\epsilon\left(x+\frac{\epsilon z}{2}\right)\bar{u}_\epsilon\left(x-\frac{\epsilon z}{2}\right)|dx \\
&\leq \left(\frac{1}{2\pi}\right)^N ||\phi||_\mathcal{A}||u_\epsilon||^2
\end{aligned} \tag{41}$$

♡

Denote by $\mathcal{A}'$ the topological dual of $\mathcal{A}$.

From Proposition 2 one derives *by compactness* that there exists a subsequence $\{u_{\epsilon_n}\}$ which converges weakly to an element of $u \in \mathcal{A}''$ and at the same time $W^{\epsilon_n}_{u_{\epsilon_n}}$ converges in the *-weak topology to an element of $\mathcal{A}''$ that we denote by μ.

Note that the convergence of u_{ϵ_n} to u does not imply weak convergence of $W^{\epsilon_n}_{u_{\epsilon_n}}$; in general one must select a further subsequence.

In the same way we can construct sequences of Husimi functions $H^{\epsilon_n}_{u_{\epsilon_n}}$ and of corresponding measures that converge weakly.

Denote by $\tilde{\mu}$ the limit measure. One has

Theorem 2 *(1) One has always* $\mu = \tilde{\mu}$
(2) $\mu \geq |u(x)|^2 \delta_0(\xi)$
(3) $\int_{R^N} |u(x)|^2 dx \leq \int_{R^{2N}} d\mu \leq liminf_{\epsilon\to 0} \int_{R^N} |u_\epsilon|^2 dx.$ ◇

Proof We provide the proof only in the case $n = 1$. Notice that

$$H^\epsilon_{u_\epsilon} = W^\epsilon_{u_\epsilon} * G_\epsilon, \qquad G_\epsilon = (\pi\epsilon)^{-\frac{1}{2}} e^{-\frac{(|x|^2+|\xi|^2)}{\epsilon}} \tag{42}$$

($*$ denotes convolution in ξ).

We must prove that if $\phi \in \mathcal{A}$ (or in a dense subset) then $\phi * G_\epsilon$ converges to ϕ in the topology of $\mathcal{A}$.

From

$$\mathcal{F}_\xi(\phi * G_\epsilon)(x,z) \;=\; \left[(\mathcal{F}_\xi\phi)(x,z)(\pi\epsilon)^{-1/2} * e^{-\frac{|x|^2}{\epsilon}}\right] e^{-\epsilon\frac{|z|^2}{4}} \tag{43}$$

it follows

$$|\phi\ * G_\epsilon - \phi|_{\mathcal{A}} \leq \int_{R^N} sup_x|\ \mathcal{F}_\xi\phi - \mathcal{F}_\xi\phi * (\pi\epsilon)^{-\frac{1}{2}} e^{-\frac{|x|^2}{\epsilon}}\ |dz$$
$$+ \int (1 - e^{\epsilon|z|^2/4})\ sup_x|\mathcal{F}_\xi\phi|\ dz \tag{44}$$

The second term converges to zero so does the first term if $\phi \in \mathcal{S}(R \times R)$. Point 1 of the theorem is proved, since $\mathcal{S}(R \times R)$ is dense in $\mathcal{A}$.

Point 3 follows from Point 1 since

$$\int_{R^2} \tilde{\mu}_u dx \leq lim\ inf \int_R |u_\epsilon|^2 dx \tag{45}$$

To prove Point 2 notice that for a compact sequence u_ϵ that converges weakly to u in $L^2(R)$ one has, for every $z \in R$

$$u_\epsilon\left(x + \frac{\epsilon z}{2}\right)\ \bar{u}_\epsilon\left(x - \frac{\epsilon z}{2}\right) \Rightarrow |u(x)|^2 \tag{46}$$

Therefore one has, weakly for subsequences in $\mathcal{S}'(R \times R)$

$$W^\epsilon_{u_\epsilon} \to |u(x)|^2 \tag{47}$$

and from this one derives $\mu_u = \|u\|^2 \delta_0(\xi)$.

Define

$$H(u,v)(\xi,x) = (2\pi)^{-1} \int e^{-i\xi\cdot y} u\left(x+\frac{y}{2}\right) \bar{v}\left(x-\frac{y}{2}\right) d\xi \tag{48}$$

Then

$$H^{\epsilon}_{u_\epsilon} \geq H^{\epsilon}_{u} + 2H^{\epsilon}_{u_\epsilon - u} \tag{49}$$

and to prove Point 2 it suffices to prove that $\tilde{W}^{\epsilon}(u, v_\epsilon)$ converges weakly (in the topology of Borel measures) if $u \in C_0^\infty(R)$ and v_ϵ converge weakly to zero in $L^2(R)$.

One has

$$H^{\epsilon}(u, v_\epsilon) * G_\epsilon \tag{50}$$

$$W^{\epsilon}(u,v_\epsilon) = (2\pi)^{-1} Re \int e^{-i\xi.y} u\left(x+\frac{\epsilon y}{2}\right) \bar{u}\left(x-\frac{\epsilon y}{2}\right) d\xi \tag{51}$$

Therefore for every $\phi \in \mathcal{S}(R^N \times R^N)$

$$< W^{\epsilon}(u,v_\epsilon), \phi >= (2\pi)^{-1} Re \int_{R^2} dydz \bar{v}_\epsilon(y) u\left(x+\frac{\epsilon y}{2}\right) (\mathcal{F}_\xi \phi)(y - \epsilon z/2, z) \tag{52}$$

If $u \in C_0^\infty(R^N)$ one has moreover

$$lim_{\epsilon\to 0} u\left(x+\frac{\epsilon y}{2}\right) (\mathcal{F}_\xi \phi)\left(y-\frac{\epsilon z}{2}, z\right) = u(x)\ \mathcal{F}_\xi \phi(y,z) \tag{53}$$

in the topology of $L^2(R^N_z, L^2(R^N_x))$.

It follows that $W^{\epsilon}(u, v_\epsilon)$ converges weakly to zero in $\mathcal{A}'$.

Similar estimates show that $H^{\epsilon}(u, v_\epsilon)$ converges weakly to zero in the sense of measures. ♡

The following remarks are useful and easily verifiable.

(a) It may occur that $\mu = 0$ (we shall presently see an example)

(b) If μ_u is the measure associated to the subsequence u_ϵ weakly convergent to u, then $\mu(. - x_0, . - \xi_{\xi_0})$ is the measure associated to to the subsequence

$$u_\epsilon(x - x_0) e^{i\frac{(\xi_0 - x)}{\epsilon}} \tag{54}$$

(c) The measure μ_u is also the limit of

$$(2\pi)^{-n} \int e^{-i\xi.z} u_\epsilon\left(x+\frac{\alpha\epsilon z}{2}\right) \bar{u}_\epsilon\left(x+\frac{\beta\epsilon z}{2}\right) dz \tag{55}$$

for all values of the parameters $\alpha, \beta \in (0,1), \alpha+\beta = 1$

(d) If the measure μ_u is associated to the sequence u_ϵ and the measure ν_v to the sequence v_ϵ, in general the measure $\mu_u + \nu_v$ *is not associated* to the sequence $(u_\epsilon + v_\epsilon)$ (for example if $u_\epsilon = v_\epsilon$ the associated measure is $4\mu_u$). This fact is a consequence of the superposition principle.

Additivity always holds when μ and ν are mutually singular.

3 Semiclassical Limit Using Wigner Functions

Example 1 Sequence of functions that concentrate in one point

$$u_\epsilon(x) \equiv \frac{1}{\epsilon^{\frac{nN\alpha}{2}}} u\left(\frac{x}{\epsilon^\alpha}\right) \tag{56}$$

One has

$$\alpha < 1 \qquad lim_{\epsilon\to 0} W^\epsilon_{u_\epsilon} = \delta_0(x)\delta_0(\xi) \int |u(y)|^2 dy \tag{57}$$

$$\alpha > 1 \qquad lim_{\epsilon\to 0} W^\epsilon_{u_\epsilon} = 0 \tag{58}$$

$$\alpha = 1 \qquad lim_{\epsilon\to 0} W^\epsilon_{u_\epsilon} = \frac{1}{(4\pi)^N} |\hat{u}(\xi)|^2 \, \delta_0(x) \tag{59}$$

♣

Example 2 (*coherent states*)

$$u_\epsilon = \frac{1}{\epsilon^{\frac{n\alpha}{2}}} u\left(\frac{x - x_0}{\epsilon^\alpha}\right) e^{i\frac{\xi_0 \cdot x}{\epsilon}} \tag{60}$$

$$0 < \alpha < 1 \qquad lim_{\epsilon\to 0} W_{\epsilon, u_\epsilon} = \|u\|_2^2 \, \delta_{x_0}(x)\delta_{\xi_0}(\xi) \tag{61}$$

$$\alpha > 1 \qquad lim_{\epsilon\to 0} W^\epsilon_{u_\epsilon} = 0 \tag{62}$$

$$\alpha = 1 \qquad lim_{\epsilon\to 0} W_{\epsilon, u_\epsilon} = (2\pi)^{-n} |\hat{u}(\xi - \xi_0)|^2 \, \delta_{x_0}(x) \tag{63}$$

♣

Example 3 (*WKB states*)

$$u_\epsilon(x) \equiv u(x) e^{ia(x)/\epsilon}, \qquad u \in L^2(R^N), \;\; u(x) \in R \qquad a \in W^{1,1}_{loc} \tag{64}$$

Notice that $u_\epsilon(x+\frac{\epsilon^\alpha z}{2})\bar{u}_\epsilon(x-\frac{\epsilon^\alpha z}{2})$ converges in $\mathcal{S}'(R^{2N})$ to $|u(x)|^2$ if $0<\alpha<1$ and to $|u(x)|^2 e^{i\nabla a(x).z}$ if $\alpha=1$.

One has therefore

$$\alpha < 1 \qquad lim_{\epsilon\to 0} W^\epsilon_{u_\epsilon} = |u(x)|^2\delta_0(\xi) \tag{65}$$

$$\alpha = 1 \qquad lim_{\epsilon\to 0} W^\epsilon_{u_\epsilon} = |u(x)|^2\delta(\xi - \nabla u(x)) \tag{66}$$

♣

Example 4 (*superposition of coherent states*)

$$u_\epsilon = \sum_{x_j\neq x_k} \beta_j u^j\left(\frac{x-x_j}{\epsilon^\alpha}\right) e^{N/4} e^{i\xi.x/\epsilon} \qquad 0<\alpha<1 \tag{67}$$

It can be verified that the limit is

$$\sum |\beta_j|^2 \delta(x-x_j)\delta(\xi-\xi_j) \tag{68}$$

♣

A detailed analysis of Wigner functions in the semiclassical limit can be found in [6].

We now give details of the use of Wigner functions in study the semiclassical limit.

Theorem 3 ([6]) *(i) Let $V \in C^1(R^N)$ and verify (12). Then for every $T>0$ there is a subsequence $f^{\hbar_n}(t)$ that converges in the weak* -topology of $\mathbf{A}'$ for $|t|<T$ to a function $f \in C_b(R^N)$ which satisfies (20) in distributional sense.*
(ii) If moreover $V \in C^{1,1}(R^N)$ and $V(x) \geq -c(1+|x|^2)$, then $f(t)$ is the unique solution of (28) and represents the evolution of f_0 under the flow defined by

$$\dot{x} = \xi, \qquad \dot{\xi} = -\nabla V \tag{69}$$

◇

Notice that under the assumptions we have made this equation does not have in general a unique solution.

It is possible to construct examples of lack of uniqueness by taking coherent states localized on different solutions.

Sketch of the proof of Theorem 3
By density it is sufficient to prove that if

$$\phi \in \mathcal{S}\left(R^N_x \times R^N_\xi\right), \qquad \mathcal{F}_\xi\phi \in C^\infty_0\left(R^N_x \times R^N_\xi\right) \tag{70}$$

and $K_\hbar$ is defined as in (20), then $< K_\hbar *_\xi f^\hbar, \phi >$ is bounced for $|t| \leq T$ and converges weakly when $\hbar \to 0$, to

$$\int_{R^{2N}} \nabla V(x) \cdot \nabla_\xi \phi(x,\xi) f(x,\xi) dx d\xi \tag{71}$$

One has

$$< K^\hbar * f, \phi >= \frac{i}{(2\pi)^N} < f^\hbar, \phi^\hbar >_{\mathcal{A}' \otimes \mathcal{A}} \tag{72}$$

where

$$\phi^\hbar(x,y) = \int_{R^N} \hbar^{-1} (\mathcal{F}_\xi \phi)(x,y)(y \cdot \nabla V(x)) e^{i\eta . y} \left(\left(V(x + \frac{\hbar z}{2} \right) \right) - \left(V \left(x - \frac{\hbar z}{2} \right) \right) dy \tag{73}$$

It follows for every $\phi \in \mathcal{A}'$

$$< \frac{\partial f^\hbar}{\partial t}, \phi^\hbar > - < f^\hbar, \xi \cdot \nabla_x \phi^\hbar > + < f^\hbar, (K_\hbar * I)\phi^\hbar >= 0 \tag{74}$$

When $\hbar$ tends to zero, the sequences $\phi^\hbar$ and $\frac{\partial f^\hbar}{\partial t}$ are convergent in the topology induced by $\mathcal{A}'$.

Taking into account that $\phi^\hbar$ converges to ϕ in the topology of $\mathcal{A}'$ one has

$$< \frac{\partial f}{\partial t}, \phi > - < f, \xi \cdot \nabla_x \phi > + < f, \nabla V \cdot \nabla_\xi \phi) = 0 \qquad \phi \in \mathcal{A}' \tag{75}$$

Therefore $f \equiv lim_{\hbar \to 0} f^\hbar$ is a weak solution of (20). This proves (i).

To prove point (ii) an integration by parts is needed in order to pass from the weak form of the solution to the classical solution. For this, it is convenient to regularize f and then undo the regularization *after having taken the limit* $\epsilon \to 0$ taking advantage from the fact that the classical solution is of Lipschitz class. This is legitimate under the assumptions made on V.

♡

This analysis of the semiclassical limit for the Schroedinger equation can be extended with minor modifications to the *Schroedinger–Poisson system* which describes the propagation of a system of N quantum mechanical particles subject to the electric field generated by their charges and possibly to an external field E_0 generated by an external charge ρ_0.

The equations which describe this quantum system are

$$i\hbar \frac{\partial \phi_j}{\partial t} = -\frac{\hbar^2}{2} \Delta \phi_j + V\phi_j, \qquad j = 1, \ldots, N \tag{76}$$

$$V = E_0 - \sum_j e_j |\phi_j(t,x)|^2 \tag{77}$$

It is possible to show that the limit $\hbar \to 0$, denoted by f (a function on classical phase space) of the Wigner function associated to the density matrix of any particle does not depend on the particle chosen and satisfies the system of classical equations (called *equations of Vlasov–Poisson*)

$$\frac{\partial f}{\partial t} + \xi \cdot \nabla f - E \cdot \nabla_\xi f = 0, \quad E(x) = \nabla_x \left(\int \frac{1}{|x-y|} \left[\rho_0(y) - \int f(y,\xi) d\xi \right] dy \right) \tag{78}$$

where x, ξ are coordinates in the classical phase space.

4 Gabor Transform

For completeness we mention here the *Gabor transform* [3], much used in signal analysis: the time modulation and frequency modulation of an acoustic signal have the same role as position and momentum in the description of a wave function.

The Gabor transform $G_{\bar{g}} f$ or two complex-valued functions f, g on R^1 is obtained from the corresponding Wigner function by a change of variables

$$W(f,g)(x,\xi) = (2\pi)^N e^{2i(x.\xi)} [G_{\bar{g}}(f)](2x, 2\xi) \tag{79}$$

Inverting this formula one obtains

$$G_{\bar{g}} f(x,\xi) = (2\pi)^{\frac{N}{2}} (f, M_\xi T_{-x} g)$$

$$M_\xi(h)(t) = e^{it\xi} h(t) \qquad T_{-t} h = h(t-x) \tag{80}$$

The Gabor transform is also called *short time Fourier transform* of f *with window* $\bar{g}$.

The function

$$\phi_{x,\xi} = M_\xi T_{-x} \phi \tag{81}$$

is called *Gabor wavelet generated by* ϕ.

The operators M_ξ and T_{-x} are called respectively *modulation operator* and *translation operator* Occasionally one uses the notation $\phi_{t,\nu}$ to stress the *time-frequency analysis*.

The role of the Gabor wavelets in signal analysis is seen in the following formula that allow to reconstruct a signal from its *Gabor spectrum*.

Let $\phi \in L^1(R^N) \cap L^2(R^N)$ with $\int \phi(x,\xi)) d^N x = 1$. The function ϕ is called *the window*.

For all $f \in L^2(R^N)$

$$f = (2\pi)^{-N} \int \int (f, \phi_{x,\xi}) \phi_{x,\xi} dx d\xi \tag{82}$$

Let $\sigma \in L^2(R^N \otimes R^N)$. The *Gabor multiplier* $G_{\sigma,\phi} : L^2(R^N) \to L^2(R^N)$ is defined by

$$(G_{\sigma,\phi} f, g) = \int\int \sigma(x,\xi) G_\phi f(x,\xi)(x,\xi)(\bar{G}_\phi g)(x,\xi) dx d\xi$$
$$= (2\pi)^{-N} \int\int (\sigma(x,\xi)(f, \phi_{x,\xi})_{L^2} \phi(x,\xi) dx d\xi \qquad (83)$$

for $f, g \in L^2(R^N)$.

Gabor operators are also called *localization operators*.

One proves the following results:

(1) if $\sigma \in L^2(R^N \times R^N$ the Gabor multiplier is a Hilbert–Schmidt operator
(2) if $\sigma \in L^2(R^N \times R^N)$ one has

$$G^*_{\sigma,\phi} G_{\bar{\sigma},\phi} = \int |\sigma(x)|^2 dx \cdot \int |\phi(x,\xi)|^2 dx d\xi \qquad (84)$$

(3) moreover

$$G_{\sigma,\phi} G_{\tau,\phi} = G_{\lambda,\phi} \quad \lambda = (2\pi)^{-N} \left(\hat{\sigma}, *\frac{1}{2}\hat{\tau} \right) \qquad (85)$$

where

$$(f *^{\frac{1}{2}} g) = \int_{C^N} f(z-w) g(w) e^{\frac{(z\bar{w}-|w|)^2}{2}} dw. \qquad (86)$$

5 Semiclassical Limit of Joint Distribution Function

Recall that in Quantum Mechanics a state can be characterized by the expectation values of the operators. Typically one considers expectation values in the state described by the wave function ϕ of product of the canonical operators $\hat{q}_k, \ \hat{p}_k, \ k = 1, \ldots, d$

$$(\phi, \Pi_{i,j;h,k} \hat{q}_i^k p_j^h \phi) \quad i, j = 1 \ldots d, \quad k.h = 1, 2, \ldots \qquad (87)$$

Notice that by using the canonical commutation relations one can restrict oneself to polynomials of this type.

To avoid proliferation of indices we shall consider from now on a system with one degree of freedom.

The evolution of the state $\phi(0) \to \phi(t)$ under the Schödinger equation is described by evolution of the correlation functions.

In Classical Mechanics correspond to measures concentrated in a point in phase space, expectation values correspond to position and momentum of the particle considered and the evolution of the pure state is described by Hamilton's equation of motion for the canonical coordinates.

In Quantum Mechanics a state which is not pure is described by a ρ and again the evolution $\rho \to \rho(t)$ corresponding to the Hamiltonian $H(q, p)$ can be described by giving the map

$$Tr(\rho q_k^m p_h^n) \to Tr(\rho(t) q_k^m p_h^n)) = Tr(\rho, e^{itH(\hat{q},\hat{p})} q_k^m p_h^M e^{-itH(\hat{q},\hat{p})}) \tag{88}$$

Correspondingly one has in Classical Mechanics the evolution of a probability distribution under the Liouville equation corresponding to the hamiltonian $H(q, p)$.

Of course in Quantum Mechanics the expressions we have given are formal, since the canonical variables $\hat{q}$, $\hat{p}$ and the hamiltonian H are unbounded operators and one must keep track of their domains.

Before giving a precise statement we note the analogy with *mean field models.*

In these models, mostly used in (Quantum) Statistical Mechanics, one considers (in one dimension)a system with N degrees of freedom and a set of M (quantum) *intensive* observables α_N^k, $k = 1, \ldots, M$ which are *space averages*

$$\alpha_N^M = N^{-1} \sum_{n=M}^{N+M} a_n$$

of *local* observables a_n.

The *local observables almost commute* at long distances: $[a_n, a_m] \simeq (n - m)^{-p}$ for some large p. Both the local and the intensive observables depend on time through a (Quantum) hamiltonian H.

The intensive observables became *classical* in the limit $N \to \infty$ (i.e. they form a commutative algebra). And under the hamiltonian H the evolution $t \to \alpha(t)$ of the observables is describes by an *effective* equation.

One is interested in the structure and the evolution in of the *fluctuations*

$$\beta_t^k = lim_{N\to\infty} \sqrt{N} \left(\frac{1}{N} \sum_{h=1}^{N} a_{h+k}(t) - \alpha^k(t) \right) \tag{89}$$

Under suitable assumptions [7] one proves that if a_n^k are quantum canonical variables their fluctuations $\beta(t)$ are at first order in $\sqrt{\hbar}$ again quantum canonical variables for a system with M degrees of freedom and they evolve according to a quadratic hamiltonian.

In this sense our analysis of the semiclassical limit $\hbar \to 0$ is analogous to the analysis for $N \to \infty$ in Quantum Statistical Mechanics of a system of N particles (with $\hbar$ taking the place of $\frac{1}{M}$.

6 Semiclassical Limit Using Coherent States

We return to the semiclassical approximation which we now understand a semi-classical limit of quantum correlation functions. Of course one should expect this convergence to hold only for a suitable class of initial states, which we take to be coherent states with joint dispersion of order $\hbar$.

We give here first a formulation of the problem in terms of the Weyl algebra and unitary evolution, due to K. Hepp [7]. We shall later sketch a formulation in terms of correlation functions for canonical variables.

We consider hamiltonian system with one degree of freedom and classical hamiltonians of the form $H^{cl}(p,q) = \frac{p^2}{2m} + V(x)$ with V sufficiently regular so that both the classical equation of motion and the quantum one have a unique solution for the initial data we will consider, at least up to a time a time T.

The classical equations of motion are

$$\dot{x}(t) = p(t) \qquad \dot{p}(t) = -\nabla V(x(t)) \tag{90}$$

The corresponding equation in Quantum Mechanics are

$$\frac{d}{dt}(\psi_t, \hat{q}_\hbar \psi_t) = (\psi_t, \hat{p}_\hbar \psi(t))$$

$$\frac{d}{dt}(\psi_t, \hat{p}_\hbar \psi_t) = -(\psi_t, \nabla V(\hat{x})\psi(t)) \tag{91}$$

where $\nabla V(\hat{x})$ is defined by the functional calculus for the self-adjoint operator $V(\hat{x})$ and we have assumed that $\hat{q}_\hbar$ $\hat{p}_\hbar$ satisfy the Heisenberg relations

$$[\hat{q}_\hbar, \hat{p}_\hbar] = -i\hbar \tag{92}$$

on suitable dense domain (by antisymmetry one has $[\hat{q}_\hbar, \hat{q}_\hbar] = [\hat{p}_\hbar \hat{p}_\hbar] = 0$).

However $(\psi_t, \nabla V(\hat{x})\psi(t)) \neq \nabla V((\psi_t, \hat{q}_\hbar \psi_t)$ unless the potential V is at most quadratic, and even if the error is small for $t = 0$ it may become incontrollable for large values of T even if $\hbar$ is very small.

Formally one recovers (91) from (90) in the limit $\hbar \to 0$ (Eherenfest theorem) when ψ is a coherent state centered around large mean values $\hbar^{-\frac{1}{2}} p_0, \hbar^{-\frac{1}{2}} x_0$.

The introduction of the following *macroscopic* representation of the Heisenberg relations

$$\hat{p} = \hbar^{-\frac{1}{2}} \hat{p}_\hbar, \quad \hat{q} = \hbar^{-\frac{1}{2}} \hat{q}_\hbar$$

$$[\hat{q}, \hat{p}] = I, \quad [\hat{q}, \hat{q}] = [\hat{p}, \hat{p}] = 0 \tag{93}$$

is suggested by the fact that the product $\hat{q}(t_1) \dots \hat{p}(t_N)$ bar should be *observed at scale* $\frac{1}{\hbar}$.

This change of scale can be achieved considering the expectation value of these observables in *semiclassical states*, in particular in coherent states, localized at points of the phase space (in the present case, R^2) and seen at a *semiclassical scale*, i.e. at a scale that differs from the atomic scale by a factor $\hbar^{-\frac{1}{2}}$.

The states that we will consider are coherent states ϕ_α centered at the point α of classical phase space.

Recall that a coherent state is given by

$$\phi_\alpha \equiv U(\alpha)\Omega \qquad U(\alpha) = e^{\alpha a^* - \alpha^* a} = e^{p\hat{q} - q\hat{p}} \tag{94}$$

where Ω is the Fock vacuum.

The operator $U(\alpha)$ acts on the annihilation operator as follows

$$U(\alpha)aU^*(\alpha) = a - \alpha, \qquad a = \hat{p} + i\hat{q} \qquad \alpha = q + ip \tag{95}$$

For any choice of monomial P in the $\hat{p}\ \hat{q}$ one has

$$(\hbar^{-\frac{1}{2}}\alpha, P[\hat{q} - (\hbar^{-\frac{1}{2}}q), \hat{p} - (\hbar^{-\frac{1}{2}}p)]\hbar^{-\frac{1}{2}}\alpha) = (\Omega, \hat{q} \dots \hat{p}\Omega)$$

and therefore

$$lim_{\hbar\to 0}(\hbar^{-\frac{1}{2}}\alpha, \hat{q}_\hbar \dots \hat{p}_\hbar \hbar^{-\frac{1}{2}}\alpha) = q \dots p \tag{96}$$

We now show that this relation is preserved under time evolution $U_\hbar(t)$ associated to the self-adjoint extension of $H_\hbar = \frac{p_\hbar^2}{2m} + V(q_\hbar)$ i.e.

$$lim_{\hbar\to 0}(h^{-\frac{1}{2}}\alpha), q_\hbar(t_1) \dots p_\hbar(t_N)(h^{-\frac{1}{2}}\alpha) = q(\alpha, t_1) \dots p(\alpha, t_N) \tag{97}$$

(for more singular Hamiltonians the statement is true for times for which the classical orbit exists).

This result must be compared with the statement (see Volume I) that *along coherent states* the quantum mechanical evolution

$$(h^{-\frac{1}{2}}\alpha, a_\hbar(t)h^{-\frac{1}{2}}\alpha)$$

and the classical evolution

$$z(\alpha, t) = (h^{-\frac{1}{2}}\alpha(t)a_\hbar h^{-\frac{1}{2}}\alpha(t))$$

differ by terms of order $\hbar^{\frac{1}{2}}$ (more exactly are in correspondence and their difference vanishes for $\hbar \to 0$).

Notice that the result can be put in a probabilistic setting (stressing the analogy with the central limit theorem) as a comparison between the expectation value of a quantum observable under the *classical* evolution of coherent states (i.e. the parameter of the coherent states evolve according to a classical equation of motion) and the quantum mechanical evolution of the expectation values in given coherent state.

As a consequence one can view, to first order in $\hbar$, the quantum mechanical evolution as quantum mechanical (central limit) gaussian oscillations around the classical evolution.

One has indeed

$$lim_{\hbar\to 0}(\hbar^{-\frac{1}{2}}\alpha, [\hat{q}(t_1)-q(\alpha,t_1)]\dots\hbar^{-\frac{1}{2}}[\hat{p}(t_1)-p(\alpha,t_N)]\hbar^{-\frac{1}{2}}\alpha) = q(\alpha,t_1)\dots p(\alpha,t_N) \quad (98)$$

where $q(\alpha,t)$ and $p(\alpha,t)$ are solutions of the linearized classical equation around $\xi(\alpha,t)$ (the classical trajectory of the *barycenter* of the coherent state)

$$\dot{q}(\alpha,t) = p(\alpha,t) \qquad \dot{p}(\alpha,t) - \nabla V(\xi(\alpha,t))q(\alpha,t)$$

$$(\phi_{\hbar^{-\frac{1}{2}}\alpha}, (\hat{Q}_1 - \hbar^{-\frac{1}{2}}\xi_1)\dots(\hat{P}_n - \hbar^{-\frac{1}{2}}\pi_n)\phi_{\hbar^{-\frac{1}{2}}\alpha}) = (\Omega, \hat{Q}_1, \dots \hat{P}_n\Omega) \quad (99)$$

From (98), multiplying by $\hbar^{s/2}$ (s is the degree of the monomial) one obtains

$$lim_{\hbar\to 0}(\phi_{\hbar^{-\frac{1}{2}}\alpha}, (q_1^{\hbar} - \xi_1), \dots (\pi_n^{\hbar} - \eta_n)\phi_{\hbar^{-\frac{1}{2}}\alpha}) = 0$$

By iteration, for polynomials of type $P^* \ P$,

$$lim_{\hbar\to 0}(\phi_{\hbar^{-\frac{1}{2}}\alpha}, (q_1^{\hbar}, \dots p_n^{\hbar})\phi_{\hbar^{-\frac{1}{2}}\alpha}) = \xi_1 \dots \pi_n \quad (100)$$

We want to prove that, if $\{\xi_m(s)\}$, $\{\eta_m(s)\}$ are solutions of Hamilton's equation with potential term V and if

$$q(\alpha,t) \equiv \{q_m(\alpha,t)\} \qquad p(\alpha,t) \equiv \{p_m(\alpha,t)\} \quad (101)$$

are the solutions of the tangent flow i.e.

$$\dot{q}(\alpha,t) = p(\alpha,t), \qquad \dot{p}(\alpha,t) = -\nabla V(\xi(\alpha,t).q(\alpha,t)$$

$$\dot{\xi}(t) = \pi(t), \qquad \dot{\pi}(t) = -\nabla V(\xi(\alpha,t))q(\alpha,t) \quad (102)$$

then (96) and (97) are satisfied at all finite times if they are satisfied by the initial conditions, i.e. for any T and all $|s| \le T$ one has

$$lim_{\hbar\to 0}\left(\phi_{h^{-\frac{1}{2}\alpha}}, (\hat{q}_1^{\hbar} - \xi_1(\alpha,s)), \dots (\hat{p}_n^{\hbar} - \eta_n(\alpha,s))\phi_{h^{-\frac{1}{2}\alpha}}\right) = q_1(\alpha,s)\dots p_n(\alpha,s) \quad (103)$$

7 Convergence of Quantum Solutions to Classical Solutions

We state the following theorem in the case of one degree of freedom; it is easy to generalize the proof to the case of an arbitrary *finite* number of degrees of freedom.

Formally the result can be extended to the case of a system with infinitely many degrees of freedom, but in that case care must be put in the choice of the representation and on the definition of the Hamiltonian.

Theorem 4 (Hepp [7]) *Let $\xi(\alpha, t)$ be a solution of the classical equation of motion for the hamiltonian $H_{class} = \frac{1}{2m}p^2 + V(x,t)$ for $|t| \leq T$. Let $V(x)$ be real and of Kato class so that the quantum Hamiltonian $H_\hbar$ is self-adjoint; we use the notation*

$$U_\hbar(t) = e^{i\frac{t}{\hbar}H_\hbar}, \qquad U(t) = e^{itH_\hbar} \tag{104}$$

Let $V(x)$ be of class $C^{2+\epsilon}$ in a neighborhood of the classical trajectory $\xi(\alpha, t)$, $\pi(\alpha, t)$ so that the cotangent flow $q(\alpha, t)$, $p(\alpha, t)$ is well defined for $|t| < T$ and of class $C^{1+\epsilon}$.

Then for all $r, s, \in R^2$ and uniformly in $|t| \leq T$

$$\begin{aligned} s - lim_{\hbar \to 0} U(\hbar^{-\frac{1}{2}}\alpha)^* U_\hbar(t)^* e^{i[rq_\hbar + sp_\hbar]} U_\hbar(t) U(\hbar^{-\frac{1}{2}}\alpha) \\ = e^{i[r\xi(\alpha,t) + s\pi(\alpha,t)]} \end{aligned} \tag{105}$$

Let $\xi(\alpha, t)$, $\pi(\alpha, t)$ be a solution of Hamilton's equation with initial data $\alpha \equiv (\xi, \pi)$ and defined in $t \in (-T, +T)$.

Let $V(x)$ be of class $C^{2+\delta}$ $\delta > 0$, in a neighborhood of $\xi(\alpha, t)$, and let

$$\int |V(x)|^2 e^{-\rho x^2} dx < \infty \tag{106}$$

for some $\rho > 0$.

Let $H_\hbar$ be a self-adjoint extension of the symmetric operator

$$-\frac{\hbar}{2}\frac{d^2}{dx^2} + V(\sqrt{\hbar}x) \tag{107}$$

and set

$$U_\hbar(t) \equiv e^{-i\frac{H_\hbar t}{\hbar}} \tag{108}$$

Let $\{p(t), q(t)\}$ be the solution of the linearized flow at $\xi(\alpha, t)$ with initial data p, q. This flow corresponds to the Hamiltonian

$$H(t) \equiv \frac{p^2}{2} + C(t)\frac{q^2}{2} \qquad C(t) \equiv \frac{d^2V}{dq^2}(\xi(\alpha, t)) \tag{109}$$

Under these assumptions, for every r, s $\in R^2$ *and uniformly in t* $\in (-T, +T)$ *one has*

(1)

$$s - lim_{\hbar\to 0} U^*\left(\frac{\alpha}{\sqrt{\hbar}}\right) U_{\hbar}^* e^{[ir(q-\frac{\xi(\alpha,t)}{\hbar})+is(p-\frac{\pi(\alpha,t)}{\sqrt{\hbar}})]} U_{\hbar}(t) U\left(\frac{\alpha}{\sqrt{\hbar}}\right)$$

$$= e^{i(rq(\alpha,t)+sp(\alpha,t))} \tag{110}$$

(2)

$$s - lim_{\hbar\to 0} U^*\left(\frac{\alpha}{\sqrt{\hbar}}\right) U_{\hbar}^*(t) e^{i(rQ^{\hbar}+sP^{\hbar})} U_{\hbar}(t) U\left(\frac{\alpha}{\sqrt{\hbar}}\right). = e^{i(r\xi(\alpha,t)+s\pi(\alpha,t))} \tag{111}$$

◇

We remark that the same result is obtained using modified (squeezed) coherent states for which the dispersion in configuration space is of order $\hbar^\alpha$ and the dispersion in momentum space is of order $\hbar^{\frac{1}{2}-\alpha}$ with $0 < \alpha < \frac{1}{2}$.

Proof of Theorem 4 The strategy of the proof is to *expand* formally the Hamiltonian around the classical orbit in powers of $\sqrt{\hbar}$ up to the second order and to consider the corresponding evolution equations (this corresponds classically to consider only the tangent flow).

If the potential is smooth the first term is a constant (as a function of $\hat{q}$) and does not contribute to the dynamics of the canonical variables.

Since the term of first order is linear in the $\hat{q}$, $\hat{p}$ the evolution corresponds to a scalar shift in the canonical variables.This provides a *rotating frame* for the canonical variables

The second order term provides in the rotating frame a linear homogeneous map that depends differentiably on time. Classically this would give the evolution of the fluctuations.

Our assumption on the hamiltonian imply that the higher order terms provide a negligible effect in the limit $\hbar \to 0$.

These remarks imply that one can obtain similar results for classical Hamiltonians with smooth coefficients.

We shall give a sketch of the proof. Details can be found in Hepp's paper [7].

Expand formally $\hbar^{-1} H_\hbar$ in powers of $\hbar$ in a $\hbar$ neighborhood of the classical orbit $\xi(\alpha, t) \equiv \xi(t)$

$$\hbar^{-1} H_\hbar = H_\hbar^0(t) + H_\hbar^1(t) + H_\hbar^2(t) + H_\hbar^3(t)$$

$$H_\hbar^0(t) = \hbar^{-1} H(\pi, \xi)$$

$$H_\hbar^1(t) = \hbar^{-1}\left[\pi_t\left(\hat{p} - \frac{\pi_t}{\sqrt{\hbar}}\right) + \frac{dV}{dx}(\xi(t))\left(\hat{q} - \frac{\xi(t)}{\sqrt{\hbar}}\right)\right]$$

$$H^2_\hbar(t) = \frac{1}{2}\left(\hat{p} - \frac{\pi_t}{\sqrt{\hbar}}\right)^2 \frac{1}{2}\left(\hat{q} - \frac{\xi(t)}{\sqrt{\hbar}}\right)^2 \tag{112}$$

Define

$$H^3_\hbar(t) = V(x) - V(\xi(t)) - (x - \xi(t)) \cdot \frac{dV}{dx}(\xi(t)) - \frac{1}{2}(x - \xi(t))^2 \frac{d^2V}{dx^2}(\xi(t)) \tag{113}$$

The term $H^3_\hbar(t)$ is, *at least formally*, an operator of order $O(\hbar^{\frac{1}{2}})$.

Since the operator is unbounded, one must qualify the meaning of this statement. We consider the restriction of the operator to functions of fast space decay (coherent states) and will have to control that this remains true under evolution.

This will be guaranteed by the fact that under the total evolution the set of coherent spates is left invariant modulo a small correction.

Notice that $H^1_\hbar$ is a liner function of $\hat{p}$ e $\hat{q}$ and therefore the propagator $U^s_\hbar(t)$ exists for every t and is unitary.

This provided a one-parameter map of gaussian coherent states differentiable in time. To see this, use on the convex closure of Hermite functions Dyson's perturbation series, or apply the result on the metaplectic group described in Volume I of these Lecture Notes.

The unitary operators $U^1_\hbar(t) \equiv e^{iH^1_\hbar t}$ provide a family of automorphisms of Weyl algebra and the evolution can be written

$$W^\hbar(t,0)^* e^{i(r\hat{q}+s\hat{p})} W^\hbar(t,0) \tag{114}$$

with

$$W^\hbar(t,s) \equiv U^*\left(\frac{\alpha}{\sqrt{\hbar}}\right) U^1_\hbar(t)^* \, U_\hbar(t-s) U^1_\hbar(s) U\left(\frac{\alpha}{\sqrt{\hbar}}\right) e^{i\int_s^t dr H^0_\hbar(r)} \tag{115}$$

This proves point (1) in the theorem.

Point (2) is proved if one shows that

$$s - lim_{\hbar \to 0} W^\hbar(t,s) \equiv W(t,s) = e^{-i\int_s^t H^0(r)dr} \tag{116}$$

where the (time ordered) integral on the right hand side is defined using spectral representation.

Since the operators in the Weyl system are uniformly bounded it is sufficient to prove this on a dense set of states, which we shall choose to be the coherent states

$$\phi_a(x) \equiv e^{-(x-a)^2/2}, \quad a \in R \tag{117}$$

We must show that for every τ, $|\tau| < T$ one can find $\hbar(\tau) > 0$ such that for any $\hbar < \hbar(\tau)$ the states

$$\phi_a^{\hbar,s} \equiv U_\hbar^1 U\left(\frac{a}{\sqrt{\hbar}}\right) W(s,0)\phi_a \tag{118}$$

belong to the domain of the operator H^1.

To show this, we note H^1 is quadratic and we use the explicit form of $W^\hbar(t,s)$ and the identity

$$W(s,0)\hat{q}W(s,0)^* = \alpha\,\hat{q} + \beta\,\hat{p} \qquad W(s,0)\hat{p}W(s,0)^* = \gamma\,\hat{q} + \delta\,\hat{p} \tag{119}$$

in which $\alpha, \beta, \gamma, \delta$ depend continuously on time. We obtain

$$\phi_a^{\hbar,s} = Cexp\left[-\frac{\alpha+i\gamma}{2(\delta-i\beta)}\left(x-\frac{\xi_s}{\sqrt{\hbar}}\right) - \frac{a}{(\alpha+i\gamma)^2} + i\frac{\pi_s}{\sqrt{\hbar}}x\right] \tag{120}$$

for any $|s| < T$. Since for a suitable $\rho > 0$

$$\int dx|V(x)|^2\, e^{-\rho x^2} < \infty \tag{121}$$

one has the inclusion in the domain of H^1 if

$$\hbar(\tau) = 2\eta_\epsilon(\tau)\rho^{-1}, \qquad \eta_\epsilon = |Re\frac{\alpha+i\,\gamma}{\delta-i\,\beta}| \tag{122}$$

We conclude that for $\hbar < \hbar(\tau)$ the product $W(t,s)W(s,r)$ is strongly differentiable in s.

From Duhamel's formula one has

$$\begin{aligned} W(t,0)\phi_a - \phi_a &= \int ds \frac{d}{ds} W_\hbar(t,s)\phi_a \\ &= \int_0^t i W_{\hbar,s}\left[\hbar^{-1}(V(\xi_s+\hbar;q) - V(\xi_s)) - \hbar^{-\frac{1}{2}}V'(\xi_s,q) - V''(\xi_0)\frac{q^2}{2}\right]\phi_a \end{aligned} \tag{123}$$

We provide now an estimate of the L^2 norm of the right hand side. Using $V \in C^{2+\delta}$ we can bound the integral over a ball of radius $O(1)$ (and therefore $\sqrt{\hbar}x \simeq \sqrt{\hbar}$)

For large values of $|x|$ we use instead the rapid decrease of $|W(s,0)\phi_0|^2$.

In $\sqrt{\hbar}x \simeq \sqrt{\hbar}$ we use the estimate

$$|\hbar^{-1}[V(\xi+\hbar\,x) - V(\xi)] - \hbar^{-1/2}x\;V'(\xi) - \frac{1}{2}x^2\;V''(\xi_s)| \le x^{2+\delta}\hbar^{\delta/2} \tag{124}$$

and derive

$$|W(t,0)\phi_\alpha - W_\hbar(t,0)\phi_\alpha| = o(\hbar^{\frac{\delta}{2}}) \tag{125}$$

With a similar estimate one completes the proof of Theorem using the identity

$$|U\left(\frac{\alpha}{\sqrt{\hbar}}\right)^* U_\hbar^* e^{i(r\, q_\hbar + s\, p_\hbar)} U_\hbar(t) U(\frac{-\alpha}{\sqrt{\hbar}})\phi - e^{i(r\, \xi_t + s\, \pi_t)}\phi|$$
$$= |W^\hbar(t,0) e^{i\hbar(r\, q + s\, p)} W^\hbar(t,0)\phi - \phi| \tag{126}$$

and

$$s - lim_{\hbar\to 0}\, W^\hbar(t,0) \equiv W(t,0), \qquad s - lim_{\hbar \to o}\, e^{i\hbar(r\, q + s\, p)} = 1 \tag{127}$$

This concludes the sketch of the proof of Theorem 4. ♡

It is easy to verify that under our hypothesis on the hamiltonian the formulas obtained for the coherent states $e^{ia\hat{x}+ib\hat{p}}\Omega$ are differentiable in the parameters $a,\ b$ and therefore provide a semiclassical approximation for expectation values of any polynomial.

A direct proof (i.e. without going through the Weyl operators and keeping track of domain problems) can be given but requires attention to the domain problems.

The proof using gaussian coherent states holds for potential that grow not more than a polynomial at infinity since at each step the decay in space of the wave function must compensate uniformly the increase of the potential.

Notice the change in time scale between the unitary groups $U_\hbar(t)$ and $U(t)$: the motion is seen as *adiabatic* at macroscopic scale.

This is in accordance with Eherenfest theorem [8].

The space-time change of scale $x \to \hbar^{-\frac{1}{2}}x, \quad t \to \frac{1}{\hbar}t$ leaves the Schrödinger equation invariant.

The space-time adiabatic change of scales $x \to \frac{x}{\epsilon}, t \to \frac{t}{\epsilon}$, which is efficiently used in solid state physics, corresponds for the Schródinger equation on macroscopic scale to an adiabatic scaling $t \to \frac{1}{\hbar^{-\frac{1}{2}}}t$,

We remark that a similar estimates one proves that a "semiclassical" limit holds in the case of a quantum particle of mass M in the limit $M \to \infty$.

For this one considers the Hamiltonian

$$H(\xi, x) = \frac{1}{2M}\xi^2 + V(x) \tag{128}$$

and sets

$$\xi_\lambda = M\sqrt{\lambda}\xi, \quad x_\lambda = \sqrt{\lambda}x \tag{129}$$

Let $\hbar^{-1} H_\lambda$ be a self-adjoint extension of H and denote by U_λ the operator

$$exp\{-i H_\lambda t\} \tag{130}$$

Then in the limit $\lambda \to 0$ Theorem 4 holds.

It is worth remarking that in the semiclassical limit superpositions of vector states

$$\phi_\hbar = \sum U(\frac{\alpha_n}{\sqrt{\hbar}})\phi_n \tag{131}$$

tend weakly to the corresponding statistical mixtures. Indeed one has for $|t| \leq T$

$$lim_{\hbar \to 0}(\phi_\hbar, U_\hbar^*(t) e^{i(r\,\hat{q}+s\,\hat{p})} U_\hbar(t)\phi_\hbar) = \sum_n |\phi_n|^2 e^{i(r\,\xi(\alpha_n,t)+s\,\eta(\alpha_n,t))} \tag{132}$$

References

1. E.M. Stein, R. Shakarki, *Princeton Lectures in Analysis II* (Princeton University Press, Princeton, 2003)
2. R.L. Hudson, Rep. Math. Phys. **6**, 249–266 (1974)
3. H. Feichtinger, T. Stohmer, *Gabor Analysis and Algorithms* (Birkhauser, Boston, 1998)
4. D. Kastler, Comm. Math. Phys. **1**, 1–15 (1965)
5. G. Loupias, S. Miracle-Sole, Ann. Inst. H. Poincar **6**, 39–60 (1967)
6. P.L. Lions, T. Paul, Rev. Mat. Iberoamericana **9**, 553–618 (1993)
7. K. Hepp, Comm. Math. Phys. **35**, 265–277 (1974)
8. P. Ehrenfest, Zeits. Phys. **45**, 455–471 (1927)

Lecture 2: Pseudo-differential Operators. Berezin, Kohn–Nirenberg, Born–Jordan Quantizations

Weyl quantization is strictly linked to Wigner transform.

If $l(q,p)$ is a linear function of the $q's$ and of the $p's$ (coordinates of the cotangent space at any point $q \in R^d$) the *Weyl quantization* is defined, in the Schrödinger representation, by

$$Op^w(e^{it(q,p)}) = e^{it(x,-i\hbar\nabla_x)} \tag{1}$$

Let $\mathcal{S}$ be the Schwartz class of functions on R^{2d}. It follows from the definition of Wigner function $W_\psi(q,p)$ that in the Weyl quantization one can associate to a function $a \in \mathcal{S}(R^{2d})$ an operator $Op^w(a)$ through the relation

$$(\psi, Op^w(a)\psi) = \int W_\psi(q,p)(\mathcal{F}a)(q,p)dpdq, \quad q,p \in R^d \quad \psi \in L^2(R^d) \tag{2}$$

where the symbol $\mathcal{F}$ stands for Fourier transform *in the second variable* (a map $-\frac{i\nabla}{\hbar} \to p$).

To motivate this relation recall that the Weyl algebra is formally defined a twisted product (twisted by a phase).

Introducing the parameter $\hbar$ to define a microscopic scale, we define the operator $Op^w_\hbar(a)$, as operator on $L^2(R^d)$, by

$$[Op^w_\hbar(a)\ \phi](x) \equiv (2\pi\hbar)^{-d} \int\int a\left(\frac{x+y}{2}, \xi\right) e^{\frac{i}{\hbar}(x-y,\xi)}\phi(y)dy\, d\xi \tag{3}$$

or equivalently

$$(Op^w_\hbar(a)\phi)(x) = \int \tilde{a}\left(\frac{x+y}{2}, x-y\right)\phi(y)dy,$$
$$\tilde{a}(\eta,\xi) \equiv \left(\frac{1}{2\hbar}\right)^N \int a(\eta,z)e^{\frac{i}{\hbar}(z,\xi)}dz \tag{4}$$

G. Dell'Antonio, *Lectures on the Mathematics of Quantum Mechanics II: Selected Topics*, Atlantis Studies in Mathematical Physics: Theory and Applications 2, DOI 10.2991/978-94-6239-115-4_2

Notice that (3) can also be written (from now on, for brevity, we omit the symbol $\hbar$ in the operator)

$$(Op^w(a)\,\phi)(x) = \left(\frac{1}{2\pi\hbar}\right)^d \int\int e^{\frac{i}{\hbar}(y-\xi),\xi)} a\left(\frac{1}{2}(x-y),\xi\right)\phi(y)d\xi\,dy \quad (5)$$

In this form it can be used *to extend the definition* (at least as quadratic form) to functions $a(x, y))$ that are not in $\mathcal{S}$.

1 Weyl Symbols

We will call the function *a Weyl symbol* of the operator $Op^w_\hbar(a)$.

Some Authors refer to the function a in (3) as *contravariant symbol* and define as *covariant symbol* the following expression

$$a^\#(z) = (2\pi\hbar)^{-d} a(Jz) \quad (6)$$

With this definition one has

$$Op^w(a) = (2\pi\hbar)^{-1}\int a^\#(z)\hat{T}(z)dz \qquad \hat{T}(z) = e^{\frac{i}{\hbar}(y\cdot\hat{q}-x\cdot\hat{p})} \quad z = x+iy \quad (7)$$

Notice that $T(z)$ is translation by z in the Weyl system. The use of covariant symbols is therefore most convenient if one works in the Heisenberg representation, regarding q, p as translation parameters (hence the name *covariant*).

It is easy to prove

$$(\phi, Op^w(a)\psi) = \left(\frac{1}{2\pi\hbar}\right)^{-d}\int a^\#(z)(\phi,\hat{T}(z)\psi)dz \qquad \forall\phi,\psi\in\mathcal{S}(R^d) \quad (8)$$

In particular in the case of coherent states centered in z

$$(\psi_z, Op^w(a)\psi_z) = \left(\frac{1}{2\pi\hbar}\right)^{-d}\int a^\#(z')e^{-\frac{|z-z'|^2}{4\hbar}-\frac{i}{2\hbar}\sigma(z',z)}dz' \quad (9)$$

Recall that

$$\psi_z = \hat{T}(z)\psi_0, \qquad \psi_0 = \left(\frac{1}{\pi\hbar}\right)^{\frac{d}{2}} e^{-\frac{x^2}{2\hbar}} \quad (10)$$

where $\hat{T}(z)$ is the operator of translation by z in the Weyl representation.

The covariant symbol $a^\#$ is therefore suited for the analysis of the semiclassical limit in the coherent states representation and in real Bergmann–Segal representation.

2 Pseudo-differential Operators

Definition 1 (*Pseudo-differential operators* [1–4]) The operators obtained by Weyl's quantization are called *pseudo-differential operators.*

They are a subclass of the Fourier Integral Operators [5] which are defined as in (4) by substituting the factor $e^{\frac{i}{\hbar}(x-y,\xi)}$ with $e^{\frac{i}{\hbar}(f(x,y),\xi)}$ where f is a regular function.

Notice that when $\hbar$ is very small, this function is fast oscillating in space.

◇

As a remark we mention that the notation *pseudo-differential* originates from that fact that if $a(q,p) = P(p)$ where P is a polynomial, the operator $Op^w(a)$ is the differential operator $P(-i\nabla)$ and if $a(q,p) = f(q)$, the operator $Op^w(a)$ acts as multiplication by the function $f(x)$.

In the case of a generic function a the operator $Op^w(a)$ is far being a simple differential operator (whence the name *pseudo-differential*).

Later we discuss other definitions of *quantization*; Weyl quantization has the advantage of being invariant under symplectic transformations (since it is defined through a symplectic form) and therefore is most suited to consider a semiclassical limit.

In the analysis of the regularity of the solutions of a P.D.E. with space dependent coefficients other quantization procedures may be more useful, e.g. the one of Kohn–Nirenberg [7] that we shall define later.

For the generalization to system with an infinite number of degrees of freedom other quantizations (e.g.the Berezin one) [6] are more suited because they stress the role of a particular element in the Hilbert space of the representation, the *vacuum.*

In a finite dimensional setting this vector is represented by function $\iota(z)$ which takes everywhere the value one and therefore satisfies $\frac{\partial \iota(z)}{\partial z_k} = 0 \ \forall k$ (is *annihilated* by all destruction operators) in the Berezin-Fock representation.

In this representation a natural role is taken by the operator $N = \sum_k z_k \frac{\partial}{\partial z_k}$, which has as eigenvalues the integer numbers and as eigenvectors the homogeneous polynomials in the z_k's.

In the Theoretical Physics literature this representation is often called the *Wick representation* and the operator N is called *number operator*.

For a detailed analysis of pseudo-differential operators, also in connection with the semiclassical limit, one can consult e.g. [1–3, 8].

Let us notice that one has

$$Op^w_\hbar(a) = \int\int e^{i[(p,x)+\hbar(q,D_x)]}\mathcal{F}a(p,q)dpdq \tag{11}$$

where $\mathcal{F}a$ is the Fourier transform of a in the second variable. In particular

$$\|Op^w_\hbar(a)\|^2_{L^2} = \int\int |(\mathcal{F}a)(p,q)|^2 dqdp \tag{12}$$

Remark that integrability of the absolute value of $\mathcal{F}a$ is a *sufficient* (but not a necessary) condition for $Op^w_\hbar \in \mathcal{B}(\mathcal{H})$.

The relation between Weyl symbols and Wigner functions associated to vectors in the Hilbert space (or to density matrices) is obtained by considering the pairing between bounded operators and bilinear forms in $\mathcal{S}$.

More explicitely on has

$$(Op^w(a)f, g) = \int a(\xi, x) W_{f,g}(\xi, x) d\xi\, dx \quad \forall f, g \in \mathcal{S} \tag{13}$$

where

$$W_{f,g}(\xi, x) \equiv \int e^{-i(\xi,p)} f\left(x + \hbar\frac{p}{2}\right) g\left(x - \hbar\frac{p}{2}\right) dp \tag{14}$$

From this one concludes that the Wigner function associated to a density matrix ρ *is the symbol of* ρ as a pseudo-differential operator.

Define in general, for $f \in \mathcal{S}'$

$$W_f(\xi, x) \equiv \int e^{-i(p,\xi)} f\left(x + \hbar\frac{p}{2}, x - \hbar\frac{p}{2}\right) dp \tag{15}$$

Notice that it is the composition of Fourier transform with a change of variable that preserves Lebesgue measure:

It follows that (11) preserves the classes $\mathcal{S}$ and $\mathcal{S}'$ and is unitary in $L^2(R^{2n})$.

One has moreover

$$Op^w_\hbar(\bar{a}) = [Op^w_\hbar(a)]^* \tag{16}$$

and therefore if the function a is real the operator $Op^w_\hbar(\bar{a})$ is symmetric.

One can prove that if the symbol a is sufficiently regular this operator is essentially self-adjoint on $\mathcal{S}(R^d)$.

One can give sufficient conditions in order that a pseudo-differential operator belong to a specific class (bounded, compact, Hilbert–Schmidt, trace class...).

We shall make use of the following theorem

Theorem 1 ([2, 3]) *Let $l_1, \dots l_k$ be independent linear function on R^{2d} and $\{l_h, l_i\} = 0$. Let $\tau : R^k \to R$ be a polynomial.*

Define

$$a(\xi, x) \equiv \tau(l_1(\xi, x), \dots l_k(\xi, x)) \tag{17}$$

Then

(i) $a(\xi, x)$ maps $\mathcal{S}$ in $\mathcal{B}(L^2(R^d))$ and is a self-adjoint operator

(ii) For every continuous function g one has

$$(g.a)(\xi, x) = g(a(\xi, x)) \tag{18}$$

◇

We leave to the reader the easy proof.

From the relation between Wigner functions and pseudo-differential operators one derives the following properties ($\mathcal{L}$ denotes a linear map).
(1) $Op^w(a)$ is a continuous map from $\mathcal{S}(R^d)$ to $\mathcal{L}(\mathcal{S}(R^d), \mathcal{S}'(R^d))$
(2) $Op^w(a)$ extends to a continuous map $\mathcal{S}'(R^d) \to \mathcal{L}(\mathcal{S}(R^d)\mathcal{S}'(R^d))$
(3) If $a(z) \in L^2(C^d)$ one has

$$|Op^w(a)|_{H.S.} = (2\pi\,\hbar)^{-\frac{n}{2}} \left[\int |a(z)|^2\, dz\right]^{1/2} \tag{19}$$

(4) If $a, b \in L^2(R^d)$, then the product $Op^w(a) \cdot Op^w(b)$ is a trace class operator and

$$Tr\,(Op^w(a)\; Op^w(b)) = (2\pi\,\hbar)^{-d} \int \bar{a}(z)\; b(z)\; dz \tag{20}$$

In order to find conditions on the symbol a under which $Op^w(a)$ is a bounded operator on $\mathcal{H}$ one can use the duality between states and operators and

$$(\psi, Op^w(a)\psi) = \int \mathcal{F}a(p,q) W_\psi(p,q) dq dp \tag{21}$$

One can verify in this way that $\|Op(a)\| \leq |\hat{a}|_1$, but $\hat{a} \in L^1$ is *not necessary* in order $Op(a)$ be a bounded operator.

Remark that using this duality one can verify that Weyl quantization is a *strict quantization* (see Volume I).

One can indeed verify that, if $\mathcal{A}_0$ is the class of functions continuous together with all derivatives, introducing explicitly the dependence on $\hbar$.
(i) *Rieffel condition.* If $a \in \mathcal{A}_0$ then $\hbar \to Op^w_\hbar(a)$ is continuous in $\hbar$.
(ii) *von Neumann condition.* If $a \in \mathcal{A}_0$

$$lim_{\hbar\to 0} \|Op^w_\hbar(a) Op^w_\hbar(b) - Op^w_\hbar(a \otimes b)\| = 0 \tag{22}$$

where $\otimes$ is convolution.
(iii) *Dirac condition.* If $a \in \mathcal{A}_0$

$$lim_{\hbar\to\infty} \|\frac{1}{2\hbar}\left[Op^w_\hbar(a)Op^w_\hbar(b) - Op^w_\hbar(b)Op^w_\hbar(a) - Op^w_\hbar(\{a,b\})\right]\| = 0 \tag{23}$$

where $\{a, b\}$ are the Poisson brackets.

If one wants to make use of the duality with Wigner function to find bounds on $Op^w_\hbar(a)$ in term of its symbol $a(x, \hbar\nabla)$ one should consider that Wigner's functions can have *strong local oscillations* at scale $\hbar$.

3 Calderon–Vaillantcourt Theorem

The corresponding quadratic forms are well defined in $\mathcal{S}$ but to obtain regular operators on $L^2(R^d)$ these oscillations (which become stronger as $\hbar \to 0$) *must be smoothed out* by using regularity properties of the symbol.

This is the content of the theorem of Calderon and Vaillantcourt.

From the proof we shall give one sees that the conditions we will put on the symbol a in order to estimate the norm of the pseudo-differential operator $Op^w(a)$ *are far from being necessary.*

We give an outline of the proof of this theorem because it is a prototype of similar proofs and points out the semiclassical aspects of Weyl's quantization.

Theorem 2 (Calderon–Vaillantcourt [2, 3, 8]) *If*

$$A_0(a) \equiv \sum_{|\alpha|+|\beta|\leq 2d+1} |D_\xi^\alpha D_x^\beta a(x,\xi)|_\infty < \infty, \qquad x, \xi \in R^d \tag{24}$$

then $Op^w(a)$ is a bonded operator on $L^2(R^d)$ and its norm satisfies

$$||Op^w(a)|| < c(d) A_0(a) \tag{25}$$

where the constant $c(d)$ depends on the dimensions of configuration space.

◇

The proof relies on the decomposition of the symbol a as

$$a(x,\xi) = \sum_{j,k} a(x,\xi)\zeta_{j,k}(x,\xi) \qquad \sum_{j,k} \zeta_{j,k} = 1 \tag{26}$$

where $\zeta_{j,k}$ are smooth function providing a covering of R^{2d} each having support in a hypercube of side $1+\delta$ centered in $\{j,k\}$ and taking value one in a cube of side $1-\delta$ with the same center.

One gives then estimates of the norm of $Op^w(\sum_\Gamma a(x,\xi)\zeta_{j,k})$, where Γ is a bounded domain in terms of the derivatives of $a(x,\xi)$ up to an order which depends on the dimension of configuration space.

These bounds rely on embeddings of Sobolev spaces $H^p(R^{2d})$ in the space of continuous functions for a suitable choice p (that depends on d).

The convergence $\Gamma \to R^d$ is controlled by the decay at infinity of the symbol $a(x,\xi)$.

A standard procedure is to require at first more decay, and prove by density the theorem in the general case.

The estimates on $Op^w(a\zeta_{j,k})$ are obtained noticing that the symbols of these operators are the product of a function that is *almost* the product the characteristic function of a set on configuration space and of a function that is *almost* the characteristic function of a set on momentum space.

The word *almost* refers to the fact that the partition is smooth, and the functions one uses tend to characteristic functions as $\hbar \to 0$.

If one chooses the side of the hypercubes to be of order $\sqrt{\hbar}$ these qualitative remarks explain why the estimates that are provided in the analysis of pseudo-differential operators have relevance for the study of the semiclassical limit.

And explains why pseudo-differential calculus is relevant if one considers a macroscopic crystal, the partition is at the scale of the elementary cell and one must analyze the properties of projection operator in a Bloch band (these pseudo-differential operators are far from being simple polynomials).

The proof of the theorem of Calderon–Vaillantcourt is based on two results of independent interest.

The first is the theorem of Cotlar–Knapp–Stein; we give the version by L. Hormander [5]; this paper is a very good reference for a detailed analysis of pseudo-differential operators.

In what follows we shall use units in which $\hbar = 1$.

Theorem 3 (Cotlar–Knapp–Stein) *If a sequence $A_1, A_2, \ldots, A_N$ of bounded operators in a Hilbert space $\mathcal{H}$ satisfies*

$$\sum_{k,j=1}^{N} \|A_j^* A_k\| \le M \qquad \sum_{k,j=1}^{N} \|A_j A_k^*\| \le M \tag{27}$$

then

$$\sum_{k=1}^{N} \|A_k\| \le M \tag{28}$$

◇

Proof The proof follows the lines of the corresponding proof for finite matrices. For each integer m

$$\|A\|^{2m} = \|(A^*A)^m\| \tag{29}$$

Also

$$(A^*A)^m = \sum_{1 \le j_1 \le j_2 \ldots \le j_m} A_{j_1}^* A_{j_2} \ldots A_{j_{2m-1}}^* A_{j_{2m}} \tag{30}$$

and

$$\|A_{j_1}^* A_{j_2} \ldots A_{j_{2m-1}}^*, A_{j_{2m}}\| \le \min\{\|A_{j_1}^* A_{j_2}\| \ldots \|A_{j_{2m-1}}^* A_{j_{2m}}\|, \\ \|A_{j_1}^*\| \|A_{j_1} A_{j_2}^*\| \ldots \|A_{j_{2m}}\|\} \tag{31}$$

Making use of the inequality for positive numbers $min\{a,\ b\} \leq \sqrt{ab}$ and taking into account the assumption $\|A_j\| \leq M$ e $\|A_j^*\| \leq M$ one has

$$\|A_{j_1}^* A_{j_2} \dots A_{j_{2m-1}j_{2m}}^*\| \leq M \|A_{j_1}^* A_{j_2}\|^{\frac{1}{2}} \dots \|A_{j_{2m-1}}^* A_{j_{2m}}\|^{\frac{1}{2}} \tag{32}$$

Performing the summation j_2, j_3, j_{2m} one obtains

$$\|A\|^{2m} \leq NM^{2m} \tag{33}$$

so that, taking logarithms, for $m \to \infty$

$$\|A\| \leq M \frac{logN}{m} \tag{34}$$

◇

It is possible [2, 5] to generalize the theorem replacing the sum by the integration over a finite measure space Y. In this case the theorem takes the form

Theorem 4 (Cotlar–Knapp–Stein, continuous version) *Let $\{Y, \mu\}$ be a finite measure space and $A(y)$ be a measurable family of operators on a Hilbert space $\mathcal{H}$ such that*

$$\int \|A(x)A(y)^*\| d\mu \leq C \qquad \int \|A(x)^* A(y)\| d\mu \leq C \tag{35}$$

Then the integral $A = \int A(x) d\mu$ is well defined under weak convergence and one has $\|A\| \leq C$.

◇

Outline of the Proof of the Theorem of Calderon–Vaillantcourt
We build a smooth partition of the identity by means of functions $\zeta_{j,k}(x, \xi)$ of class C^∞ such that

$$\zeta_{j,k}(x, \xi) = \zeta_{0,0}(x - j, \xi - k), \qquad \sum_{j,k \in Z} \zeta(x - j, \xi - k) = 1 \quad x,\ \xi \in R^d \tag{36}$$

We choose $\zeta_{0,0}(x, \xi)$ to have value one if $|x|^2 + |\xi|^2 \leq 1$ and zero if $|x|^2 + |\xi|^2 \geq 2$.

Define

$$a_{j,k} = \zeta_{j,k} a \qquad A_{j,k} = Op^w(a_{j,k}) \tag{37}$$

We must verify that the corresponding operators are bounded and that their sum converges in the weak (or strong) topology. We shall see that these requirements can be satisfied provided the symbol a is sufficiently regular as a function of x and ξ.

The regularity conditions do not depend on the value of the indices j, k since the functions $\zeta_{j,k}$ differ from each other by translations.

It follows from the definitions that $\sum A_{j,k}$ converges to $Op^w(a)$ in the weak topology of the functions from $\mathcal{L}(\mathcal{S}(R^d))$ to $\mathcal{L}(\mathcal{S}'(R^d))$. We are interested in conditions under which convergence is in $\mathcal{B}(L^2(R^d))$.

For this it is sufficient to prove that there exists an integer $K(d)$ such that, for any finite part Γ of the lattice with integer coordinates

$$\| \sum_{j,k\in\Gamma} A\| \leq Csup_{|\alpha|\leq K(d),|\beta|\leq K(d),(x,\xi)\in R^d} |\partial_\xi^\alpha \, \partial_x^\beta a(x,\xi)| \tag{38}$$

This provides conditions on the symbol and at the same time provides bounds for the operator norm.

From Theorem 3 it follows that it is sufficient to obtain bounds on the norm of

$$A_\gamma^* \cdot A'_\gamma \tag{39}$$

for any choice of the index $\gamma \equiv \{j, k\}$.

Define $a_{\gamma,\gamma'}$ by

$$Op^w(a_{\gamma,\gamma'}) = A_\gamma^* \cdot A'_\gamma \tag{40}$$

One derives

$$\begin{gathered} a_{\gamma,\gamma'} = e^{\frac{i}{2}\sigma(D_x,D_\xi;D_y,D_\eta)}(\bar{a}_\gamma(x,\xi).a_{\gamma'}(y,\eta))_{y=x,\eta=\xi} \\ \forall\gamma\in Z^{2d} \quad a_\gamma \in L^2(R^d) \end{gathered} \tag{41}$$

where σ is the standard symplectic form.

Notice that we have used estimates on Sobolev embeddings to obtain an estimate of the norm of $Op^w(a)$ in term of Sobolev norms of the symbol a; recall that the operator norm of $Op(a)$ is the L^2 norm of its Fourier transform. Remark that a_γ has support in R^{2d} of radius $\sqrt{2}$.

The partition of phase space serves the purpose of localizing the estimates; the number of elements in the Cottlar–Kneipp–Stein procedure depends on the dimension $2d$ of phase space.

Remark that $\sum_{k\in Z^{2d}} A(\gamma)$ converges to $A = Op^w a$ in the topology of linear bound operators from $\mathcal{S}(R^d)$ to $\mathcal{S}'(R^d)$.

Therefore it is sufficient to prove, for any bounded subset $\Gamma \subset Z^{2d}$,

$$\| \sum_{\gamma\in\Gamma} A(\gamma)\|_{L^2(R^d)} \leq C(d)sup_{|\alpha|\leq 2d+1,|\beta|\leq 2d+1,(x,\xi)\in R^{2d}} |\partial_\xi^\alpha \, \partial_x^\beta a(x,\xi)| \tag{42}$$

We must have a control over $\|A_\gamma^* \cdot A_{\gamma'}\|$ and therefore of the norm of the operator with symbol $a_{\gamma,\gamma'}$.

We use Sobolev-type estimates. If B is a real quadratic form on R^{2n}, for every $R > 0$ and integer $M \geq 1$ there exists a constant $C(R, M)$ such that

$$|(e^{iB(x,D)}u(x)| \leq C(R, M)(1 + |x - x_0|^2)^{-d} sup_{|\alpha| \leq 2M+d+1, x \in B(x_0,R)} |\partial_x^\alpha u(x)| \quad (43)$$

for every function $u \in C_0^\infty(B(x_0, R))$ and every $x_0 \in R^{2d}$.

This is a classical Sobolev inequality for $x_0 = 0, \quad M = 0$; it holds $x_0 \neq 0$ since the operator commutes with translation and it is satisfied for every M since

$$\mathcal{F}_{x \to \zeta}[(1 + q|x|^2)^M e^{iB(D)}u](\zeta) = e^{iB(\zeta)} \sum_{|\alpha|+|\beta| \leq 2M} C_{\alpha,\beta} \mathcal{F}_{x \to \zeta}[x^\beta \, \partial^\alpha u](\zeta) \quad (44)$$

where $\mathcal{F}_{x \to \zeta}$ denotes total Fourier transform and the constants $c_{\alpha,\beta}$ depend only on the dimension n and on the quadratic form B.

This ends our sketch of the proof of the theorem of Calderon–Valliantcourt.

♡

4 Classes of Pseudo-differential Operators. Regularity Properties

To characterize other classes of pseudo-differential operators we introduce two further definitions.

Definition 2 We shall denote by *tempered weight* on R^d a continuous positive function $m(x)$ for which there exist positive constants $C_0, \; N_0$ such that

$$\forall x, \; y \in R^d \qquad m(x) \leq C_0 m(y)(1 + |y - x|)^{N_0} \quad (45)$$

◇

Definition 3 If Ω is open in R^d, $\rho \in [0, 1]$ and m is a tempered weight, we denote *symbol of weight* (m, ρ) in Ω a function $a \in C^\infty(\Omega)$ such that

$$\forall x \in \Omega \qquad |\partial^\alpha a(x)| \leq C_\alpha \cdot m(x)(1 + |x|)^{-\rho|\alpha|} \quad (46)$$

We shall denote by $\Sigma_{m,\rho}$ the space of symbols of weight (m, ρ); in particular $\Sigma_\rho \equiv \Sigma_{\iota,\rho}$ where ι is the function identically equal to one. ◇

With these notations one can prove (following the lines of the proof of the Theorem of Calderon–Vaillantcourt).

Theorem 5 *(1) If $a \in \Sigma_{\iota,0}$, there exists $T(d) \in R$ such that*

$$\|Op^w(a)\|_{Tr} \leq T(d) \sum_{|\alpha|+|\beta|\leq d+2} \int\int |\partial_x^\alpha \partial_\eta^\beta a(x,\eta)| dx d\eta \tag{47}$$

(α and β are multi-indices).
(2) If $a \in \Sigma_{m,0}$ and $lim_{|x|+|\eta|\to\infty} a(x,\eta) = 0$ then the closure of $Op^w(a)$ is a compact operator on $L^2(R^d)$. ◇

The proof is obtained exploiting the duality with Wigner's functions taking into account that both trace class operators and Hilbert–Schmidt operators are sum of one-dimensional projection operators and the eigenvalues converge respectively in l^1 and l^2 norm, and that a compact operator is norm-limit of Hilbert–Schmidt operators.

A *more stringent* condition which is easier to prove (making use of the duality with Wigner's functions) and provides an estimate of the trace norm is given by the following theorem

Theorem 6 *Let $a \in \Sigma_{\iota,0}$ be such that for all multi-indices α, β*

$$\partial_x^\alpha \partial_\eta^\beta a \in L^1(R^{2d}) \tag{48}$$

Then $Op^w(a)$ is trace class and one has

$$trOp^w(a) = \int\int a(x,\eta) dx d\eta \tag{49}$$

◇

Since Hilbert–Schmidt operators form a Hilbert space, it is easier to verify convergence and then to find conditions on the symbol such that the resulting operator be of Hilbert–Schmidt class. A first result is the following

Theorem 7 *Let $a \in \Sigma_{m,0}$, $b \in \mathcal{S}(R^{2n})$. Then*

$$tr[Op^w(a) \cdot Op^w(b)] = \int\int a(x,\xi) b(x,\xi) dx d\xi \tag{50}$$

◇

Proof If $B \equiv Op^w(b)$ is a rank one operator $B = \psi \otimes \phi \quad \psi, \ \phi \in \mathcal{S}$ one has

$$tr(A \cdot B) = (\phi, A\psi) \qquad A = Op^w(a) \tag{51}$$

From the definition of $Op^w(a)$ it follows

$$(\phi, A\psi) = \int \int a(x,p) \left[\int e^{ip\zeta} \bar{\phi}(x + \frac{\zeta}{2}) \psi(x - \frac{\zeta}{2}) d\zeta \right] dxdp \tag{52}$$

and (50) is proven in this particular case.

The proof is the same if B has finite rank, and using the regularity of $a(x,p)$, $b(x \cdot p)$ one achieves the proof of (50).

♡

The bilinear form

$$A,\ B \to Tr(A^* B) \equiv < A, B > \tag{53}$$

can be extended to

$$\mathcal{L}(\mathcal{S}(R^{2d}), \mathcal{S}'(R^{2d})) \times \mathcal{L}(\mathcal{S}'(R^{2d}), \mathcal{S}(R^{2d})) \tag{54}$$

with the property $< A, B > = < B, A >^*$.

This duality can be used to extend the definition of the symbol $\sigma^w(a)$ to an operator-valued tempered distribution A by

$$Tr(A \cdot Op^w(b)) = 2\pi^{-n}(b, \sigma^w(a)) \tag{55}$$

and the duality can extended to symbols belonging to Sobolev classes dual with respect to $L^2(R^d)$. Remark that one has

$$a \in L^2(R^{2d}) \leftrightarrow Op^w(a) \in H.S. \tag{56}$$

but $|a|_\infty < \infty$ does not imply that $a(D,x)$ be bounded.

For example if $a(\xi, x) = e^{i(\xi, x)}$ one has $(a(D,x))f(x) = \int f(y)dy \cdot \delta(x)$.

It is convenient to introduce a further definition.

Definition 4 (*Class O(M)*) A function a on $C^d \equiv R^{2d}$ belongs to $O(M)$ if and only if $f \in C^\infty(R^d)$ and for every multi-index $m\ :\ |m| = M$ one has

$$|\frac{\partial^m}{\partial z^m} a(z)| \le C|z^M|, \qquad \forall z \in C^d \tag{57}$$

We shall denote by Σ_M the collection of functions in $O(M)$.

◇

Following the lines of the proof of Theorem 7 one proves

Theorem 8 *(i) If $a \in O(0)$, then $Op^w(a)$is a bounded operator*
(ii) If $a \in O(M)$, $M \leq -2d$ then $Op^w(a)$ is trace-class and

$$TrOp^w(a) = (2\pi\,\hbar)^{-d} \int |A(z)|\, dz \tag{58}$$

(iii) If $a \in O(M)$ is real, then $Op^w(a)$ is essentially self-adjoint on $C_0^\infty(R^d)$.

◇

5 Product of Operator Versus Products of Symbols

The next step is to establish the correspondence between the product of symbols and the product of the corresponding operators. We can inquire e.g. whether, given two symbols a e b, there exists a symbol c such that $Op^w(c) = Op^w(a) \cdot Op^w(b)$.

The answer is in general no. To obtain a (partially) positive answer it will be necessary to *enlarge the class of symbols considered* and add symbols that depend *explicitly* on the small parameter $\hbar$.

In their dependence on $\hbar$ they must admit an expansion to an order M such that the remainder has the regularity properties that imply that the corresponding operator is a bounded operator with suitable estimates for its norm.

This possibility to control the residual term is *an important advantage* of the (strict) quantization with pseudo-differential operators as compared to formal power series quantization. We limit ourselves to consider pseudo-differential operators with symbols in $O(M)$.

Definition 5 ($\hbar$ *admissible symbol*) A $\hbar-$ admissible symbol of weight M is a C^∞ map from $\hbar \in]0, \hbar_0]$ to Σ_M such that there exists a collection of functions $a_j(z) \in O(M)$ with the property that, for every integer N and for every multi-index γ with $|\gamma| = N$ there exists a constant C_N such that

$$sup_z \left[\left(\frac{1}{1+|z|^2} \right)^{d/2} |\frac{\partial^\gamma}{\partial z^\gamma} a(z,\ \hbar) - \sum_1^N \hbar^j a_j(z)| < c_N \hbar^{N+1} \right. \tag{59}$$

◇

Definition 6 ($\hbar$-*admissible operator*) An $\hbar$-admissible operator of weight M is a C^∞ map

$$A_\hbar \ : \hbar \in]0, h_0 \Rightarrow \mathcal{L}(\mathcal{S}(R^d), L^2(R^d)) \tag{60}$$

for which there exists a sequence of symbols $a_j \in \Sigma_M$ and a sequence $R_N \in \mathcal{L}(L^2(R^d))$ such that for all $\phi \in \mathcal{S}$

$$A_\hbar = \sum_1^N h^j Op_h^w a_j + R_N(\hbar), \quad sup_{0<\hbar\leq\hbar_0} |R_N(h)\phi|_2 < \infty \quad \forall \phi \in L^2(R^d) \tag{61}$$

The function $a_0(z)$ is called *principal symbol* of the $\hbar$-admissible operator $A_\hbar$; it will be denoted $\sigma_P(A_\hbar)$.

The function $a_1(z)$ is called *sub-principal symbol* of the $\hbar$-admissible operator $A_\hbar$; it will be denoted $\sigma_{SP}(A_\hbar)$.

◇

Definition 7 (*class $\hat{O}^{s.c.}$ operators*) We shall denote $\hat{O}^{sc}_M$ the set in $\mathcal{L}(\mathcal{S}(X))$ (the collection of all bounded operators in $\mathcal{S}(X)$) that is obtained associating to each function in Σ_M the operator obtained by Weyl quantization.

This class of operators is sometimes called $\hbar$-admissible.

◇

The following theorem states that the $\hbar$-admissible operators form an algebra:

Theorem 9 *For any pair $a \in O(M)$ and $b \in O(P)$ there exists unique a semiclassical observable $\hat{C} \in O^{sc}_{M+P}$ such that*

$$Op^w(a) \cdot Op^w(b) = \hat{C} \tag{62}$$

The semiclassical observable has the representation

$$\hat{C} = \sum \hbar^j Op^w(c_j) \tag{63}$$

$$c_j = 2^{-j} \sum_{|\alpha|+|\beta|=j} \frac{(1)^\beta}{\alpha!\ \beta!} (D^\beta_x\ D^\alpha_\xi a)\ (D^\beta_x\ D^\alpha_\xi b)(x, \xi) \tag{64}$$

Moreover

$$\frac{i}{\hbar}[Op^w(a), Op^w(b)] \in \hat{O}_{sc}(M + P) \tag{65}$$

with principal symbol the Poisson bracket $\{a, b\}$.

◇

Sketch of the Proof

The proof follow the same lines as the proof of the Theorem of Calderon–Vaillantcourt and makes use of the definition of pseudo-differential operator, the duality with Wigner's functions and the explicit form of the phase factor in Weyl product.

Notice that the structure of Weyl algebra implies that if L is a linear form on R^{2d} and $a \in O(M)$ one has for any linear operator L

$$L(x, \hbar\nabla)Op^w_\hbar(a) = Op^w_\hbar(b), \qquad b = L \cdot a + \frac{\hbar}{2i}\{L, a\} \tag{66}$$

where $\{.,.\}$ denotes Poisson brackets.

This remark is useful to write in a more convenient form the product of the phase factors that enter in the definition of the product $Op^w(a) \cdot Op^w(a_1)$.

Recall that by definition

$$(Op^w_\hbar a\phi)(x) = \hbar^{-n} \int \int a(y,z) e^{\frac{2\pi i}{\hbar}(\frac{x+z}{2},\xi)} \phi(y) dy d\xi \quad (67)$$

and that $Op^w_\hbar(b)$ is given by a similar expression.

The integral kernel $K_{Op^w_\hbar(a).Op^w_\hbar(b)}$ of the operator $Op^w_\hbar(a),\ Op^w_\hbar(b)$ is then given by

$$K_{Op^w_\hbar(a)\cdot Op^w_\hbar(b}(x,y)$$
$$= \hbar^{2n} \int \int e^{\frac{i}{\hbar}[(x-z,\xi)+(z-y,\eta)]} a\left(\frac{1}{2}(x+z),\xi\right) b\left(\frac{1}{2}(y+z),\eta)\right) dz d\xi d\eta \quad (68)$$

In general there is no symbol c such that $C_\hbar \equiv Op^w_\hbar(a) Op^w_\hbar(b) = Op^w(c)$

One can verify, making use of (67), that if $a_\hbar,\ b_\hbar \in O(M)$ the operator $C_\hbar$ is $\hbar$-admissible, i.e. for some $N \in \mathcal{Z}$ it can be written as

$$C_\hbar = \sum_{n=0,\dots N} c_n(\hbar) + \hbar^{N+1} R_{N+1}(h) \quad (69)$$

where $a_n \in O(M)$ and $sup_{h\in[0,\hbar_0]} \|R_{N+1}(h)\|_{\mathcal{L}(L^2(R^d))}$.

♡

With the new definition Theorem 9 *can be extended to all semiclassical observables.*

For each $A \in O^{sc}(M),\ B \in O^{sc}(N)$ there exists *a unique semiclassical observable* $C \in \hat{O}^{sc}(N{+}M)$ such $\hat{A} \cdot \hat{B} = \hat{C}$.

Moreover the usual composition and inversion rules apply.

6 Correspondence Between Commutators and Poisson Brackets; Time Evolution

From the analysis given above one derives the following relations.

Let $A_\hbar$ and $B_\hbar$ be two $\hbar$-admissible operators and denote by $\sigma_P(A)$ the principal symbol of A and by $\sigma_{SP}(A)$ its sub-principal symbol. Then

(1)

$$\sigma_P(A_\hbar \cdot B_\hbar) = \sigma_P(A_\hbar) \cdot \sigma_P(B_\hbar) \quad (70)$$

(2)

$$\sigma_{SP}(A_\hbar \cdot B_\hbar) = \sigma_P(A_\hbar) \cdot \sigma_P(B_\hbar) + \sigma_{SP}(A_\hbar) \cdot \sigma_P(B_\hbar) + \frac{\hbar}{2i}\{\sigma_P(A_\hbar) \cdot \sigma_P(B_\hbar)\} \quad (71)$$

These relations give the correspondence between the commutator of two quantum variables and the Poisson brackets of the corresponding classical variables.

The introduction of semiclassical observables is also useful in the study of time evolution. One has [2–4].

Theorem 10 *Let $H \in O^{sc}(2)$ be a classical hamiltonian satisfying*

$$|\partial_z^\gamma H_j(z)| < c_\gamma, \qquad \gamma + j \geq 2 \tag{72}$$

$$\hbar^{-2}(H - H_0 - \hbar H_1) \in \hat{O}_{sc}(0) \tag{73}$$

Let $a \in O(m)$, $m \in \mathcal{Z}$. Then
(i) For any sufficiently small value of $\hbar$, $\hat{H}$ is essentially selfadjoint with natural domain $S(X)$. Therefore $exp\{-i\hbar^{-1}\hat{H}t\}$ is well defined and unitary for each value of t and continuous in t in the strong topology.
(ii)

$$\forall t \in R, \quad Op^w(a(t)) \equiv e^{i\frac{t}{\hbar}\hat{H}} Op^w(a) e^{-i\frac{t}{\hbar}\hat{H}} \in \hat{O}_{sc}(m) \tag{74}$$

Moreover

$$a(t) = \sum_{k \geq 0} \hbar^k a_k(t) \quad a_k(t) \in \quad O_{sc}(m) \tag{75}$$

uniformly over compacts. ◇

Proof (outline) Under the conditions stated the classical flow $z \to z(t)$ exists globally. From the properties of the tangent flow it easy to deduce that $a(z(t)) \in O(m)$ uniformly over compacts in t.

With $U_H(t) = exp\{-i\frac{t}{\hbar}\hat{H}\}$ Heisenberg equations give

$$\begin{aligned} &\frac{d}{ds}(U_H(-s)Op^w(a(z(t-s)))U_H(s) \\ &\quad = U_H(-s)(\frac{i}{\hbar}[H, Op^w(a(z(t-s)))] - Op^w(\{H, a_0\})])U_H(s) \end{aligned} \tag{76}$$

From the product rule one derives then that the principal symbol of

$$\frac{i}{\hbar}[H, Op^w a_0(t-s)] - Op^w(\{H_0, a_0(t-s)\}) \tag{77}$$

vanishes. Therefore the right hand side of (76) is of order one in $\hbar$ and the thesis of the follows by a formal iteration as an expansion in $\hbar$ through Duhamel series.

Using the estimates one proves convergence of the series.

♡

Remark that if H is a polynomial at most of second order in $x, i\frac{d}{dx}$ one has

$$(Op^w(a))(t) = Op^w(a(z(t))) \tag{78}$$

where ψ_H^t is the classical solution of Hamilton's equations. Indeed in this case one has

$$\frac{i}{\hbar}[\hat{H}, Op(b)] = Op(h, b) \tag{79}$$

In particular if $W(z)$ is an element of Weyl's algebra

$$W(z)Op(b)W(-z) = Op(b_z) \qquad b_z(z') = b(z' - z) \tag{80}$$

(this is a corollary of Eherenfest theorem). Relation (80) does not hold in general if H is not a polynomial of order ≤ 2.

Still, under the assumptions of Theorem 10 a relation of type (80) holds in the limit $\hbar \to 0$ in a weak sense, i.e. as an identity for the matrix elements between *semiclassical states* (e.g. coherent states). We have remarked this in our analysis of the semiclassical limit in volume I of these Lecture Notes.

Theorem 10 can be extended to Hamiltonians which are not in $O(2)$ (for example to Hamiltonians of type $H = \frac{p^2}{2} + V(q)$ with V bounded below) if the classical hamiltonian flow is defined globally in time.

Weyl quantization can be extended to distributions in $\mathcal{S}'$; in this case the operator $\hat{A}$ is bounded from $\mathcal{S}$ to $\mathcal{S}'$ and the correspondence it induces is a bijection.

This follows from an analogue of a Theorem of L.Schwartz which states that every bilinear map from $\mathcal{S}(X)$ to $L^2(X)$ continuous in the $L^2(X)$ topology can be extended as a continuous map from $\mathcal{S}(X)$ to $\mathcal{S}'(X)$.

A way to achieve this extension exploits the properties of Weyl symbol $Op^W(\Pi_{u,v})$ of the rank-one operator $\Pi_{u,v}$ defined, for $u, \ v \in \mathcal{S}(X)$, by

$$\Pi_{u,v}\psi = (\psi, u)v \tag{81}$$

One has then

$$(Op^w(a)u, v) = (2\pi\hbar)^{-1} \int a(x, \xi)\pi_{u,v}(x, \xi)dxd\xi \tag{82}$$

since by definition

$$< Op^w(a)u, v >= Tr(\Pi_{u,v}Op^w(a)) = \int \Pi_{u,v}(x, \xi)A(x, \ \xi)dx\, d\xi \tag{83}$$

The function $\pi_{u,v}(x, \xi)$ is the Wigner function of the pair $u, \ v$. Remark that

$$(Op^w(a)u, v) = (2\pi\ \hbar)^{-n} \int a(z)\pi_{u,v}(z)dz \tag{84}$$

The definition of pseudo-differential operator on a Hilbert space $\mathcal{H}$ can be extended to the case in which the symbol $a(q, p)$ is itself an operator on a Hilbert space $\mathcal{K}$.

A typical case, occasionally used in information theory, is the one in which the *phase space* is substituted with the (linear) space of the (Hilbert) space of Hilbert–Schmidt operators with the commutator as symplectic form.

This linear space is itself a Hilbert space with scalar product $< A, B >= Tr(A^B)$ (in information theory the Hilbert space $\mathcal{K}$ on which the Hilbert–Schmidt operators act is usually chosen to be finite-dimensional).

Another case, which has gained relevance in the Mathematics of Solid State Physics, is the treatment of adiabatic perturbation theory *through the Weyl formalism* [10].

This procedure is useful e.g. in the study of the dynamics of the atoms in crystals but also in the study of a system composed of N nuclei of mass m_N with charge Z and of NZ electrons of mass m_e.

In the latter case one chooses the ratio $\epsilon \equiv \frac{m_e}{m_N}$ as small parameter in a multi-scale approach. We shall come back to this problem in Lecture 5.

7 Berezin Quantization

A quantization which associates to a positive function a positive operator in the Berezin quantization defined by means of *coherent states* i.e. substituting the Wigner function with its Husimi transform.

This quantization does not preserve polynomial relations, the product rules are more complicated than in Weyl quantization and the equivalent to Eherenfest theorem does not hold.

Recall that a coherent state "centered in the point" $(y,\ \eta)$ of phase space is by definition

$$\phi_{y,\eta} \equiv e^{\frac{i}{\hbar}(\eta,x)+i(y\,D_x)}\phi_0(x) \tag{85}$$

where $\phi_0(x)$ is the ground state of the harmonic oscillator for a system with d degrees of freedom.

$$\phi_0 \equiv (\pi\ \hbar)^{-n} e^{-\frac{|x|^2}{2\hbar}} \tag{86}$$

Definition 8 The *Berezin quantization* of the classic observable a is the map $a \to Op^B(a)$ given by

$$Op^B_\hbar(a)\phi \equiv (2\pi\ \hbar)^{-d} \int\int a(y,\eta)(\psi, \bar{\phi}_{y,\eta})\phi_{y,\eta} dy\, d\eta \tag{87}$$

◇

One can prove, either directly or through its relation with Weyl quantization to construct $Op^B_\hbar(a)$, that the Berezin quantization has the following properties:
(1) If $a \geq 0$ then $Op^B_\hbar(a) \geq 0$
(2) The Weyl symbol a^B of the operator $Op^B_\hbar(a)$ is

$$a^B(x, \xi) = (\pi \hbar)^{-d} \int \int a(y, \eta) e^{-\frac{1}{\hbar}[(x-y)^2+(\xi-\eta)^2]} dy \, d\eta \tag{88}$$

(3) For every $a \in O(0)$ (bounded with all its derivatives) one has

$$||Op^B_\hbar(a) - Op^w_\hbar(a)|| = O(\hbar) \tag{89}$$

We have noticed that the Berezin quantization is dual to the operation that associates to a vector ψ in the Hilbert space $\mathcal{H}$ a *positive measure* μ_ψ in phase space, the *Husimi measure*.

On the contrary Weyl's quantization is dual to the operation which associates to ψ the Wigner function W_ψ, which is real but not positive in general.

We recall the

Definition 9 (*Husimi measure*) The Husimi's measure μ_ϕ associated to the vector ϕ is defined by

$$d\mu_\psi = \tilde{\rho}(q, p) dq \, dp \qquad \tilde{\rho}(q, p) \equiv |(\phi_{q,p}, \psi)|^2 \tag{90}$$

From this one derives that Husimi measure is a positive Radon measure. Its relation with Berezin quantization is given by

$$\int a \, d\mu_\psi = (Op^B_\hbar(a)\psi, \psi), \quad a \in \mathcal{S} \tag{91}$$

◇

Although it gives *a map between positive functions and positive operators* the Berezin quantization is less suitable for a description of the evolution of quantum observables. In particular Eherenfest's and Egorov's theorems do no hold and the semiclassical propagation theorem has a more complicated form.

The same is true for the formula that gives the Berezin symbol of an operator which is the product of two operators $Op^B_\hbar(a) Op^B_\hbar(b)$ where $a, \ b$ are functions on phase space.

Berezin representation is connected the Bargman–Segal representation of the Weyl system (in the same way as Weyl representation has its origin in the Weyl–Schroedinger representation).

Recall that the Bargman–Segal representation is set in the space of function over C^d which are holomorphic in the sector $Im z_k \geq 0, \ k = 1, \ldots, d$ and square integrable with respect to the gaussian probability measure

$$d\mu_r(z) = \left(\frac{r}{\pi}\right)^d e^{-r|z|^2} dz, \quad r > 0 \tag{92}$$

We shall denote this space $\mathcal{H}_r$. In the formulation of the semiclassical limit the parameter r plays the role h^{-1}.

8 Toeplitz Operators

In the Berezin representation an important is played by the *Toeplitz operators*.

For $g \in L^2(d\mu_r)$ the Toeplitz operator $T_g^{(r)}$ is defined on a dense subspace of $\mathcal{H}_r$ by

$$(T_g^{(r)} f)(z) = \int g(w) f(w) e^{i(z.w)} d\mu_r(w) \tag{93}$$

In part I of these Lecture Notes we introduced the *reproducing kernel* $e^{i(z.w)}$ within the discussion of the Bargman–Segal representation.

Remark that if $g\, f \in L^2(d\mu_r)$ then $T_g^{(r)} f \in \mathcal{H}_r$. The map $g \to T_g^{(r)}$ (Berezin quantization) is a complete strict deformation (the deformation parameter is r^{-1}).

Under the Bargman–Segal isometry $B_r : L^2(R^n, dx) \to \mathcal{H}_r$ the Weyl–Schroedinger representation is mapped onto the Bargman–Segal complex representation and the quantized operators $\hat{z}_k$ are mapped into Toeplitz operators.

For these Toeplitz operators are valid the same "deformation estimates" which hold in Berezin quantization (and are useful when studying the semiclassical limit)

$$\| T_f^{(r)} T_g^{(r)} - T_{fg}^{(r)} + \frac{1}{r} T^{(r)}_{\sum_j (\frac{\partial f}{\partial z_j} \frac{\partial g}{\partial \bar{z}_j})} \|_r \le C(f, g) r^{-2} \tag{94}$$

The interested reader can consult [6, 7] for the Berezin quantization and its relation with Toeplitz operators.

Let us remark that Berezin quantization is rarely used in non relativistic Quantum Mechanics to describe the dynamics of particles; as mentioned, Eherenfest's and Egorov's Theorems do not hold and the formulation of theorems about semiclassical evolution is less simple.

On the contrary Berezin quantization is much used in Relativistic Quantum Field Theory (under the name of *Wick quantization*) since it leads naturally to the definition of *vacuum state* Ω (in the case of finite number of degrees of freedom in the Schrödinger representation it is constant function) as the state which is *annihilated* by $\frac{\partial}{\partial z_k}$, $\forall k$) and of the $\sum_k z_k \partial z_k$ (*number operator*).

In the infinite dimensional case it is useful to choose as *vacuum* a gaussian state or equivalently use instead of the Lebesgue measure a Gaussian measure (which is defined in R^∞)

In turn this permits the definition of *normal ordered polynomials* (or *Wick ordered polynomials*) in the variables z_h, $\frac{\partial}{\partial z_k}$; the normal order is defined by the prescription that all the operators $\frac{\partial}{\partial z_k}$ stand to the right of all operators z_k.

The Berezin quantization is much used in Quantum Optics where the coherent states play a dominant role (coherent states are in some way "classical states" of the quantized electromagnetic field).

9 Kohn–Nirenberg Quantization

For completeness we describe briefly the quantization prescription of Kohn-Nirenberg [7], often introduced in the study of inhomogeneous elliptic equation and of the regularity of their solutions. It is seldom used in Quantum Mechanics.

By definition

$$(\sigma_{K.N.}(D,x)f)(x) \equiv \int \sigma(\xi,x)e^{i(x-y)}f(y)dyd\xi \times \int\int \hat{\sigma}(p,q)(e^{i(q,x)}e^{i(p,D)}f)(x)dp\,dq \tag{95}$$

In the particular case $\sigma(\xi,x) = \sum a_k(x)\xi^k$ one has

$$\sigma_{K.N}(D,x) = \sum_k a_k(x)D^k \tag{96}$$

In the Kohn–Nirenberg quantization the relation between an operator and its *symbol* is

$$Op^{KN}(a)\phi(x) = \left(\frac{1}{2\pi\hbar}\right)^{\frac{d}{2}} \int_{R^d} e^{\frac{i}{\hbar}px}a^{K,N}(x,p)\hat{\phi}(p)dp \tag{97}$$

where we have indicated with $\hat{\phi}$ the Fourier transform of ϕ.

This is the definition of pseudo-differential operator that is found in most books of Partial Differential Equations. In this theory one proves that the pseudo-differential operators are singled out by the fact that they satisfy the weak maximum principle.

In general if the K.N. symbol $a^{K,N}(q,p)$ is real the operator $Op^{K,N}(a)$ is not (essentially) self-adjoint. The quantization of Kohn–Nirenberg is usually employed in micro-local analysis and also in the time-frequency analysis since in these fields it leads to simpler formulations [2, 9].

In general this quantization is most useful when considering equations in which the differential operators appear as low order polynomials (usually second or fourth); in this case it is not interesting to study operators of the form $L(x,\nabla)$ for a generic smooth function L.

If the K.N. operators do not depend polynomially in the differential operators, their reduction to spectral subspaces is not easy. For this reason Weyl quantization is preferred in solid state physics when one wants to analyze operators which refer to *Bloch bands*.

10 Shubin Quantization

The quantizations of Weyl and of Berezin are particular cases of a more general form of quantization, parametrized by a parameter $\tau \in [0, 1]$, as pointed out by Shubin [4].

In this more general form to the function $a \in \mathcal{S}(R^{2d})$ one associates the continuous family of operators $Op^{S,\tau}(a)$ on $\mathcal{S}(R^d)$ defined by

$$(Op^{S,\tau}(a)\phi)(x) = \left(\frac{1}{2\pi\hbar}\right)^d \int\int a((\tau x + (1-\tau)y), \xi) f(y) e^{\frac{i}{\hbar}(x-y,\xi)} dy d\xi \quad (98)$$

It is easy to verify that the choice $\tau = \frac{1}{2}$ corresponds to Weyl's quantization, $\tau = 0$ to Kohn–Niremberg's and $\tau = 1$ to Berezin's. Notice that Eherenfest theorem holds only for $\tau = \frac{1}{2}$.

It is easy to verify that only for $\tau = \frac{1}{2}$ the relation between the operator and its symbol is *covariant* under linear symplectic transformations. In general if $s \in Sp(2d, R)$ is a linear symplectic transformation, there exists a unitary operator S such that

$$S^{-1}(s) Op^w(a) S(s) = Op^w(a \circ s) \quad (99)$$

$S(s)$ belongs to a representation of the metaplectic group generated by quadratic form in the canonical variables.

For all values of the parameter τ one has

$$\mathcal{F} Op^{S,\tau} \mathcal{F}^{-1} = Op^{S,1-\tau}(a \circ J^{-1}) \quad (100)$$

where J is the standard symplectic matrix and $\mathcal{F}$ denoted Fourier transform.

One can consider also *Wigner functions* associated to Shubin's τ-quantization. In particular

$$W_\tau(\phi, \psi)(x \cdot p) = \left(\frac{1}{2\pi\hbar}\right)^d \int_{R^d} e^{-\frac{i}{\hbar}py} \phi(x + \tau y)\psi(x - (1-\tau)y) dy \quad (101)$$

Independently of the value of the parameter τ one has

$$\int_{R^d} W_\tau(x, p) dp = |\phi(x)|^2, \qquad \int_{R^d} W_\tau(x, p) dx = |\hat{\phi}(p)|^2 \quad (102)$$

The relation between W_τ and $Op_\tau(a)$ is

$$(Op_\tau(a)\psi, \phi)_{L^2} = (a, W_\tau(\psi, \phi)) \quad (103)$$

For all τ

$$\Pi_\phi(x, y)) = [2\pi\hbar]^{-d} W_\tau \phi(x, p) \tag{104}$$

where Π_ϕ is the projection operator on the vector ϕ.

11 Born–Jordarn Quantization

We end this lecture with the quantization introduced by Born and Jordan [11] to give a prescription for associating operators to functions over classical phase space of the form $\sum_k f_k(x)P_k(p)$ where $f_k(x),\ x \in R^d$ are sufficiently regular function and $P_k(p)$ are polynomials in the momenta $\{p_j\}$.

Notice that all Hamiltonians introduced in non relativistic Quantum Mechanics have this structure. The correspondence proposed by Born and Jordan is

$$f(x)p_j^n \to \frac{1}{n+1}\sum_{k=0}^{n} \hat{p}_j^{n-k} f(\hat{(x)})\hat{p}_j^k \tag{105}$$

where $\hat{x}_j$ (in the Schroedinger representation) is multiplication by x_j and $\hat{p}_j = -i\hbar\frac{\partial}{\partial x_j.}$

For comparison, Weyl quantization corresponds to the prescription

$$f(x)p_j^k \to \frac{1}{2^k}\sum_{m=0}^{k} \frac{m!}{m!(k-m)!}\hat{p}_j^{n-m} f(\hat{x})\hat{p}_j^m \tag{106}$$

One has

$$Op^{BJ}(a) = \left(\frac{1}{2\pi}\right)^d \int Op_\tau(a)d\tau \tag{107}$$

Weyl's prescription *coincides with that of Born and Jordan* if the monomial is of rank at most two in the momentum.

Therefore the quantization of Born and Jordan coincides with Weyl's for Hamiltonians that are of polynomial type in the position coordinates (the Hamiltonians that are of common use in Quantum Mechanics).

Also for the quantization of the magnetic hamiltonian the B–J quantization coincides with the Weyl quantization.

One can verify that the symbol of $a_{B,J}$ of operator A in Born–Jordan quantization is given by

$$a_W = \left(\frac{1}{2\pi}\right)^d a * \mathcal{F}_\sigma \Theta \tag{108}$$

where $\mathcal{F}_\sigma$ is the symplectic Fourier transform and the function Θ is given by

$$\Theta(z) = \frac{sin\frac{px}{\hbar}}{\frac{px}{\hbar}} \tag{109}$$

This implies that the symbols a_w and $a_{B,J}$ are related by

$$a_w = \left(\frac{1}{2\pi}\right)^d a_{B,J} * \mathcal{F}_\sigma \Theta \tag{110}$$

therefore $a_{B,J}$ is not determined by a_w.

Through (110) one can define the equivalent of the Wigner function (Born–Jordan functions) in phase space. They are not positive but the negative part for elementary elements is somewhat reduced.

The quantization of Born and Jordan is related to the Shubin quantization by the formula

$$Op^{BJ}(a)\phi = \left(\frac{1}{2\pi\hbar}\right)^d \int_0^1 Op^{S,\tau}(a)\phi d\tau \tag{111}$$

on a suitable domain (in general the operator one obtains is unbounded). From the relation between $Op^{S,\tau}(a)$ and $Op_{1-\tau}(\bar{a})$ one derives

$$Op^{BJ}(a)^* = Op^{BJ}(\bar{a}) \tag{112}$$

Therefore the operator $Op^{BJ}(a)$ is formally self-adjoint if and only if a is a real function.

The relation between a *magnetic Born–Jordan quantization* and the quantization given by the magnetic Weyl algebra is still unexplored.

References

1. D. Robert, *Autour de l'approximation semiclassique* (Birkhauser, Boston, 1987)
2. A. Grigis, J. Sjostrand, *Microlocal Analysis for Differential Operators*, vol. 196, London Mathematical Society Lecture Note (Cambdrige University Press, Cambdrige, 1994)
3. A. Martinez, *An Introduction to Semiclassical and Microlocal Analysis*, Universitext (Springer, New York, 2002)
4. M. Shubin, *Pseudo-differential Operator and Spectral Theory* (Springer, Berlin, 2001)
5. L. Hörmander, Comm. Pure Appl. Math. **32**, 359–443 (1979)
6. C.A. Berger, A. Coburn, Trans. Am. Math. Soc. **301**, 813–829 (1987)
7. J.J. Kohn, L. Nirenberg, Comm. Pure Appl. Math. **18**, 269–305 (1965)
8. G. Gerard, A. Martinez e, J. Sjostrand, Comm. Math. Phys. **142**, 217–244 (1991)
9. P. Markovic, N. Mauser, Math. Models Methods Appl. Sci. **3**, 109–124 (1993)
10. S. Teufel, *Adiabatic Perturbation Theory in Quantum Dynamics*, vol. 1821, Lecture Notes in Mathematics (Springer, Berlin, 2003)
11. M. Born, P. Jordan, Zeits. für Physik **34**, 128–139 (1925)

Lecture 3: Compact and Schatten Class Operators. Compactness Criteria. Bouquet of Inequalities

Compactness is a property that is very frequently used in the theory of Schroedinger operators. For example, as we shall see, in scattering theory compactness of the resolvent operator plays an important role. In this chapter we shall give a collection of definitions and results that pertain to the problem of compactness and some useful inequalities.

Definition 1 (*Compact Operator*) A closable operator A on a Hilbert space $\mathcal{H}$ is *compact* if the set $\{A\phi \;\phi \in D(A),\;\; |\phi| = 1\}$ is pre-compact in H (i.e. its closure is compact). ◇

Recall that a closed subset Y of a topological space X is compact if from any bounded sequence in Y one can extract a convergent subsequence. The unit ball in $\mathcal{H}$ is compact in the weak topology. It follows that A is compact iff for any sequence $\{\phi_n\}$ which converges weakly in $\mathcal{H}$ the sequence $\{A\;\phi_n\}$ converges strongly.

From the definition one derives that the set of compact operators is closed in the uniform topology and that it is a bilateral ideal in $B(\mathcal{H})$. One proves easily that if A is compact also A^* is compact.

Definition 2 (*Finite Rank Operator*) An operator is *of finite rank* if its range is finite-dimensional, i.e. there exist $N < \infty$ vectors ϕ_n in $\mathcal{H}$ and N linear functionals γ_n such that $A\psi = \sum_1^N \gamma_n(\psi)\phi_n$ for any $\psi \in H$. ◇

Since any closed set in R^N is compact, every finite rank operator is compact.

Theorem 1 *Every compact operator is norm-limit of finite rank operators.* ◇

Proof Let $\mathcal{H}$ a separable infinite-dimensional Hilbert space (the proof in the non-separable case is slightly more elaborated, makes use of Zorn's lemma and of the fact that the norm topology is separable).

Let $\{\phi_n\}$ be an orthonormal basis in $\mathcal{H}$ and denote by $\mathcal{H}_N$ the subspace spanned by $\{\phi_k,\;\; k = 1, \ldots N\}$. Define

G. Dell'Antonio, *Lectures on the Mathematics of Quantum Mechanics II: Selected Topics*, Atlantis Studies in Mathematical Physics: Theory and Applications 2, DOI 10.2991/978-94-6239-115-4_3

$$\lambda_N \equiv sup_{\{\phi \in \mathcal{H}_N^{\perp}, |\phi|=1\}} |A\ \phi| \tag{1}$$

The numerical sequence λ_N is monotone decreasing; let λ be the limit. By construction $\lambda_N = |A - A_N|$ where A_N is the restriction of A to $\mathcal{H}_N$. The theorem is proved if $\lambda = 0$.

Suppose that $\lambda > 0$; then $|A\phi| \geq \lambda|\phi|$ and the image under A of the unit ball contains a ball of finite radius.

This contradicts the fact that A is compact. ♡

A particularly useful result is

Theorem 2 *The self-adjoint operator A is compact iff its spectrum is pure point, the eigenvalues different from zero have finite multiplicity and zero is the only possible accumulation point.* ◇

Proof If σ_{cont} is not empty it contains an interval $I \equiv (\lambda_0 - \epsilon, \lambda_0 + \epsilon)$. Without loss of generality we can assume $\lambda_0 = 0$.

Denote with Π the orthogonal projection one the subspace associated to the continuous spectrum in I; by Weyl's lemma this subspace has infinite dimension.

By construction if $\phi \in \Pi\ \mathcal{H}$ then $|A\ \phi| \geq \epsilon|\phi|$. Therefore the image under A of the unit ball in $\mathcal{H}$ contains a finite ball in a subspace of infinite dimension and cannot be compact.

In the same way one proves that the eigenvalues different from zero have finite multiplicity. ♡

Let $\mathcal{H}$ be a separable Hilbert space and $\{\phi_n\}$ an ortho-normal complete basis. Let A be a positive operator.

Set

$$Tr(A) \equiv lim_{N\to\infty} \sum_{n=1}^{N} (\phi_n, A\phi_n) \tag{2}$$

The sequence is not decreasing and therefore the limit exists (may be $+\infty$). One easily verifies that the function Tr (for the moment defined only for positive operators) is invariant under unitary transformations.

If $U(t)$ is a one-parameter group of unitary operators, $\frac{d}{dt} Tr_{\Phi}\ U(t)AU(t)_{t=0} = 0$. Choosing as basis the eigenvectors of A one has $Tr(A) = \sum_{n=1}^{\infty} a_n$ where a_n are the eigenvalues.

Definition 3 The operator $A \in \mathcal{B}(\mathcal{H})$ is of class *Hilbert–Schmidt* if there exist an orthonormal basis $\{\phi_n\}$, $n \in Z$ in $\mathcal{H}$ such that

$$\sum_{n\geq 1} \|A\phi_n\|^2 < +\infty \tag{3}$$

One proves that this condition is independent of the basis chosen. ◇

An equivalent definition is

Definition 4 The operator A is of class Hilbert–Schmidt if A^*A is of trace class with the following definition. ◇

Definition 5 The operator A is of trace class if $Tr\sqrt{A^*A} < \infty$.

One can prove that the operator A is trace class iff there a decomposition $A = A_1A_2$ where A_1, A_2 are of class Hilbert-Schmidt. ◇

Every bounded self-adjoint operator A can be written as $A = A_+ - A_-$ where $A_\pm$ are positive operators; A is of trace class iff both $A_\pm$ are of trace class. In this case one has $TrA = TrA_+TrA_-$.

The function Tr can be extended to a class of bounded operators. Recall that any bounded closed operator A can be written as sum over the complex field of two self-adjoint operators

$$A = \frac{A + A^*}{2} + \frac{A - A^*}{2} \equiv Re\, A + i\, ImA \tag{4}$$

Therefore the function Tr is defined for any closed bounded operator whose real and imaginary parts are of trace class.

The function Tr has the following properties
(i) $Tr(A + B) = Tr\, A + Tr\, B$
(ii) $Tr(\lambda\, A) = \lambda Tr\, A$
(iii) $0 < A \leq B \;\; \Rightarrow Tr\, A \leq Tr\, B$

Theorem 3 *The collection J_1 of the trace class operators is a bilateral *-ideal in $B(\mathcal{H})$ and a Banach space with norm $\|A\| = Tr|A|$ where $|A| = \sqrt{A^*A}$.* ◇

Proof We must prove
(a) J_1 is a vector space
(b) $A \in J_1, \;\; b \in B(\mathcal{H}) \;\; \Rightarrow A\, B \in J_1, \quad B\, A \in J_1$
(c) $A \in J_1 \;\; \Rightarrow A^* \in J_1$
(d) The space J_1 is closed for the topology given by the norm $\|A\|_1 = Tr|A|$.
We call it *trace topology*.

The first two statements follow from the definition. Notice that every closed bounded operator maps compact sets in compact sets and that if A, B are closed and bounded one has $(AB)^* = B^*A^*$.

We now prove (d). If $A \in J_1$, consider $A_N = A - \sum_1^N(\phi_n,)A\phi_n$ where $\{\phi_k\}$ are the eigenvectors of $|A| \equiv \sqrt{A^*A}$.

Let a_k be the eigenvalues of $|A|$ in decreasing order. Let $\mathcal{H}_N$ be the subspace spanned by the first N eigenvectors; by construction $A_N\phi = 0, \phi \in \mathcal{H}_N$.

If A is positive, $a_n \geq 0$ and $lim_{N\to\infty} \sum_{k>N} a_k = 0$ since the series converges. If A is not positive, consider the polar decomposition $A = U_A|A|$ where U_A is a partial isometry from the closure of the range of $|A|$ to the closure of the range of A and is such that $Ker|A| \subset Ker\; U_A$.

Let $A_N = U|A|_N$. Then $lim_{N\to\infty} Tr(A_N) = 0$ and therefore $lim_{Nto\infty} Tr(|A_N|) = 0$. $Tr(|A|)$ defines a norm

$$Tr(|A+B|) \leq Tr|A| + Tr|B| \tag{5}$$

and J_1 is closed in this topology. Remark that $A+B = U_{A+B}|A+B|$, $A = U_A|A|$, $B = u_B|B|$. Therefore

$$\sum_n (\phi_n, |A+B|\phi_n) = \sum_n [(\phi_n, U^*_{A+B} U_A |A|\phi_n) + (\phi_n, U^*_{A+B} U_B |B|\phi_n)] \tag{6}$$

$$(\phi_n, U^*\ V\ |A|\phi_n) = (|A|^{1/2} V^*\ u\phi_n, |A|^{1/2}\phi_n) \tag{7}$$

and

$$\left|\sum_n (\phi_n, U^*\ V\ |A|\phi_n)\right| \leq \left(\sum_n |\ |A|^{1/2} V^* U\phi_n|^2\right)^{1/2} \left(\sum_n |\ |A|^{1/2}\phi_n|^2\right)^{1/2} \tag{8}$$

Partial isometries map orthonormal complete bases to orthonormal bases which are in general not complete. Therefore the right hand side in (8) is no bigger than $\sum_n (\phi_n, \|A\|\phi_n)$. This inequality, together with the same inequality for B concludes the proof. ♡

Theorem 4 *The finite rank operators are dense in* J_1. ◇

Proof Any trace class operator can be written as sum over the complex field of positive trace class operators. For positive trace class operators the non zero eigenvalues have 0 as accumulation point. ♡

We have proved that the function Tr defined on J_1 is positive, order preserving, and has the following properties
(i) $Tr(A\ B) = Tr(B\ A) \qquad A, B \in J_1$
(ii) $Tr(U\ A\ U^*) = TrA \qquad$ if U is unitary.

If $A \in J_1$, $B \in B(H)$ also AB and BA are in J_1 (J_1) is an ideal in $\mathcal{BH}$.

Moreover the following identity holds $Tr(AB) = Tr(BA)$. Therefore Tr is defined on a product of bounded operators when at least one of the factors is of trace class.

We remark explicitly that *only for positive operators in* J_1 one has $TrA = \sum_n (\phi_n, A\phi_n)$ where $\{\phi_n\}$ is an orthonormal complete basis.

We have seen in Volume I that the trace class operators have an important role in Quantum Mechanics because those of trace one represent states of the system. In that context we have noticed that $B(H)$ is the dual of J_1 and that the states represented by $J_{1,+,1}$ are *normal states*.

In fact it can be proved that Tr is completely additive. We denote with J_2 the class of Hilbert–Schmidt operators. It is easy to verify that J_2 is a *bilateral ideal of $B(\mathcal{H})$. Let $A, B \in J_2$ and let $\{\phi_n\}$be an orthonormal basis in $\mathcal{H}$. Then A^*B is trace class and

$$Tr(A^* \, B \, B^* \, A)^{\frac{1}{2}} = Tr[(A; A^*)(B \, B^*)]^{\frac{1}{2}} \leq |A| Tr|B| \tag{9}$$

Define

$$< A, B >_2 = Tr(A^* \, B) \equiv \sum_n (A\phi_n, B\phi_n) \tag{10}$$

(it is easy to see that this definition does not depend on the basis).

The quadratic form $< ., \; . >$ defines J_2 a non-degenerate scalar product and therefore a pre-hilbert structure. It is not difficult to verify that with this scalar product J_2 has the structure of a complete Hilbert space.

Moreover

$$| < A, B > | \leq (TrA^* \, A)^{1/2} \, (TrB^* \, B)^{1/2} = \|A\|_1 \, \|B\|_1 \tag{11}$$

Setting $\|A\|_2 = (Tr(A^*A))^{1/2}$ one has

$$\{A \; : \|A\|_1 \leq 1\} \; \subset \{A \; : \|A\|_2 \leq 1\} \; \subset \{A \; : |A| \leq 1\} \tag{12}$$

and $\|A\|_1 \geq \|A\|_2 \geq |A|$.

Therefore the topology of J_2 is intermediate between that of J_1 and the uniform topology of $B(\mathcal{H})$. Proceeding as we have done for J_1 one can prove that the hermitian part of J_2, denoted J_2^{her}, satisfies

$$J_2^{her} \equiv \{A \in B(H), \;\; A^* = A, \; \sum_k a_k^2 < \infty\} \tag{13}$$

where $\{a_n\}$ are the eigenvalues of A.

It is convenient to keep in mind the following *inclusion and density scheme.*

Denote by $\mathcal{F}$ the finite rank operators and by $\mathcal{K}$ the compact operators. Then
(1) $\mathcal{F}$ is dense in J_1 in the topology $\|.\|_1$.
(2) J_1 is dense in J_2 in the topology $\|.\|_2$.
(3) J_2 is dense $\mathcal{K}$ in the uniform operator topology.
(4) $\mathcal{K}$ is dense in $B(\mathcal{H})$ in the strong operator topology.

Moreover $\mathcal{F} \subset J_1 \subset J_2 \subset \mathcal{K} \subset B(\mathcal{H})$ and all inclusions are strict if $\mathcal{H}$ has infinite dimension.

The elements of J_1 and J_2 are particular cases of Schatten class operators.

1 Schatten Classes

Definition 6 (*Schatten Classes*) Let $1 \leq p < \infty$. An operator A is a *Schatten operator of class p* if $Tr(|A^*A|^{\frac{p}{2}}) < \infty$. ◇

We denote the space of all Schatten operators of class p by $\mathcal{S}^p(\mathcal{H})$. It is a Banach space with norm $\|A\|_p = (tr(|A|^p))^{\frac{1}{p}}$ and it has properties (in particular interpolation properties) similar to Lebesgue's spaces $L^p(\mu)$ where μ is a Lebesgue measure.

In particular $\mathcal{S}^1(\mathcal{H})$ are the trace class operators and $\mathcal{S}^2(\mathcal{H})$ are the Hilbert–Schmidt operators.

Let $\mu_n(A)$ be the eigenvalues of $|A|$ taken in decreasing order. The operator A belongs to $\mathcal{S}^p(\mathcal{H})$ iff $\sum_n \mu_n(A)^p < \infty$. One has
(1) $\mathcal{S}^p(\mathcal{H})$ is a *-ideal of $\mathcal{B}(\mathcal{H})$.
(2) $\mathcal{S}^p(\mathcal{H})$ is complete with respect to $\|A\|_p$.
(3) $p \le q \Rightarrow \mathcal{S}^p(\mathcal{H}) \subset \mathcal{S}^q(\mathcal{H})$ and $\|A\| \le \|A\|_p \le \|A\|_q$.
(4) Hoelder inequality holds for $0 \le p, q, r < \infty$:

$$\frac{1}{q} + \frac{1}{p} = \frac{1}{r}, \quad A \in \mathcal{S}^p(\mathcal{H})\ B \in \mathcal{L}^q(\mathcal{H}) \Rightarrow AB \in \mathcal{S}^r(\mathcal{H}), \quad \|AB\|_r \le \|A\|_p \|B\|_q \tag{14}$$

(5) Let $1 < p, q < \infty$ satisfy $\frac{1}{q} + \frac{1}{p} = 1$. Then for $A \in \mathcal{S}^p(\mathcal{H})$ and $B \in \mathcal{S}^q(\mathcal{H})$ one has $Tr(AB) = Tr(BA)$. Moreover for $A \in \mathcal{S}^1(\mathcal{H})$ and $B \in \mathcal{B}(\mathcal{H})$ one has $Tr(AB) = Tr(BA)$.

As one sees from the properties listed above, the spaces $\mathcal{S}^p(\mathcal{H})$ are non-commutative analogues of $L^p(X, \mu)$ spaces defined for a measure space X with finite measure μ.

These properties are also shared by the space $\mathcal{S}^p$ defined on a von Neumann algebra $\mathcal{M}$ with a trace state ω by defining $\mathcal{L}^p(\mathcal{M})$ as those elements in $\mathcal{M}$ for which

$$\|a\|_p = (\omega(a^* a))^{\frac{p}{2}} < \infty \tag{15}$$

The corresponding non-commutative integration theory has been developed, among others, by D. Gross, I. Segal, E. Nelson. We shall treat it briefly in Lecture 16

One can prove the following theorem (Lidskii identity [1])

Theorem 5 (Lidskii) *For every trace-class operator $A \in \mathcal{B}(\mathcal{H})$ and for every orthonormal base $\{\phi_n\}$ of $\mathcal{H}$ the following holds*

$$\sum_{i \ge 1} (\phi_n, A\phi_n) = sum_{i \ge 1} \lambda_j(A) \tag{16}$$

where $\{\lambda_i\}$ are the eigenvalues of A. ◇

Notice that Lidskii's theorem is a fundamental theorem for the spectral analysis of non-self-adjoint operators.

In general it is difficult to determine these eigenvalues but one can write a *trace formula* of the type

$$Trf_\mu(A) = \sum_{j \ge 1} f_\mu(\lambda_j(A)) \tag{17}$$

where μ is a parameter (real or complex) and estimate $\{\lambda_i\}$ with a Tauberian-type procedure.

2 General Traces

The definition of Schatten class can be generalized to the case $0 < p < 1$.

The Schatten classes $\mathcal{S}^p$ for $0 < p < 1$ are composed by all the operators such that

$$\sum_{j=0}^{\infty} s_j(A)^p < \infty \tag{18}$$

where s_j are the eigenvalues of $(A^*A)^{\frac{1}{2}}$ arranged in increasing order, counting multiplicities.

For $p < 1$ the space $\mathcal{S}^p$ is not a Banach space but rather a *quasi-Banach space* ($\|A\|\|B\| \leq c\|AB\|,\ c > 1$).

For $p > 1$ one defines also the Schatten classes $\mathcal{S}^{p,\infty}$ consisting of all A for which

$$\sup_j (j+1)^{\frac{1}{p}} s_j(A) < \infty \tag{19}$$

Also this class in a Banach space.

For $p = 1$ this norm *is not* a Banach norm. To have a Banach space one must introduce a norm, *the Dixmier norm*

$$sup_{k\geq 2} \frac{\sum_{j=0}^{k} s_j(A)}{logk} \tag{20}$$

and the corresponding *Dixmier class* $\mathcal{S}^{Dixm}$.

All spaces $\mathcal{S}^p,\ 0 < p \leq \infty$ and $\mathcal{S}^{p,\infty}$, $1 \leq p \leq \infty$ are ideals for $\mathcal{B}(\mathcal{H})$. One can also verify that $A, B \in \mathcal{S}^{2,\infty}$ implies $AB \in \mathcal{S}^{1,\infty}$.

For the positive operators which belong to the Dixmier class one can define a *trace* Tr_{Dixm} (Dixmier trace) by

$$Tr_{Dixm}(A) = lim_{k\to\infty} \frac{logk}{\sum_{j=1}^{k} s_j}(A) \tag{21}$$

This trace plays a relevant role in the study of von Neumann algebras which are *type 2* factors. In particular note that if $A \in \mathcal{S}^{Dixm}$ then $TrA = 0$ whenever A is trace-class.

3 General L^p Spaces

One can further generalize the definition of trace.

Denote by $\mathcal{L}_+(\mathcal{H})$ the cone of positive operators on the separable Hilbert space $\mathcal{H}$.

One can define *trace* any function with values in $[0, \infty]$ $\mathcal{L}_+(\mathcal{H})$ which is positive, additive and homogeneous. A trace is *normal* if it is *completely additive*.

It is possible to prove that every normal trace is proportional to the trace we have studied.

We shall now study the structure of some $\mathcal{L}^p(\mathcal{H})$ spaces in the representation of the Hilbert space $\mathcal{H}$ as $L^2(X, d\mu)$ for some locally compact space X and regular measure μ.

In this case the Schatten class operators have a representation as integral kernels.

Theorem 6 *Let $\mathcal{H} \equiv L^2(X, d\mu)$. Then $A \in J_2$ iff there exists a measurable function*

$$a(x, y) \in L^2(X \times X, d\mu \times d\mu) \tag{22}$$

such that, for every $f \in H$

$$(A f)(x) = \int a(x, y) f(y) d\mu(y) \tag{23}$$

Moreover one has

$$\|A\|_2^2 = \int |a(x, y|^2 d\mu(x) d\mu(y) \tag{24}$$

◇

Proof To prove sufficiency, let $a(x, y) \in L^2(X \times X, d\mu \times d\mu)$ and set, for any function $f \in \mathcal{H} \equiv L^2(X, d\mu)$

$$(Af)(x) \equiv \int a(x, y) f(y) d\mu(y) \tag{25}$$

Then for any $g \in \mathcal{H}$ Schwartz inequality gives

$$(g, Af) = \int \bar{g}(x) a(x, y) f(y) d\mu(y) d\mu(y) \leq |a|_2 \, |f| \, |g| \tag{26}$$

Therefore A is bounded $\|A\|_2 \leq |a|_2$.

Let $\{\phi_n\}$ an orthonormal basis of $\mathcal{H}$, then $\phi_n \otimes \phi_m$ is an orthonormal basis in $\mathcal{H} \otimes \mathcal{H}$. Therefore there exists $c_{n,m} \in C$ for which

$$a(x, y) = \sum_{n,m} c_{n,m} \bar{\phi}_n(x) \phi_m(y), \quad \sum_{n,m} |c_{n,m}|^2 = |a|_2^2 < \infty \tag{27}$$

Setting

$$a_N(x,y) = \sum_{n,m\le N} c_{n,m}\bar{\phi}_n(x)\phi_m(y), \quad (A_N f)(x) = \sum_{n,m} a_N(x,y) f(y) d\mu(y) \tag{28}$$

one has

$$lim_{N\to\infty}|a_N - a|_2 = 0, \qquad lim_{N\to\infty}\|A_N - A\|_2 = 0 \tag{29}$$

and therefore A is compact (as norm limit of compact operators). Moreover

$$Tr(A^* A) = \sum_N |A\phi_n|^2 = \sum_{n,m} |c_{n,m}|^2 = |b|_2^2 < \infty \tag{30}$$

and therefore A is of Hilbert–Schmidt class.

To prove that the condition is necessary one makes use of the fact that J_2 is the closure of $\mathcal{F}$ in the $\|.\|_2$ norm.

By definition every finite rank operator is represented by an integral kernel. Choosing a sequence $A_n \in \mathcal{F}$ that converges to A it easy to see that the corresponding integral kernels a_n converge in the topology of $L^2(X \times X, d\mu \times d\mu)$.

Let $a(x,y)$ be the limit integral kernel. For any $f \in L^2(X, d\mu)$ one has $(Af) = \int a(x,y) f(y) d\mu(y)$ and $\|A\|_2 = |a|_2$. ♡

If $A \in J_1$ and $A > 0$ one can prove $\|A\|_1 = \int a(x,x)\mu(dx)$.

But in general if A is not positive *is not true* that $|\int a(x,x)d\mu(x)| \le \infty \Rightarrow A \in J_1$.

Example The operator $-\frac{d^2}{dx^2}$ on $L^2((0,\pi), dx)$ with Dirichlet boundary conditions has discrete spectrum with simple eigenvalues n^2 and corresponding eigenfunctions $\sqrt{2/\pi}\ sen\ n\ x, \quad n \ge 1$.

The resolvent R_λ is represented by the integral kernel

$$R_\lambda(x,y) = \frac{2c}{\pi}\sum_1^\infty \frac{1}{n^2+\lambda} sen\ n\ x.\ sen\ n\ y \tag{31}$$

and has eigenvalues $(\lambda + n^2)^{-1}$. Therefore for $\lambda \notin (-\infty, -1]$ the operator R_λ is of trace class.

One can extend this result to all λ which are not in the spectrum using the resolvent identity and the fact that J_1 is a bilateral ideal of $\mathcal{B}(\mathcal{H})$.

An easy consequence of Theorem 5 is the following proposition which we state without proof.

Proposition 1 *Let A be a linear operator on $L^2(X, d\mu)$. The following statements are equivalent to each other*
(a) A is a Hilbert–Schmidt operator
(b) There exists $\xi(x) \in L^2(X, d\mu)$ such that $f \in D(A) \Rightarrow |(Af)(x)| \le |f|\ \xi(x)$.

(c) There exists a kernel $K(x, y) \in L^2(X, d\mu)$ such that, for any $f \in L^2(X, d\mu)$ and for almost all $x \in X$ one has $(Af)(x) = \int K(x, y)f(y)dy$. ◇

Notice that an operator A defined by an integral kernel may be bounded also when the kernel is singular. For example, the identity operator has integral kernel $K(x, y) = \delta(x - y)$.

On the other hand, the kernel $K(x, y) = h(x)\ \delta(y) \quad h \in C^\infty$ corresponds to the operator $(Kf)(x) = h(x)\ + f(0)$ with domain the functions in $L^2(X, d\mu)$ which are continuous at the origin.

This operator is not closable since the map $f(.) \rightarrow f(0)$ is not continuous $L^2(X, d\mu)$.

4 Carleman Operators

Before discussing more in detail the compact operators we briefly mention the *Carleman operators*. They are frequently encountered in Quantum Mechanics because they intervene naturally in the inversion of differential operators.

Definition 7 A linear map T from $\mathcal{H}$ to $L^2(X, d\mu)$ is a *Carleman operator* if there exists a measurable function $K_T(x)$ with values in $\mathcal{H}$ such that for any $f \in D(A)$

$$(Tf)(x) = (K_T(x), f) \tag{32}$$

holds for almost all x. The measurable function K_T is called *Carleman kernel* associated to T. ◇

Theorem 7 *The map T is a Carleman operator iff there exists a positive measurable function $g(x)$ such that for any $f \in D(T)$ one has, for μ-almost all $x \in X$,*

$$|(Tf)(x)| \leq g(x)|f|_2 \tag{33}$$

◇

Proof The condition is necessary: let T be a Carleman operator and let $K(x)$ be its kernel. The inequality is satisfied by taking $g(x) \equiv |K(x)|$.

The condition is sufficient: let $\rho(x)$ be a positive bounded measurable function such that $g(x)\ \rho(x) \in L^2(X, d\mu)$. Then $|(\rho\ T\ f)(x)| \leq |f|_2 g(x)\ \rho(x)$ and therefore according to proposition 1 $\rho\ T$ is a Hilbert–Schmidt operator.

It follows that there exist a measurable function $\tilde{K}(x)$ with value in the Hilbert–Schmidt operators such that for almost all x $(\rho\ T\ f)(x) =< \tilde{K}(x), T >$.

Setting $K(x) = \rho^{-1}\ \tilde{K}(x)$ one has $(Tf)(x) = (K(x), f)$. ♡

Often the integral kernels that one encounters in the study of Schroedinger equation have the form $K(x, y) = K_1(x, y)\ K_2(x, y)$.

Let T be an operator represented by this integral kernel and let K_1, K_2 satisfy, for almost all $x, y \in R^N$

$$\int |K_1(x,y)^2|\mu(dy) < C_1, \quad \int |K_2(y,x)^2|\mu(dy) < C_2 \tag{34}$$

Since $\|T\| \equiv sup_{\phi,|\phi|=1}|(\phi, T\phi)|$ it is easy to see that T is bounded in $L^2(R^N\ d\mu)$ and its norm satisfies $\|T\| \leq (C_1\ C_2)^{1/2}$. Moreover the adjoint T^* has integral kernel $\bar{K}(x,y)$. Choosing

$$K_1(x,y) \equiv |K(x,y)|^{1/2} \qquad K_2(x,y) \equiv Sign\ K(x,y)\ |K|^{1/2}(x,y) \tag{35}$$

one derives the following important result

Theorem 8 *If the integral kernel of T is such that a.e.*

$$\int |K(x,y)|\mu(dy) \leq C_1 \qquad \int |K(x,y)|\mu\ (dx) \leq C_2 \tag{36}$$

then $\|T\| \leq \sqrt{C_1\ C_2}$. ◇

5 Criteria for Compactness

We give now a useful criterion which gives a *sufficient condition* for the compactness of an operator.

Theorem 9 *Let A be operator with integral kernel* $K(x,y) = K_1(x,y)\ K_2(x,y)$ *where* $K_{1,2}$ *are measurable. Let* X_n^1, X_n^2 *two increasing sequences of measurable subsets of X, and X be their common limit.*

The operator A is compact if for n

$$\int_{X_n^1 \times X_n^2} \int |K(x,y)|^2 \mu(dx)\ \mu(dy) < \infty \tag{37}$$

and moreover for any $\epsilon > 0$ *there exists an integer* $N(\epsilon)$ *such that the following inequalities are satisfied*

(a) $\int_X |K_1(x,y)|\mu(dy) < \epsilon$ *a.e. in* $X - X_{1,N(\epsilon)}$

(b) $\int_X |K_2(x,y)|\mu(dx) < \epsilon$ *a.e. in* $X - X_{2,N(\epsilon)}$

(c) $\int_{X-X_{1,N(\epsilon}} |K_1(x,y)|\mu(dy) < \epsilon$

(d) $\int_{X-X_{2,N(\epsilon)}} |K_2(x,y)|\mu(dy) < \epsilon$ ♡

Outline of the proof Consider the operators A_n with integral kernel given by $K^n(x,y) = K(x,y)$ if $x, y \in X_n \times X_n$, and zero otherwise. The preceding theorems imply that A_n is of Hilbert–Schmidt class.

It follows from (a), (b), (c), (d) that $|S_n| < \epsilon$ where the integral kernel of S_n is the restriction of K to $X_{1,N(\epsilon)} \times X^{\perp}_{2,N(\epsilon)}$. Hence A is norm limit of Hilbert- Schmidt operators and therefore compact. ◇

Example 1 Let $X \equiv R^d$, let μ be Lebesgue measure and assume that T has integral kernel

$$K(x,y) = f_1(x)\, f_2(y)\, f_3(x-y) \tag{38}$$

with f_1, f_2 bounded measurable, $lim_{|x|\to\infty} f_k(x) = 0$ and $f_3 \in L^1(R^d)$. Set

$$X_{(1,n)} = X_{(1,n)} \equiv \{x \,:\, x \in R^d, \quad |x| < n\} \tag{39}$$

$$K_1 = f_1(x)|f_3(x-y)|^{1/2}, \quad K_2 = f_2(y)|f_3(x-y)|^{1/2} sign\ (f_3(x-y)) \tag{40}$$

Then A is compact. ♣

It can be shown that if one assumes a suitable decay at infinity of the functions f_1 and f_2 to prove the result it is sufficient that $f_3 \in L^1_{loc}$.

The result is applicable therefore for $f_3(z) = \frac{1}{|z|}$.

Example 2 Let $X = R^d$ and μ Lebesgue measure. Let A have integral kernel

$$K(x,y) = |x-y|^{\alpha-d} H(x,y) \qquad x \neq y \qquad K(x,x) \equiv 0 \qquad \alpha > 0 \tag{41}$$

where H bounded measurable. Set $K_n(x,y) = |x-y|^{\alpha-d} H(x,y)$ if $|x-y| \geq n^{-1}$, zero otherwise. Then for every n, A_n is of Hilbert–Schmidt class and the sequence A_n converges to A in norm. Therefore A is compact. ♣

Given the importance of the compactness property in Quantum Mechanics we give yet another compactness criterion.

Theorem 10 *Let A be a positive operator. The following properties are equivalent*
(i) $(A-\mu_0)^{-1}$ *is compact when we choose* $\mu_0 \in \rho(A)$.
(ii) $(A-\mu)^{-1}$ *is compact for every* $\mu \in \rho(A)$.
(iii) $\{\phi \in D(A), \ |A\,\phi| \leq I, \ \ |A\phi| \leq b\}$ *is a compact set for every* $b > 0$.
(iv) $\{\phi \in D(A), \ \ |\phi| \leq I, \ \ (\phi, A\phi) \leq b\}$ *is a compact set for every* $b > 0$.
(v) A has discrete spectrum and, denoting by a_n *the eigenvalues taken in decreasing order* $lim_{n\to\infty} a_n = 0$. ◇

Proof (i) $\leftrightarrow$ (ii). We use the resolvent identity

$$(A-\mu)^{-1} = (A-\mu_0)^{-1} + (A-\mu_0)^{-1}(\mu-\mu_0)(A-\mu)^{-1} \tag{42}$$

The first term on the right-hand side is compact by assumption. Also the second is compact because the compact operators are an ideal in $B(H)$.
$(i) \to (v)$. By definition

$(v) \rightarrow (iv)$. Let $Q(A)$ the domain of the close positive quadratic form $\phi \rightarrow (\phi, A\phi) \in R^+$ and let

$$\mathcal{F}_b^A \equiv \{\psi \in Q, \quad |\psi| = 1, \quad (\psi, A\psi) \le b\} \tag{43}$$

The set $\mathcal{F}_b^A$ is closed because the form is closed. To show that it is compact we prove that for any $\epsilon > 0$ it is possible to cover $\mathcal{F}_b^A$ with a finite number of ball in $\mathcal{H}$ of radius ϵ.

We choose N such that $a_N \le b\epsilon$. Then (v) implies

$$\sum_{n>N} |(\psi, \phi_n)|^2 \le \epsilon \tag{44}$$

Therefore every $\psi \in \mathcal{F}_b^A$ is at a distance less than $\sqrt{\epsilon}$ from the intersection of the ball of radius ϵ with the subspace spanned by the first N eigenfunctions.

Since this is compact set (it is a closed bounded subset in a finite-dimensional space) it can be covered with a finite number of balls of radius $\sqrt{\epsilon}$. Therefore also $\mathcal{F}_b^A$ has this property.

$(iv) \rightarrow (iii)$. By assumption (iv) holds for A and therefore also for A^2. It follows that $\mathcal{F}_b^{A^2}$ is compact. On the other hand, $(\psi, A^2\psi) = |A\psi)|^2$.

$(iii) \rightarrow (i)$. Let $\mathcal{M} \equiv \psi : \exists \phi \in H, \ \psi = (A+1)^{-1}\phi, \ |\phi| \le 1\}$. Then

$$|\psi| \le |\phi| \le 1 \quad |A\psi| = |A(A+1)^{-1}\phi| \le |\phi| \le 1 \tag{45}$$

Therefore the set $\mathcal{G}_b^A \equiv \{\psi \in H, \ |\psi| \le 1, \ |(A+1)\psi)| \le b\}$ is closed and contained in $\mathcal{F}_b^B$. It follows that $(1+A)^{-1}$is compact. ♡

From this theorem one sees that it is convenient to have criteria to decide whether a subset of $\mathcal{H}$ is compact.

In the realization of $\mathcal{H}$ as $L^2(\Omega, dx)$, $\Omega \subset R^N$, these criteria rely on inequalities among norms in suitable function spaces (often related to Sobolev immersion theorems).

We shall collect in the Appendix to this chapter a collection of inequalities that are useful in studying the solutions of the Schroedinger equation and in estimating their regularity.

We also give some other *compactness criteria* that are derivable from general inequalities.

As an example consider the operator $H \equiv L^2(-\pi, \pi)$ defined by

$$A \equiv -\frac{d^2}{dx^2}, \quad D(A) = \{\phi \in C^\infty, \quad \phi(-\pi) = \phi(\pi)\} \tag{46}$$

and define

$$S \equiv \{\phi \in L^2\ (-\pi, \pi), \quad |\phi|_2 \le 1, \quad |\frac{d\phi}{dx}|_2 \le 1\} \tag{47}$$

(i.e. the domain of the quadratic form associated to A).

Denoting with c_n the Fourier coefficients of a function, it is easy to see that S is characterized by $\sum |c_n|^2 < \infty, \sum n^2\, |c_n|^2 < \infty$.

In these notations one sees immediately that S is compact in the topology of H and therefore $(\bar{A}+I)^{-1}$ is compact. In the same way one proves compactness of the closure of $-\frac{d^2}{dx^2}+x^2$ defined on $C_0^\infty(R)$.

Remark that the same *is not true* for the closure of the operator $-\frac{d^2}{dx^2}$ defined on $C_0^\infty(R)$. Indeed the closure of this operator is a self-adjoint operator with continuous spectrum.

Rellich Compactness Criterion

Let F and G be two continuous positive functions on R^d which satisfy

$$lim_{|x|\to\infty}F(x) = +\infty, \qquad lim_{|p|\to\infty}G(p) = +\infty \tag{48}$$

The set

$$S \equiv \{f \; : \; \int F(x)\; |f(x)|^2 dx \le 1, \quad \int G(p)\,\hat{f}(p)|^2 dp \le 1\} \tag{49}$$

is compact in $L^2(R^d)$. ◇

Proof The set S is closed. Without loss of generality we can assume

$$F(x) \le x^2, \;\; G(p) \le p^2 \tag{50}$$

Indeed if this equation is not satisfied the set S is closed and contained in the set of functions that satisfy the equation.

The set S is dense in $L^2(R^d)$. Denote by $\hat{G}$ the operator that acts as $G(p)$ in the Fourier transformed space. If $V(x)$ is bounded and has compact support then $[V(x)(\hat{G})^{-1}](x,y)$ is compact. Indeed for every value of $\epsilon > 0$ the kernel of $V(x)[\epsilon(\hat{p})^d + \hat{G} + 1)^{-1}]$ belongs to $L^2(R^d) \otimes L^2(R^d)$ and $[\epsilon p^d + G(p) + 1)^{-1}]$ converges to $[G(p)+1)^{-1}]$ in L^∞.

Therefore $V[\hat{G}+1)^{-1}]$ is compact since it is the norm-limit of compact operators.

For $\alpha > 0$ define $V_\alpha \equiv min\{F(x), \alpha+1\} - \alpha - 1$. Since $lim_{|x|\to\infty}F(x) = +\infty$, V_α has compact support an therefore $V_\alpha[\hat{G}+1)^{-1}]$ is compact. From the min-max principle

$$\lambda_n(A) \ge \lambda_n(\hat{G} + V_\alpha(x) + \alpha + 1) \tag{51}$$

and therefore for each $\alpha > 0$ there exists $m(\alpha)$ such that $\lambda_{m(\alpha)}(A) \ge \alpha$. Since α is arbitrary, $lim_{n\to\infty}\lambda_n = \infty$. ♡

Example 3 Let $V \in L^1_{loc}(R^d), \quad V(x) \ge 0, \quad lim\; V(x)_{|x|\to\infty} \to 0$.

Then $H \equiv -\Delta + V$ defined as sum of quadratic forms has compact resolvent. ◇

Proof Since both $-\Delta$ and V are positive operators $(\phi, H\phi) \le b \Rightarrow (\phi, -\Delta\phi) \le b \;\; (\phi, V\phi) \le b$ for every ϕ.

Therefore the set

$$\mathcal{F}_{H,b} \equiv \{\phi \in D(H),\ |\phi| \leq 1, \quad (\phi, H\phi) \leq b\ \} \tag{52}$$

is closed and contained in $\{\ \phi\ :\ |\phi| \leq 1,\ \ \int p^2|\hat{\phi}(p)|^2 dp,\ \ \int V(x)|\phi(x)|^2 dx \leq b\}$.
This set is compact by the Rellich criterion. ♡

Example 4 Let $d \geq 3$ and set

$$V = V_1 + V_2 \quad V_2 \in L^{d/2}(R^d) + L^{\infty}(R^d)$$

$$lim_{|x|\to\infty} V_1(x) = 0, \quad V_1 \in L^1_{loc}(R^d), \quad V_1 \geq 0. \tag{53}$$

Then $H \equiv -\Delta + V$ defined as sum of quadratic forms has compact resolvent. ◇

Proof V_2 is form-small with respect to Δ and therefore also with respect to $-\Delta + V_1$.
If $A \geq 0$ has compact resolvent and B is form-small with respect to A, then $C \equiv A + B$ as sum of quadratic forms has compact resolvent.

Define $q_A(\phi) = (\phi, A\phi)$ and let $Q(a)$ the domain of the form q_A i.e. the closure of $D(A)$ in the topology induced by the (strictly positive) form $q_A\phi)+!\phi|^2$. For any $\phi \in Q(C) \cap Q(A)$ one has $q_B(\phi) \leq \alpha[q_A(\phi) + b|\phi|^2]$ where $\alpha < 1, \beta > 0$.

It follows

$$q_C(\phi) \geq (1 - \alpha)(q_A(\phi) - \beta|\phi|^2 \tag{54}$$

From the min-max principle $\lambda_n(C) \geq (1 - \alpha)\lambda_n(A) - \beta$.
Therefore $\lambda_n \to \infty$ implies $\lambda(A)_n \to \infty$. ♡

A further compactness criterion which is frequently used is

Riesz Compactness Criterion
Let $1 \leq p < \infty$ and let S be a subset of the unit ball in $L^p(R^d)$.
The closure L^p of S is compact iff the following conditions hold
(a) $\forall \epsilon > 0$ there exists a compact $K \subset R^d$ such that $\int_{R^d - K} |f(x)|^p dx < \epsilon^p$ for each $f \in S$).
(b) $\forall \epsilon > 0$ there exists $\delta > 0$ such that if $f \in S$ and $|y| < \delta$ then $\int |f(x-y) - f(x)|^p dx < \epsilon$. ◇

Proof Necessity If S is compact, given $\alpha > 0$ let $f_1, \ldots f_N$ such that the balls of radius $\frac{\alpha}{3}$ centered on the unit ball of the subspace spanned by the f_k cover $\bar{S}$.
There exist therefore K and δ such that conditions (a) and (b) be satisfied for $f_1, \ldots f_N$ and $\epsilon = \frac{\alpha}{3}$. To extend this inequality to the entire set S notice that for every $g \in L^p$ one has

$$lim_{K\to R^d} \int_{R^d - K} |g(x)|^p dx = 0, \quad lim_{y\to 0} \|g_y - g\|_2 = 0, \quad g_y(x) \equiv g(x-y) \tag{55}$$

A standard argument shows then that (a) and (b) hold in S.

Sufficiency Let S satisfy (a) and (b). For any compact $\Omega \subset R^d$ and positive constants α, β the Ascoli–Arzelá theorem gives the compactness of

$$T(\Omega,\ \alpha,\ \beta) \equiv \{f \in C_0^\infty,\ \ supp\, f \in \Omega,\ \ |f|_\infty \leq \alpha,\ \ |\nabla f|_2 \leq \beta\} \tag{56}$$

Therefore, given $\epsilon > 0$, it is sufficient to find Ω, α, β such that for every $f \in S$ there exists $g \in T(\Omega, \alpha, \beta)$ with $|f - g|_p < \epsilon$.

Indeed in this case since $T(\Omega, \alpha, \beta)$ can be covered by a finite number of balls of radius ϵ and S can be covered by a finite number of balls of radius 2ϵ.

To find Ω, α, β with the desired properties, given $\epsilon > 0$ choose K, δ so that for $f \in S$,

$$\int_{R^d - K} |f(x)|^p dx < \frac{\epsilon^p}{4}, \qquad |y| < \delta \Rightarrow \||f_y - f\|_p \leq \frac{\epsilon}{4} \tag{57}$$

Let η be a positive C^∞ function with support in $y : |y| < \delta$, $\int \eta(x)dx$. Let ξ be the indicator function of the set $K' \equiv \{y : dist(y, K) < \delta\}$. Then a possible choice for $\{\Omega, \alpha\beta\}$ is

$$\Omega \equiv \{y\ :\ dist(y, K) \leq 2\delta\}, \quad \alpha = |\eta|_q, \quad \beta = |\nabla\eta|_p \ \ p^{-1} + q^{-1} = 1 \tag{58}$$

This follows from the following inequalities (the first is Hölder's inequality)

$$|f * g|_\infty \leq |f|_p |g|_q, \quad p^{-1} + q^{-1} = 1, \qquad (|f| * g|)_1 \leq |f|_1 |g|_1 \tag{59}$$

and (by interpolation) $|f * g|_s \leq |f|_q\ |g|_p, \qquad p^{-1} + q^{-1} = 1 + s^{-1}$.

We must prove that $|f - g|_p < \epsilon$. From the definitons it follows $\int_{R^d - K'} |f - g|^p dx \leq \int_{R^d - K'} |f(x)|^p dx < \frac{\epsilon}{4}^p$ and therefore

$$\|\xi f_y - \xi\ \|_p \leq \||f_y - f\|_p + \|(1 - \xi)\||f_p + \|[1 - \xi_y]f_y\|_p \leq \frac{3}{4}\epsilon \tag{60}$$

$$\|(\eta * \xi)f - \xi f\|_p \leq \int \eta(y)\|\xi f(. - y) - \xi f(.)\|_p dy \leq \frac{3\epsilon}{4} \tag{61}$$

From this one derives $\|g - f\|_p \leq \|g - \xi f\|_p + \|(1 - \xi)f\|_p < \epsilon$. ♡

6 Appendix: Inequalities

We give in this appendix a collection of inequalities that are frequently used in the theory of Schroedinger operators. A detailed account can be found in the review paper [2] and in the books [3, 4].

Some of these inequalities can be obtained in an elementary way making use of the Fourier transform. For other the proof requires more sophisticated techniques.

We give an example of an inequality which can be obtained by elementary means. In R^d one has

$$|f|_\infty \leq \int |\hat{f}(p)|dp \leq (p^2+1)^\alpha \hat{f}(p)|(p^2+1)^{-\alpha}$$
$$\leq \left[\int (p^2+1)^{2\alpha}|\hat{f}(p)|^2 dp]^{1/2} \int (p^2+1)^{-2\alpha} dp\right]^{1/2} \tag{62}$$

i.e. for $4\alpha > d$ one has $|f|_\infty \leq C|(p^2+1)^{2\alpha}\hat{f}|_2$.

This means that for any d the space $H^{\frac{d}{2}+\epsilon}$ is compact in L^∞.

Among the inequalities a relevant role is played by the *Jensen inequalities*

Recall that a real valued function f defined on a convex subset C of a real vector space E is called *convex* if

$$\forall x,\ y \in C,\ \ \forall \theta \in (0,1) \qquad f((1-\theta)x+\theta y) \leq (1-\theta)f(x)+\theta f(y) \tag{63}$$

If the inequality is strict, the function is *strictly convex.*

Jensen inequality I
Let f be convex on a convex set C and let $p_1 \dots p_n$ be positive numbers with $\sum_k p_k = 1$, then

$$f(p_1x_1+\cdots p_nx_n) \leq p_1f(x_1)+\cdots+p_nf(x_n) \tag{64}$$

If the function is strictly convex, equality holds only if $x_1 = \cdots = x_n$. ♢

Jensen Inequality II
Let μ a probability measure on the Borel subsets of an open interval I of R and let $\bar{\mu}$ be its baricenter. If f is a convex measurable function with $-\infty < \int_I f d\mu < \infty$ then

$$f(\bar{\mu}) \leq \int_I f d\mu \tag{65}$$

If f is strictly convex equality holds iff $\mu(\bar{\mu}) = 1$, i.e. if the measure is concentrated in $\bar{\mu}$. ♢

Proof It is easy to see that if f is real and convex in the interval I then for each point $x_0 \in I$ there exists an affine function $a(x)$ such that $a(x_0) = f(x_0)$ and for all $x \in I$ one has $a(x) \leq f(x)$ (which implies $\int_I a d\mu \leq \int_I f d\mu$).

If f is strictly convex then $f(x) > a(x)$ if $x \neq \bar{\mu}$ and therefore the equality holds iff the measure μ is concentrated in $\bar{\mu}$. ♡

Jensen inequality III
Let μ be a probability measure on the Borel sets of the real Banach space E, and call $\bar{\mu}$ its baricenter.

If f is continuous and convex with $-\infty < \int_E f d\mu < \infty$ then

$$f(\bar{\mu}) \leq \int_E f d\mu \tag{66}$$

If f is strictly convex equality holds if the measure is concentrated in the point $\bar{\mu}$. ◇

Proof The proof follows the lines of the proof of Jensen II.

To construct an affine comparison functional we use the separation theorem for disjoint convex sets in a Banach space. Choose as affine functional an element of $\Phi \in E^*$.

Jensen's inequality proves this inequality for the integration along the direction of the affine functional Φ

The proof follows by induction over a complete set of elements in E^*. ◇

6.1 Lebesgue Decomposition Theorem

We are going to use often the decomposition of measure in a part that is continuous with respect to Lebesgue measure and in a singular part.

Lebesgue decomposition theorem Let $\{\Omega, \Sigma, \mu\}$ be a measure space ν a measure on Σ with $\nu(\Omega) < \infty$.

There exist a non-negative measurable function $f \in L^1(\mu)$ and a measurable set $B \in \Sigma$ with $\mu(B) = 0$ such that

$$\nu(A) = \int_A f d\mu + \nu(A \cap B), \quad \forall A \in \Sigma \tag{67}$$

◇

Define a measure ν_B by $\nu_B(A) = \nu(A \cap B)$. The measures μ and ν_B are mutually singular.

If we decompose Ω as $B \cup (\Omega/B)$ one has $\mu(B) = 0$ and $\nu_B(\Omega/B) = 0$ (the measures μ and ν_B have disjoint supports). ♣

Proof of Lebesgue decomposition theorem Set $\rho(A) \equiv \mu(A) + \nu(A)$. Take $g \in L^2_\pi$ and set $L(g) = \int g d\nu$. From Schwartz's inequality

$$|L(g)| \leq (\nu(\Omega))^{\frac{1}{2}} \|g\|_{L^2_\rho} \tag{68}$$

According to the Riesz representability theorem there exists $h \in L^2_\pi$ such that $L(g) = (g, h)$. It follows $\int_\Omega g(1-h)d\nu = \int_\Omega ghd\mu$.

Choosing for g the indicator function ξ_A of the set A one has $\nu(A) = L(I_A) = \int_A hd\mu + \int_A h\, \nu$. The function h is a.e. defined.

Denote by N, G, G_N, B the collection of points in which h takes value respectively in $(-\infty, 0), [0, 1), [0, 1-\frac{1}{N}), [1, \infty)$.

It is easy to see that $\nu(N) = 0$, $\mu(B) = 0$. Set $f = \frac{h(x)}{1-h(x)}$ for $x \in G_N$ and zero elsewhere.

Then

$$\nu(A \cap G_n) = \int_\Omega \frac{1-h}{1-h} \xi_{A \cap G_n} d\nu = \int_\omega f \xi_{A \cap G_n} d\mu \tag{69}$$

By monotone convergence one has $\nu(A \cap G) = \int_A f d\mu$. It follows

$$\nu(A) = \int_A f d\mu + \nu(A \cap B) \tag{70}$$

Taking $A = \Omega$ one has $f \in L^1(\mu)$. ♡

As a corollary to the Lebesgue decomposition theorem one has

Radon-Nikodym Theorem Let $\{\Omega, \Sigma, \mu\}$ be a measure space and ν a measure on Σ with $\nu(\Omega) < \infty$.

The measure ν is absolutely continuous with respect to μ iff there exists a non negative function $f \in L^1_\mu$ such that for any $A \in \Sigma$ one has $\nu(A) = \int_A f d\mu$. ◇

6.2 *Further Inequalities*

We will give now a list of *inequalities* that are more frequently used. We shall prove the simplest ones, and give references for the others.

Hölder inequality

If $1 < p < \infty$ and $f \in L^p$, $g \in L^{p'}$ with $\frac{1}{p} + \frac{1}{p'} = 1$ then $fg \in L^1$ and

$$\int |fg| d\mu \leq \|f\|_p \|g\|_{p'} \tag{71}$$

The equality sign holds if $\|f\|_p \|g\|_{p'} = 0$ or if a.e. $g = \lambda |f|^{p-1} sign f$.

We shall call *conjugate to p* the exponent p' defined by $\frac{1}{p} + \frac{1}{p'} = 1$. ◇

We shall not prove this inequality [2, 4]. We only quote the following two corollaries

Corollary 1 *If $f \in L^p$ one has*

$$\|f\|_p = max\{|\int fg d\mu| : \|g\|_{p'} = 1\} \tag{72}$$

Conversely a measurable function f belongs to L^p, $1 \leq p < \infty$ iff $fg \in L^1$ for every function $g \in L^{p'}$. ◇

Corollary 2 *Let f be a non-negative function on $(\Omega_1, \Sigma_1, \mu_1) \times (\Omega_2, \Sigma_2, \mu_2)$ and let $0 < p \leq q < \infty$.*

Then

$$\left(\int_{\Omega_1}\left[\int_{\Omega_2} f(x,y)^p d\mu_2(y)\right]^{\frac{q}{p}} d\mu_1(x)\right)^{\frac{1}{q}} \leq \left(\int_{\Omega_2}\left[\int_{\Omega_1} f(x,y)^q d\mu_1(x)\right]^{\frac{p}{q}} d\mu_2(y)\right)^{\frac{1}{p}} \tag{73}$$

◇

The proof is obtained from Corollary 1 by using Fubini's theorem to exchange the order of integration.

Sobolev inequalities

Let f a C^1 function on R^d $d > 1$ with compact support.

For $1 \leq p < d$ the following inequality holds

$$\|f\|_{\frac{pd}{d-p}} \leq \frac{p(d-1)}{d(d-p)}\left[\Pi_{j=1}^d \|\frac{\partial f}{\partial x_j}\|_p\right]^{\frac{1}{d}} \leq \frac{p(d-1)}{2d(d-p)}\left[\Pi_{j=1}^d \|\frac{\partial f}{\partial x_j}\|_p^p\right]^{\frac{1}{p}} \tag{74}$$

◇

Proof A repeated application of the fundamental theorem of calculus gives

$$\|f\|_{\frac{d}{d-1}} \leq \frac{1}{2}\left(\Pi_{j=1}^d \|\frac{\partial f}{\partial x_j}\|_1\right)^{\frac{1}{d}} \leq \frac{1}{2d}\left(\sum_{j=1}^d \|\frac{\partial f}{\partial x_j}\|_1\right) \tag{75}$$

This proves the inequality for $p = 1$.

Consider next the case $1 < p \leq d$. For any s and any $1 \leq j \leq d$ one has

$$|f(x)|^s \leq s\int_{-\infty}^{x_j} |f(t, x^j)|^{s-1}\left|\frac{\partial f}{\partial x_j}\right| dt \tag{76}$$

(we have used the notation x^j for the remaining coordinates).

A similar inequality is obtained integration between x and ∞. Therefore

$$|f(x)| \leq [\frac{s}{2}\int_{-\infty}^{\infty} |f(t, x^j)|^{s-1}\left|\frac{\partial f}{\partial x_j}\right| dt]^{\frac{1}{s}} \tag{77}$$

and then

$$\|f\|^s_{\frac{sd}{d-1}} \leq \frac{s}{2}[\Pi_{j=1}^d \||f|^{s-1}\left|\frac{\partial f}{\partial x_j}\right|\|_1]^{\frac{1}{d}}. \tag{78}$$

Using Hölder inequality one derives

$$\|f\|^{s}_{\frac{sd}{s-1}} \leq \frac{s}{2}\left[\Pi_{j=1}^{d}\|\frac{\partial f}{\partial x_j}\|_p\right]^{\frac{1}{d}} \tag{79}$$

The choice $s = \frac{p(d-1)}{d-p}$ (and therefore $(s-1)p' = \frac{sd}{d-1} = \frac{pd}{d-p}$) concludes the proof. ♡

Schur's Test Let $k(x, y)$ be non negative measurable on a product space $(X, \Sigma, \mu) \times (Y, \Xi, \nu)$ and let $1 < p < \infty$.

Assume the existence of measurable strictly positive functions g on (X, Σ, μ) and h on (Y, Ξ, ν) and of two constants a and b such that a.e.

$$\int_Y k(x,y)(h(y))^{p'}d\nu(y) \leq (ag(x))^{p'} \quad \int_Y k(x,y)(g(x))^p d\nu(x) \leq (bh(y))^p \tag{80}$$

Then if $f \in L^p(Y)$ one has

(a) $T(f) \equiv \int_Y k(x,y)f(y)d\nu(y)$ exists almost all values of x

(b) $T(f) \in L^p(X)$ and $\|T(f)\| \leq ab\|f\|_p$ ◇

Proof The proof uses Hoelder's inequality.

Remark that it is sufficient to prove that if g is a non-negative function in $L^p(Y)$ and h is non-negative in $L^{p'}(X)$ then

$$\int_X \int_y h(x)k(x,y)g(y)d\nu(y)d\mu(x) \leq ab\|h\|_{p'}\|g\|_p \tag{81}$$

Making use of this inequality and applying twice Hoelder's inequality one completes the proof. ♡

6.3 Interpolation Inequalities

We give now *interpolation formulas* between Banach spaces; their proofs makes use of classical results from complex analysis, which have an interest of their own.

The strategy is to construct Banach space that admits as closed subspaces the two Banach spaces B_0, B_1 of interest and then to construct a family of Banach spaces parametrized by point $z \in S$ where S is the strip $Re\ z \in [0, 1]$ in the complex plane.

This spaces are defined in such a way that the norms are analytic in z in the open strips, continuous up to the boundary and at the boundary coincide with the norms of the two Banach spaces B_0, B_1.

This procedure allows the use of theorems and inequalities in the theory of complex variables, among them the *Hadamard's three-lines inequality* which is the prototype of inequalities for functions analytic in the strip $S \equiv \{z = x + iy, 0 < x < 1\ y \in R\}$ (we shall denote by $\bar{S}$ its closure).

Hadamard's inequality
Let f be continuous and bounded in $\bar{S}$ and analytic in S. Define $M_x = sup\{|f(x+iy)|, y \in R\}$. Then

$$\forall x \in [0,1] \qquad M_x \leq M_0^x M_1^{1-x} \tag{82}$$

♢

Proof Choose $a_0 > b_0 \quad a_1 > b_1$ and set $g(z) = a_0^{x-1} a_1^{-x} f(z)$. We prove $\forall z \in S \;\; |g(z)| \leq 1$.

From this follows $|f(x+iy| \leq a_0^{1-x} a_1^x$ and since we can choose $a_0 - b_0, \; a_1 - b_1$ arbitrary small the thesis of the theorem follows,

To prove $\forall z \in S \;\; |g(z)| \leq 1$ we use the maximum modulus principle for analytic functions.

To avoid a possible difficulty in the control of the function $G(z)$ for $|Imz| \to \infty$ we study the function $h_\epsilon(z) = g(z)e^{\epsilon z^2}$. This function vanishes when $|Imz| \to \infty$ and therefore the maximum of its modulus in $\bar{S}$ is reached for Imz finite.

From the maximum modulus principle we derive $|h(z)_\epsilon| \leq e^\epsilon$. Since ϵ is arbitrary, $|g(z)| \leq 1 \; \forall z \in \bar{S}$. ♡

Before introducing the Riesz-Thorin interpolation theorem, one of the most used criteria for a-priori estimates, we give some definitions.

Definition 8 (*Compatible pairs*) Let A_0 with norm $\|.\|_{A_0}$ and A_1 with norm $\|.\|_{A_1}$ be linear subspaces of a Banach space$(V, \|.\|_V)$ and assume the maps $(A_j, \|.\|_{A_j}) \to (V \;\; \|.\|_V)$ are continuous, $j = 1, 2$.

We will say that the pair $(A_0, \|.\|_{A_0}), (A_1, \|.\|_{A_1})$ is *a compatible pair*.

A Banach space $(A, \|.\|_A)$ contained in $A_0 + A_1$ and which contains $A_0 \cap A_1$ is called *intermediate space* if the maps

$$(A_0 \cap A_1, \|.\|_{A_0 \cap A_1}) \to (A, \|.\|_A) \to (A_0 + A_1, \|.\|_{A_0 + A_1}) \tag{83}$$

are continuous. Recall that the topology on $A_0 \cap A_1$ and on $A_0 + A_1$ are defined by

$$\|a\|_{A_0 \cap A_1} = max[\|a\|_{A_0}, \|a\|_{A_1}] \qquad \|a\|_{A_0 + A_1} = inf[\|a\|_{A_0}, \|a\|_{A_1}]$$

$$a = a_0 + a_1, \quad a_j \in A_j \tag{84}$$

Notice that In the applications we use $(L^p, \|.\|_p)$ and $(L^q, \|.\|_q)$, with $1 \leq p, q \leq +\infty$ as compatible pair and $(L^r, \|.\|_r) \; r \in (p, q)$ as intermediate spaces.

In order to apply Hadamard's lemma, we introduce in the strip S a suitable space of functions. Let $(A_0, \|.\|_0)$ and $(A_1, \|.\|_1)$ be a compatible pair and denote by $L_1 = \{iy, y \in R\}$ and $L_2 = \{1 + iy \; y \in R\}$ the two boundaries of S.

Denote by $\mathcal{F}(A_0, A_1)$ the complex vector space of those functions f in $\bar{S}$ that take value in $A_0 + A_1$ and are such that

(1) f is continuous in $\bar{S}$.
(2) for each $\Phi \in (A_0 + A_1)^*$ the function $\Phi(f)$ is analytic in S
(3) f is continuous and bounded from B_j to A_j, $j = 0, 1$.

The space $\mathcal{F}$ is a Banach space with norm $\|F\|_{\mathcal{F}(A_0,A_1)} = max_{j=0,1}(supz \in S \{\|F(z)\|_{A_j}, z \in B_j\})$.

Proposition 2 *If* $F \in \mathcal{F}(A_0, A_1)$ *and* $z \in S$ *then* $\|F(z)\|_{A_0+A_1} \leq \|F\|_{\mathcal{F}}$. ◇

Proof There exists $\Phi \in (A_0 + A_1)^*$ of unit norm such that $\Phi(F(z)) = \|F(z)\|_{A_0+A_1}$.
Therefore $\Phi(F)$satisfies the conditions under which Hadamard's inequality holds, and then $\Phi(F(z)) \leq \|F\|_{\mathcal{F}}$. ♡

Notice that the map $F \to F(\theta), \theta = Rez \quad 0 < \theta < 1$ is continuous from $\mathcal{F}$ a $A_0 + A_1$. Denote with A_θ its image with the quotient norm

$$\|a\|_\theta = inf\{\|F\|_{\mathcal{F}} : \; F_\theta = a\} \tag{85}$$

With this notation $(A_\theta, \|.\|_\theta)$ becomes *an intermediate space* and one has

Theorem 11 *Let* (A_0, A_1) *and* (B_0, B_1) *be compatible pairs. Let T be a linear map from* (A_0, A_1) *to* $(B_0 + B_1)$ *which maps* A_j *to* B_j *and satisfies* $\|T(a)\|_{B_j} \leq M_j \|a\|_{A_j}$ *if* $a \in A_j \;\; j = 1, 2$.
Assume moreover $0 < \theta < 1$. *Then* $T(A_\theta) \subset B_\theta$ *and one has, if* $a \in A_\theta$

$$\|T(a)\|_\theta \leq M_0^{1-\theta} M_1^\theta \|a\|_\theta \tag{86}$$

◇

Proof Let a be a non-zero element of A_θ and choose $\epsilon > 0$. There exists $F \in \mathcal{F}(A_0, A_1)$ such that $F(\theta) = a$ and $\|T(a)\|B_\theta < (1 + \epsilon)\|a\|_\theta$.
By definition the function $T(F(z))$ belongs to $\mathcal{F}(B_0, B_1)$ and one has

$$\|T(F(z))\|_{B_j} \leq (1 + \epsilon)M_j \|F(z)\|_{A_j} \qquad z \in L_j \tag{87}$$

It follows $T(a) = T(F(\theta)) \in B_\theta$.
Setting $G(z) \equiv M_0^{z-1} M_1^{-z} T(F(z))$ one concludes that $G \in \mathcal{F}(B_0, B_1)$ and $\|G(z)\|_{B_j} \leq \|F(z)\|_{A_j}$ per $z \in L_j$.
Hence

$$\|G(\theta)\|_\theta = M_0^{\theta-1} M_1^{-\theta} \|T(a)\|_\theta) \leq (1 + \epsilon)M_j \|a\|_\theta \tag{88}$$

It follows $\|T(a)\|_\theta \leq M_0^{1-\theta} M_1^\theta \|a\|_\theta)$. Since ϵ was arbitrary the thesis of the theorem follows. ♡

We can now state and prove the interpolation formula of Riesz-Thorin.

Riesz-Thorin interpolation theorem
Let (Ω, Σ, μ) and (Ψ, Ξ, ν) be regular measure space. Let $1 \leq p_0,\ p_1,\ q_0, q_1 \leq \infty$.

Let T a linear map from $L^{p_0}(\Omega, \Sigma\mu) + L^{p_1}(\Omega, \Sigma, \mu)$ to $L^{q_0}(\Psi, \Xi, \nu) + L^{q_1}(\Psi, \Xi, \nu)$.

Suppose moreover that T map continuously $L^{p_j}(\Omega, \Sigma, \mu$ on $L^{q_j}(\Psi, \Xi, \nu)$ with norms $M_j, j = 1, 2$.

Let $0 < \theta < 1$ and define $p(\theta)$ and $q(\theta)$ as

$$\frac{1}{p(\theta)} = \frac{1-\theta}{p_0} + \frac{\theta}{p_1} \qquad \frac{1}{q(\theta)} = \frac{1-\theta}{q_0} + \frac{\theta}{q_1} \tag{89}$$

Then T maps continuously $L^p(\Omega, \Sigma, \mu)$ on $L^q(\Psi, \Xi, \nu)$ with norm at most equal to $M_0^{1-\theta} M_1^{\theta}$. ◇

Proof The theorem holds if $p_0 = p_1$. If $p_0 \neq p_1$ for $z \in \bar{S}$ define $\frac{1}{p(z)} \equiv \frac{1-z}{p_0} + \frac{z}{p_1}$. Notice that if $z \in L_j$ one has $Re(\frac{1}{p(z)}) = \frac{1}{p_j}$, j=1,2.

Consider a finite measurable partition of Ω in subsets E_k and consider the simple function (weighted sum of indicator functions)

$$f = \sum_1^K r_k e^{i\alpha_k} \xi(E_k), \qquad \|f\|_{p(\theta)} = 1 \tag{90}$$

where $\xi(E_k)$ is the indicator function of the set E_k and the constants r_k are chosen so that $\|f\|_{p(\theta)} = 1$.

Define

$$F(z) = \sum_1^K r_k^{\frac{p(\theta)}{p(z)}} e^{i\alpha_k} \xi(E_k) \tag{91}$$

so that $F(\theta) = f$. If $z \in L_j$ one has

$$|F(z)| = \sum_1^K r_k^{\frac{p(\theta)}{p_j}} \xi(E_k), \qquad \|F(z)\|_{p_j} = \|f\|_p^{\frac{p(\theta)}{p_j}} = 1 \tag{92}$$

Therefore the function F is analytic in S, bounded and continuous in $\bar{S}$ in the topology of $A_0 + A_1$. It follows $\|f\|_\theta \leq 1$. Therefore $\|f\|_\theta \leq \|f\|_{p(\theta)}$ for any simple function f.

The result still holds, via approximation, for any $f \in L^p(\Omega, \Sigma.\mu)$ and therefore $\|f\|_\theta \leq \|f\|_{p(\theta)}$.

We shall now prove that $\|f\|_\theta \geq \|f\|_{p(\theta)}$. We make use of the duality between L^p and $L^{p'}$. Let f a non zero function on $(A_0, A_1)_\theta$.

If $\epsilon > 0$ there exists a function $F \in \mathcal{F}(A_0, A_1)$ such that $F(\theta) = f$ and $\|F\|_{\mathcal{F}} \leq (1+\epsilon)\|f\|_\theta$. Set $B_j = L^{p_j'(\theta)}$.

Then (B_0, B_1) is a compatible pair, $L^{p'(\theta)}(\Omega, \Sigma, \mu) \subset (B_0, B_1)_\theta$ and $\|g\|_\theta \leq \|g\|_{p'(\theta)}$ for $g \in L^{p'(\theta)}(\Omega, \Sigma, \mu)$.

If g is a simple function there exists $G \in \mathcal{F}(B_0, B_1)$ such that $G(\theta) = g$ and $\|G\|_{\mathcal{F}} \leq (1+\epsilon)\|g\|_{p'(\theta)}$.

Setting $I(z) = \int F(z)G(z)d\mu$ this function is bounded continuous in $\bar{S}$ and analytic in S. Moreover if $z \in L_j$ Hoelder's inequality gives

$$|I(z)| \leq \int |F(z)||G(z)|d\mu \leq \|F(z)\|_{p_j(\theta)}.\|G(z)\|_{p'_j(\theta)}$$

$$\leq (1+\epsilon)^2\|f\|_\theta\|g\|_\theta \leq (1+\epsilon)^2\|f\|_\theta\|g\|_{p'(\theta)} \tag{93}$$

From Hadamard's inequality one derives

$$|I(\theta)| = |\int gfd\mu| \leq (1+\epsilon)^2\|f\|_\theta\|g\|_{p'} \tag{94}$$

This inequality holds for every ϵ if g belongs to a dense subset of $L^{p'(\theta)}$ and therefore for all $f \in L^{p(\theta)}$.

It follows $f \in L^{p(\theta)}$ and $\|f\|_\theta = \|f\|_{p(\theta)}$. ♡

The last inequality for which we give a proof is Young's inequality. It refers to a locally compact metrizable group and the measure is Haar measure. In the applications it is usually R^d with Lebesgue measure or finite products of R^d with the product measure.

The same theorems are useful in other cases,e.g. for Z^d with the counting measure or $Z_2 \equiv \{1, 0\}$ with addition rule mod. two and measure $\mu(\{1\} = \mu\{-1\} = \frac{1}{2}$.

6.4 *Young Inequalities*

Let G be an abelian metrizable group σ-compact (countable union of compact sets) and assume $1 < p, q < \infty$ and $\frac{1}{p} + \frac{1}{q} = 1 + \frac{1}{r} > 1$.

Denote by $*$ the convolution product. If $g \in L^p(G), f \in L^q(G)$ then $f * g \in L^r(G)$ and one has

$$\|f * g\|_r \leq \|f\|_q\|g\|_p \tag{95}$$

◇

Proof If $f \in L^1(G) + L^{p'}(G)$ the operator $T_g : f \to f * g$ maps L^1 in L^p with norm $\leq \|g\|$. By duality it also maps $L^{p'}$ in L^∞ with the same norm.

Choosing $\theta = \frac{p}{q'} = \frac{q}{r}$ one has

$$\frac{1-\theta}{1} + \frac{\theta}{p'} = \frac{1}{q}, \qquad \frac{1-\theta}{p} + \frac{\theta}{\infty} = \frac{1}{q'} \tag{96}$$

and therefore we can use the Riesz-Thorin interpolation formula with con $p_0 = 1, p_1 = p'$ $q_0 = p$ $q_1 = \infty$. ♡

We give now, together with references, a collection of inequalities which are commonly used.

Hölder-Young inequality [5]
Set $1 \leq p, q, r \leq \infty, p^{-1} + q^{-1} = 1 + r^{-1}$. Then

$$|f * g|_r \leq |f|_q \, |g|_p \tag{97}$$

Moreover the same inequality holds for the weak L^p spaces. ◇

We recall here the definition of *weak* L^p space (in notation L^p_w)

$$f \in L^p_w(M, \mu) \Leftrightarrow \exists c > 0 \; : \; \mu(\{x : f(x) > t\} < c\, t^{-p} \;\; \forall t > O \tag{98}$$

$$|f|_p^W \equiv sup_t t^p \mu \; (\{ f(x) > t^{-1}\})^{-p} \tag{99}$$

Notice that this *is not a norm* because it does not satisfy the triangular inequality. One has $L^p \subset L^p_w$ with strict inclusion unless M is a finite collection of atoms.

If $f \in L^p_w$ then there exists a constant C such that $\int_{|t|<N} \mu(\{x \; : \; f(x) > t\})t^{p-1}dt \leq C \; logN$. ♣

Young inequality II
Let $p, q, r \geq 1$ such that $\frac{1}{p} + \frac{1}{q} + \frac{1}{r} = 2$.

As usual denote by p' the exponent dual to p. Then

$$|\int_{R^d}\int_{R^d} f(x)g(x-y)h(y)dx\,dy| \leq C_p C_q C_r \|f\|_p \|f\|_q \|h\|_r \tag{100}$$

where $C_p^2 = \frac{p^{\frac{1}{p}}}{p'^{\frac{1}{p'}}}$ and $C_p C_q C_r \leq 1$. ◇

Notice that when $s = \infty$ this inequality reduces to Hölder's and if $p = r = 2$ one obtains another variant of Hölder's inequality.

Hardy-Littlewood-Sobolev inequality
Let $1 < p, t < \infty, 0 < \lambda < d$ and $\frac{1}{p} + \frac{1}{t} + \frac{\lambda}{d} = 2$. The following inequality holds

$$|\int\int f(x)|x-y|^{-\lambda} g(y)dxdy| \leq N_{p,t,\lambda} \|f\|_p \|g\|_t \tag{101}$$

where $N_{p,t,\lambda}$ is a constant which can be given explicitly for some values of p, t, λ. If $p = t = \frac{2d}{2d-\lambda}$

$$N_{p,t,\lambda} = \pi^{\frac{\lambda}{2}} \frac{\Gamma(\frac{d}{2} - \frac{1}{2})}{\Gamma(d - \frac{\lambda}{2})} \left(\frac{\Gamma(\frac{d}{2}}{\Gamma(d)} \right)^{\frac{\lambda}{d}-1} \tag{102}$$

where for $\alpha > 0$ the function Γ is defined by $\Gamma(\alpha) = \int_0^\infty t^{\alpha-1} e^{-t} dt$. ◇

A generalization of the latter inequality is

Young's weak inequality

$$|\int \int f(x)h(x-y)g(y)dxdy| \leq N_{p,t,\lambda} \|f\|_p \|g\|_t \|h\|_q^\omega \tag{103}$$

where we have denoted by $\|g\|_q^\omega$ the norm

$$h\|_q^\omega = \left(\frac{1}{B_d}\right)^{\frac{1}{q}} sup_{\alpha>0} Vol\{x \in R^d, \|h(x)\| > \alpha\}^{\frac{1}{q}} \tag{104}$$

◇

Haussdorf-Young inequality
Let $p' \geq 2$. Then

$$\|\hat{f}|_p \leq (2\pi)^{\frac{d}{p'}} C_p^d \|f\|_{p'}. \quad C_p^2 = p^{\frac{1}{p}} (p')^{-\frac{1}{p'}} \tag{105}$$

and the equality sign holds iff the function is gaussian. This inequality shows that the Fourier transform is linear continuous from da $L^{p'}(R^d)$ to $L^p(R^d)$. ◇

6.5 Sobolev-Type Inequalities

Other inequalities compare the norm of a function with the norm of its gradient.

Generalized Sobolev inequality [6]
Let $d \geq 3, 0 \leq b \leq 1, p = \frac{2d}{2b+2d-2}$. Then

$$K_{n,p} |\nabla f|_2 \geq ||x|^{-b} f|_p \tag{106}$$

where

$$K_{n,p} = \omega_{d-1}^{-\frac{1}{2r}} (d-2)^{\frac{1}{2r-1}} \left(\frac{r-1}{r}\right)^{\frac{r-1}{2r}} \frac{\gamma(2r)}{\Gamma(r+1)\Gamma(r)})^{\frac{1}{2r}}, \tag{107}$$

$$r = \frac{p}{p-2} = \frac{d}{2(1-b)}, \quad \omega_{d-1} = \frac{2\pi^{d/2}}{\Gamma(d/2)} \tag{108}$$

(if $b = 0$ one has $p = \frac{2d}{d-2} \equiv d^*$).

If $1 - d/2 \leq b < 0$ the inequality holds for functions of the radial variable $|x|$. ◇

The generalized Sobolev inequality can be derived [7] by the following Sobolev inequality in R^1

$$|f'|_2^2 + |f|_2^2 \geq M_p^{-1}|f|_p^2, \quad M_p = 2^{\frac{1}{r}-2}\left(\frac{r-1}{r}\right)^{\frac{r-1}{r}}\left(\frac{\Gamma(2r)}{\Gamma(r)\Gamma(r+1)}\right)^{\frac{1}{r}} \tag{109}$$

where $r = \frac{p}{p-2}$.

Before giving further inequalities we introduce some notation.

Definition 9 Let Ω be an open regular subset of R^N. Define

$$W^{1,p}(\Omega) \equiv \{u \in L^p(\Omega), \exists g_1..g_N \in L^p, \quad \int_\Omega u\frac{\partial\phi}{\partial x_k} = -\int_\Omega g_k\, \phi \;\; \forall\phi \in C_0^\infty(\Omega)\} \tag{110}$$

where $C_0^\infty$is the space of C^∞ functions in a neighborhood of $\partial\Omega$ (or outside a compact if Ω is unbounded).

One often uses the notation $H^{1,p}(\Omega)$ instead of $W^{1,p}(\Omega)$. ◇

One can prove that $W^{1,p}(\Omega)$ is reflexive for $1 < p < \infty$, is a Banach space for $1 \leq p \leq \infty$ (with norm $|u|_{1,p}$) and is separable for $0 \leq p < \infty$. Let $|u|_{1,p}$ denote the norm of $u \in W^{1,p}(\Omega)$. One has

$$|u|_{1,p} = |u|_p + \sum_k \left|\frac{\partial u}{\partial x_k}\right|_p \tag{111}$$

where the derivatives are in the sense of distributions. Notice that a frequently used notation is $H^1(\Omega) \equiv W^{1,p}(\Omega)$.

If Ω is bounded the following compact inclusions hold

(i) For $q \in [N, \infty).W^{1,p}(\Omega)$ is compact in $L^q(\Omega)$

(ii) If $q \in [p, p^*]$ then $W^{1,p}(\Omega) \subset L^q(\Omega)$

(iii)

$$p < N \Rightarrow W^{1,p}(\Omega) \subset_c L^q(\Omega) \quad \forall q \in [1, p^*) \tag{112}$$

(iv)

$$p = N \Rightarrow W^{1,p}(\Omega) \subset_c L^q(\Omega) \quad \forall q \in [1, \infty) \tag{113}$$

(v)

$$p > N \Rightarrow W^{1,p}(\Omega) \subset_c C(\bar{\Omega}) \tag{114}$$

(vi) Moreover, if $\Omega \subset R^1$ is bounded

$$W^{1,p}(\Omega) \subset_c L^q(\Omega), \qquad 1 \leq q < \infty \tag{115}$$

Sobolev-Gagliardo-Nirenberg inequality

$$W^{1,p}(\Omega) \subset L^{p^*}(\Omega), \qquad \frac{1}{p*} = \frac{1}{p} - \frac{1}{N} \tag{116}$$

and

$$|u|_{p^*} \leq C(N,p)|\nabla u|_p \tag{117}$$

Remark that p^* is a natural exponent as seen setting $u_\lambda(x) \equiv u(\lambda x)$. ◇

Morrey's inequality
If $p > N$ then $W^{1,p}(\Omega) \subset L^\infty(\Omega)$ and

$$|u(x) - u(y)| \leq C(p,N)|x-y|^\alpha|\nabla u|_p \quad \forall x,\ y \in R^N, \quad \alpha = 1 - \frac{N}{p} \tag{118}$$

◇

If Ω is bounded the following compact inclusions hold
(i) For every $q \in [N,\infty)$, $W^{1,p}(\Omega)$ is immersed continuously and compactly in $L^q(\Omega)$
(ii) If $q \in [p,p^*]$ then $W^{1,p}(\Omega) \subset L^q(\Omega)$
iii) $p < N \Rightarrow W^{1,p}(\Omega) \subset_c L^q(\Omega) \quad \forall q \in [1,p^*)$
(iv) $p = N \Rightarrow W^{1,p}(\Omega) \subset_c L^q(\Omega) \quad \forall q \in [1,\infty)$
(v) $p > N \Rightarrow W^{1,p}(\Omega) \subset_c C(\bar{\Omega})$
(vi) If Ω is a bounded subset of R^1 then $W^{1,p}(\Omega) \subset_c L^q(\Omega), \qquad 1 \leq q < \infty$ ◇

Poincaré inequality
Let Ω be compact and $u \in W_0^{1,p}(\Omega), \;\; 1 \leq p < \infty$. Then $|u|_p \leq C|\nabla u|_p$ ◇

Nash inequality

$$u \in H^1 \cap L^1(R^n) \Rightarrow |u|_2^{2+1/N} \leq C_N|\nabla\ u|_2|u|_2^{\frac{2}{N}} \tag{119}$$

◇

Logarithmic Sobolev inequality
If $u \in H^1(R^N)$ there exists $a > 0$ *independent from* N such that

$$\frac{a^2}{\pi}\int |\nabla\, u|^2 dx \geq \int |u(x)|^2\; log(\frac{|u(x)|^2}{|u|_2^2})dx + C(1 + loga)|u|_2^2 \tag{120}$$

◇

To conclude we recall an inequality we have already used in Volume I

Hardy inequality
If $\phi \in L^2(R^3)$ then

$$\int_{R^3} \frac{1}{4|x|^2}|\phi(x)|^2 d^3x \leq \int_{R^3} |\nabla\phi|^2 d^3x \tag{121}$$

Equivalently

$$(\phi, |\hat{p}|^2\phi) \geq (\phi, \frac{1}{4|x|^2}\phi) \tag{122}$$

◇

Hardy's inequality can be generalized to cover the case in which a magnetic field is present. This generalization is useful to provide a-priori estimates which are useful in the study of the properties of crystalline solids in magnetic fields.

Hardy magnetic inequality
If $n \geq 3$ one has

$$\int \frac{|f(x)|^2}{|x^2|}d^nx \leq \frac{4}{(n-2)^2}\int |\nabla_A f(x)|^2 d^nx \qquad (\nabla_A f)(x) \equiv (\nabla + i.e.A(x))f(x) \tag{123}$$

◇

Proof For $f \in C^\infty$ and $\alpha \in R^+$ one has

$$0 \leq \int |\nabla_A f + \alpha\frac{x}{|x'2|}f|^2 = \int |\nabla_A f|^2 + \alpha^2\int \frac{|f^2|}{|x^2|}dx + 2\alpha Re[\int \bar{f}(x)\frac{x}{|x^2|}.\nabla_A f dx] \tag{124}$$

Using Leibnitz rule

$$2\alpha Re[\int \bar{f}(x)\frac{x}{|x^2|}.\nabla_A f dx] = -\alpha\int |f(x)|^2 div(\frac{x}{|x^2|})dx = -(n-2)\alpha\int \frac{f(x)^2}{|x^2|}dx \tag{125}$$

Therefore

$$\int |\nabla_A f(x)|^2 dx \geq [-\alpha^2 + (n-2)\alpha]\int \frac{|f(x)|^2}{|x|^2}dx \tag{126}$$

Notice now that for $n \geq 3$ one has

$$max_{\alpha \in R^{+}}[-\alpha^2 + (n-2)\alpha] = \frac{(n-2)^2}{4} \tag{127}$$

This proves the equality if $f \in C^{\infty}$.

The proof for the other functions is obtained by a density argument. ♡

References

1. V.B. Lidskii, Am. Math. Soc. Trans **47**, 43–46 (1965)
2. M. Beckner, Inequalities. Ann. Math. **102**, 159–182 (1975)
3. L. Bers, F. John, M. Schecter, *Partial Differential Equations* (1964)
4. D.J. Garling, *Inequalities* (Cambrige University Press, Cambrige, 2007)
5. A. Brascamp, E. Lieb, Adv. Math. **20**, 151–173 (1976)
6. K. Okikiolo, *Aspects of the Theory*. Academic Press (1971)
7. E. Lieb, Ann. Math. **118**, 349–374 (1983)

Lecture 4: Periodic Potentials. Wigner–Seitz Cell and Brillouen Zone. Bloch and Wannier Functions

In this lecture we will give some basic elements of the theory of Schroedinger equation with periodic potentials. This theory is considered of relevant interest for Solid State Physics, i.e. the Physics of crystalline solids and their interaction with the electromagnetic field. This is the result of some rude approximations.

Experimental data suggest that to a high degree of precision the nuclei in crystals occupy fixed positions in each periodic cell of a crystal lattice in R^3; the number of atoms and their positions depend on the material under consideration. This should be interpreted as follows: the mass of the nuclei is far larger than the one of the electrons, and therefore the wave function of the nuclei is much more localized in space and the nuclei *move more slowly*.

It is convenient, in a first approximation, to regard the nuclei as fixed points (Born-Oppenheimer approximation).

As a second step, one may consider the motion of the atom in an *effective* field due to the interaction among themselves and with the electrons.

Experimental data suggest that, in this approximation and if the temperature is not too high, the nuclei form a crystalline lattice $\mathcal{L}$.

There is so far no complete explanation for this property, although some attempts have been made to prove that this configuration corresponds to the minimal energy of a system of many atoms interacting among themselves and with the electrons through Coulomb forces.

In this lecture we shall *postulate* that the atomic nuclei form a regular periodic lattice and the electrons move in this lattice subject to the interaction among themselves and with the nuclei.

More important and drastic is the assumption we will make that the interaction among electrons is negligible and so is the interaction with the (quantized) electromagnetic field generated by the nuclei and by the electrons.

The lattice structure allows the definition of an *elementary cell* (Wigner-Satz cell).

For simplicity we assume that the interaction does not depend of the spin of the electron and the presence of spin only doubles the number of eigenvalues.

G. Dell'Antonio, *Lectures on the Mathematics of Quantum Mechanics II: Selected Topics*, Atlantis Studies in Mathematical Physics: Theory and Applications 2, DOI 10.2991/978-94-6239-115-4_4

Under this assumption, the *spin-orbit coupling* can be neglected. If this coupling has an effect on the structure of the wave function this assumption can be easily removed.

Under these simplifying assumptions the electrons are described by a Schroedinger equation with a periodic potential and possibly with an external electromagnetic field.

Finer approximations can be made to take into account interactions between electrons and nuclear dynamics. Notice in particular that the nuclei are much heavier than the electrons but their mass is not infinite. Therefore their wave function is not localized at a point, the variation in time of the function of is not negligible at a time-scale much longer than the one used to describe the motion of the electrons.

In a second approximation the motion of the nuclei (and therefore the variation of the potential that the electrons feel) can therefore be considered as *adiabatic*.

On this longer time scale the motion of the nuclei can be approximated by the motion of a material point subject to the average action of the electrons and, under suitable conditions, can be described by an *effective* differential equation.

We shall outline in the next lecture the first steps of this *adiabatic* (or multi-scale) approximation which in this context takes the name *Born-Oppenheimer approximation*. In the approximation we are considering in this lecture the wave function of a single electron is described by a Schroedinger equation in an external periodic field (originated from the presence of nuclei and from external fields).

This formulation hides a crucial assumption: the crystal is infinitely extended. Since physical crystal do not have this property, this approximation is valid if the size of the crystal is very large as compared with the size of one cell.

1 Fermi Surface, Fermi Energy

Since we have already made the one-body approximation, this last approximation can be relaxed if surface effect are relevant. For example one may consider that the crystal occupies a half-space and consider currents on the boundary due to an external field (Hall effect). A remnant of the fact that physical crystals are finite is the (artificial) introduction of a Fermi surface and of *Fermi energy*.

Notice that the electrons are identical particles which satisfy the Fermi-Dirac statistics and therefore the wave function of a system of N spineless electrons in R^3 is a square-integrable function $\phi(x_1, \sigma_1; \ldots; x_N, \sigma_N)$ $x_k \in R^3$ $\sigma_k = 1, 2$ which is antisymmetric for transposition of indices (the index x_k *and* the spin index σ_k). When considering a system of N electrons we must keep into account this antisymmetry.

In a *macroscopic* crystal the number of electrons is very large and the electron wave function should be anti-symmetrized with respect to a large number of variables.

To avoid discussing the dependence of the dynamics on the specific size and shape of the (physical) crystal it is conventional to take the *infinite volume limit*.

In this limit a formulation in terms of wave functions *is no longer possible*.

A way out is a more algebraic formulation, that relays partly on the formalism of second quantization and Quantum Statistical Mechanics. We shall not discuss here this formulation.

We shall come back to a formulation through the use of C^* algebras when we will discuss briefly the case in which the lattice is substituted with a random structure, still in the one-body approximation.

In the one-particle approximation the wave function of the electron in an infinite crystal is not normalizable.

If we are interested in the properties of the crystal at equilibrium we can follow an alternative strategy and consider the wave function of a single electron in periodic potential as being normalized to one *in a single cell.* This allows the definition of *density*, the number of electrons in a given cell. Due to translation invariance the density is a constant.

Therefore the wave function is not normalized in the whole space and we will make use of generalized eigenfunctions of a Schrödinger operator in a periodic potential.

We summarize the interaction of the electron with the nuclei and the external fields by introducing in the hamiltonian a potential V in the (one) particle Schrödinger equation that describer the dynamics of a single electron. For simplicity consider a cubic lattice with edges of length one.

Let the cell of the lattice be generated by the vectors $\eta_1, \ldots, \eta_d$ applied to the origin of the coordinates.

Let the N_{el} electrons be contained in a cube Ω_N of edges $2N\eta_k$ centered in the origin. Denote by V_N the volume of Ω_N. Then $\rho(N) \equiv \frac{N}{V_N}$ is the *density* of the N-particle system.

For the moment we neglect the spin; if the Hamiltonian does not lead to spin-orbit coupling the resulting correction consists only in doubling the multiplicity of some eigenvalues.

The free Hamiltonian is a Laplacian in Ω_N. In order to define it as a self-adjoint operator we must choose boundary conditions.

In the limit $N \to \infty$ the volume of a neighborhood of the boundary becomes negligible with respect to the volume of the bulk, and we may expect that in the limit the results be independent from the specific choice of boundary conditions.

This can be proved in the absence of a potential when the infinite volume limit $Vol(\Omega_N) \to \infty$ is taken in the van Hove sense: when N increases one consider cubes of increasing size. In the presence of a potential the same can be proved under suitable assumptions.

We shall choose to work with periodic b.c. The spectrum of $-\Delta$ in Ω_N with periodic boundary conditions on the boundary $\partial\Omega_N$ is pure point with $O(\epsilon_n^N)$ eigenvalues (taking multiplicity into account).

Since the electrons are fermions, the lowest energy state is the Slater determinant made of the first N eigenfunctions in the box Ω_N. Its energy is $\sum_{k=1}^{N} \epsilon_k^N \equiv E(N)$.

We require that $\rho \equiv lim_{N\to\infty} \frac{N_{el}}{Vol(\Omega_N)}$ exists; we call this limit *density of the infinite-volume system.* We require also that the following limits exist

$$\bar{E} \equiv lim_{N\to\infty}\frac{E(N_{el})}{Vol(\Omega_N)} \quad E_F = lim_{N\to\infty}\frac{E(N)}{N_{el}} = \rho\bar{E} \tag{1}$$

We call E_F the *Fermi potential* of the infinite system. These limits exists under very general conditions on the periodic potential V.

The proof uses, as in the case of the proof of the infinite volume limit in Quantum Statistical Mechanics, decoupling techniques for disjoint regions and the fact that for any regular bounded region Ω the sum of the first M eigenvalues with arbitrary boundary condition is contained between the corresponding sum for Neumann and Dirichlet boundary conditions.

In the case $V = 0$ the Schrödinger equation in Ω_N with periodic boundary conditions can be solved by separation of variables. The spectrum is pure point and the spectral distribution converges to a uniform distribution on the positive real line with multiplicity $2d$, i.e. to the spectral density of the Laplacian in R^d.

It is easy to see that each eigenfunction has a difference of phase at opposite sides of the unit cell of the type $e^{i\frac{2\pi M}{N}}$ i.e. restricted to the unit cell they are eigenfunction of some Laplacian defined with the corresponding b.c.

It possible to show, for sufficiently regular periodic potentials, that the counting measure μ_N (the normalized sum of delta measures on the point of the spectrum, counting multiplicity) converges weakly to a measure absolutely continuous with respect to Lebesgue measure, with a density which is zero outside disjoint intervals (*bands*). For a one-dimensional system the spectral multiplicity is one but in dimension greater than one the multiplicity can vary within a single band.

The heuristic arguments outlined above suggest that this counting measure coincides in the limit $N \to \infty$ with the spectral measure of the operator $-\Delta + V$ in R^d. In the case $d = 1$ one recovers the spectrum of the operator $-\Delta + V$. In the case $d \geq 2$ it is more difficult to make this simple argument into a formal proof.

On the basis of this expectation one *assumes* that the system be well described, for $N \to \infty$ by the Bloch-Floquet theory of a single electron in a periodic potential [1]. In particular one *expects* that this theory give correct results for quantities of interest, such as electric conductivity and polarizability, and explain some important effects, like the quantum Hall effect.

These considerations on the limit when $V \to \infty$ are used to determine the values of the parameters that enter the theory of Bloch-Floquet.

In the case of a crystalline solid one assumes that the total system be electrically neutral: for the system restricted to a finite region one assumes therefore that the number of electron present is such as to balance the charge of the nuclei. This determines the number of electrons and therefore their *density* ρ.

When $V \to \infty$ the density ρ is kept constant. The choice of the numerical value for the Fermi energy has the same empirical origins. We shall come back in the next lecture to the description of macroscopic crystals.

2 Periodic Potentials. Wigner-Satz Cell. Brillouin Zone. The Theory of Bloch-Floquet-Zak

We shall present now the Bloch-Floquet-Zak theory [1] for the Schroedinger equation for a single electron in a periodic potential, neglecting the spin. If one neglects the interaction among electrons the solution for a system of N electrons will be the anti-symmetrized product of single-electron solutions.

Consider the Schroedinger equation (in suitable units)

$$i\frac{\partial\phi}{\partial t} = -\Delta\phi(x;t) + V(x)\phi(x;t), \quad x \in R^d \tag{2}$$

where the potential $V(x)$ is periodic, i.e. there exists a minimal basis $\{a_i \in R^d\}$, $i = 1, \ldots, n$ such that

$$V(x + n_i a_i) = V(x) \quad \forall n_i \in \mathbf{N} \tag{3}$$

We will consider only the case $d = 2, d = 3$.

A *basis* is a collection of linearly independent vectors such that any element $x \in R^d$ can be uniquely written as $x = \sum_i n_i a_i, \quad n_i \in \mathbf{N}$. For a cubic lattice a_i are orthogonal unit vectors.

Each cell determines a *lattice* i.e. a subset $\Gamma \in R^d$ that has the following properties

(1) Γ has no accumulation points.
(2) Γ is an additive subgroup of R^n.

A lattice may have *several minimal bases*. It determines however a *unique cell* $\mathcal{W}$ called *Wigner-Satz cell.*

The cell associated to the lattice $\mathcal{L}$) is defined as follows

$$\mathcal{W} \equiv \{x \in R^n \; : \; d(x,0) < d(x,y), \forall y \in \mathcal{L} - \{0\}\} \tag{4}$$

($d(x, y)$ is the distance between x and y).

The Wigner–Seitz cell is in general a regular polyhedron. Define the *dual lattice* as

$$\Lambda^* \equiv \{k \in R^n \; : \; k.a \in 2\pi Z \quad \forall a \in \mathcal{L}\} \tag{5}$$

The Wigner–Seitz cell of the dual lattice is called *Brillouin zone* and is *uniquely defined*. We shall denote it with the symbol $\mathcal{B}$.

In the study of periodic potentials it is convenient to consider the space $L^2(R^n)$ as direct sum of Hilbert spaces isomorphic to $L^2(\mathcal{W})$. Every function $\phi(x) \in L^2(R^n)$ is in fact equivalent (as element of $L^2(R^n)$) to the disjoint union of the translates of its restriction to $\mathcal{W}$ by the vectors of the lattice.

This suggests the use of a formalism (Bloch-Floquet-Zak) in which the direct sum is substituted by an integral over the dual cell, in analogy with the formalism of inverse discrete Fourier transform which leads from l_2 to $L^2(0, 2\pi)$. We are therefore led to consider the space

$$\mathcal{H} = \int_{\mathcal{W}}^{\oplus} L^2(\mathcal{B})\, d\mu = \int_{\mathcal{B}}^{\oplus} L^2(\mathcal{W})\, d\mu \tag{6}$$

where μ is Lebesgue measure. Notice the symmetry between $\mathcal{B}$ and $\mathcal{W}$ in (6).

3 Decompositions

The first decomposition considers properties of the functions in *momentum space* (here called *quasi-momentum*) i.e. points in the Brillouin zone. The second decomposition considers properties in *configuration space* i.e. points in the Wigner cell.

To see the interest of the notation (6) notice that if a self-adjoint operator H on $\mathcal{H}$ commutes with a group of unitary operators $U(g)$ that form a continuous representation U_G of a Lie group G then one can write $\mathcal{H}$ as direct integral, on the spectrum σ of a maximal commutative set of generators of U_G, of Hilbert spaces $\mathcal{K}_s,\ s \in \sigma$ each of them isomorphic to the same Hilbert space $\mathcal{K}$

$$\mathcal{H} = \int_{\sigma}^{\oplus} \mathcal{K}_s d\mu \tag{7}$$

where μ in the Haar measure on the group G. For this decomposition one has

$$H = \int_{\sigma} H_s d\mu, \qquad H_s = K \tag{8}$$

where K is a self-adjoint operator on $\mathcal{K}$.

Definition 1 Let $\{M, d\mu\}$ be a measure space. A bounded operator A on $\mathcal{H} \equiv \int_M^{\oplus} \mathcal{H}_s d\mu$ is said to be *decomposable* if the exists an operator-valued function $A(m)$ with domain dense in $L^\infty(M, d\mu; \mathcal{B}(\mathcal{H}_s))$ such that for any $\phi \in \mathcal{H}$ one has

$$A\phi)(m) = A(m)\phi(m) \tag{9}$$

If this is the case, we write $A = \int_M^{\oplus} A(m) d\mu(m)$.

The operators $A(m)$ are the *fibers* of A. ◇

Conversely to each function $A(m) \in L^\infty(M, d\mu; \mathcal{B}(\mathcal{H}'))$ is associated a unique operator $A \in \mathcal{B}(\mathcal{H})$ such that (9) holds.

This provides an isometric isomorphism

$$L^\infty(M, d\mu; \mathcal{B}(\mathcal{H}')) \Leftrightarrow \mathcal{B}\left(\int_M^{\oplus} \mathcal{H}'\, d\mu\right) \tag{10}$$

One can show that the decomposable operators are characterized by the property of commuting with those decomposable operators that act on each fiber as a multiple of the identity. Since the operators which we shall introduce on each fiber are not bounded in general, we extend the definition of decomposability to the case of unbounded self-adjoint operators.

Definition 2 On a regular measure space $\{M, \mu\}$ the function A with values in the self-adjoint operators on a Hilbert space $\mathcal{H}$ is called *measurable* iff the function $(A + iI)^{-1}$ is measurable. ◊

Given such function, an operator A on $\mathcal{H} = \int_M^{\oplus} \mathcal{H}' d\mu$ is said to be continuously decomposable if for almost all $m \in M$ there exists an operator $A(m)$ with domain $D(A(m)$

$$D(A) \equiv \{\phi \in \mathcal{H} \ : \ \phi(m) \in D(A(m)) \ a.e. \ \int_M \|A(m)\phi(m)\|^2_{\mathcal{H}'} d\mu < \infty\} \qquad (11)$$

with $(A\phi)(m) = A(m)\phi(m)$. We have used in (11) the notation almost everywhere to indicate that (11) holds for a set of full measure in M. We shall use the notation $A = \int_M^{\oplus} A(m)d\mu$.

The properties of decomposable operators are summarized in the following theorem

Theorem 1 ([1]) *Let $A = \int_M^{\oplus} A(m)d\mu$ where $A(m)$ is measurable and self-adjoint for a.e. m. then the following is true*
(a) The operator A is self-adjoint
(b) The self-adjoint operator A on $\mathcal{H}$ can be written $\int_M^{\oplus} A(m)d\mu$ iff $(A + iI)^{-1}$ is bounded and decomposable.
(c) For any bounded Borel function F on R one has $F(A) = \int_M^{\oplus} F(A(m))d\mu$
(d) λ belongs to the spectrum of A iff for any $\epsilon > 0$ the measure of $\sigma(A(m)) \cap (\lambda - \epsilon, \ \lambda + \epsilon)$ is strictly positive.
(e) λ is an eigenvalue of A iff it is strictly positive the measure of the set of m for which λ is an eigenvalue of $A(m)$.
(f) If every $A(m)$ has absolutely continuous spectrum then A has absolutely continuous spectrum.
(g) Let B admit the representation $B = \int_M^{\oplus} B(m)d\mu$ with $B(m)$ self-adjoint.

If B is A-bounded with bound a then each $B(m)$ is $A(m)$-bounded and the bound satisfies $a(m) \leq a$ for a.e. m.

Moreover if $a < 1$ then $A + B$ defined as $\int_M^{\oplus} [A(m) + B(m)]d\mu$ is essentially self-adjoint on $D(A)$. ◊

For the proof of this theorem we refer to [1, 2]. We only remark that part (f) of Theorem 1 states that *sufficient condition* for A to have absolutely continuous spectrum is that a.e. operator $A(m)$ has absolutely continuous spectrum. This condition *is far from being necessary.*

The following theorem is frequently used

Theorem 2 *Let* $M = [0, 1]$ *and* μ *be Lebesgue measure. Let* $\mathcal{H} = \int_{[0,1]} \mathcal{H}_m dm$ *where* $\mathcal{H}_m$ *is an infinite-dimensional separable Hilbert space and let* $A = \int_{[0,1]} A(m)dm$ *where* $A(m)$ *is self-adjoint for a.e. m.*

Suppose that for a.e value of m the spectrum of $A(m)$ *is pure point with a complete basis of eigenvectors* $\{\phi_n(m), \quad n = 1, 2, \ldots, \}$ *and eigenvalues* $E_n(m)$.

Suppose moreover that for no value of n the function $E_n(m)$ *is constant, that almost every function* $\phi_n(m)$ *is real analytic (as a function of m) in (0, 1), continuous in [0, 1] and analytic in a complex neighborhood of [0, 1]. Then the spectrum A is absolutely continuous.* ◇

Proof Let $\mathcal{H}_n = \{\phi \in \mathcal{H} \, , \phi(m) = f(m) \; \phi_n(m)\} \;\; f \in L^2(M, d\mu)$.

The subspaces $\mathcal{H}_n$ are mutually orthogonal for different value of the index n. Moreover one has $\mathcal{H} = \oplus \mathcal{H}_n$, $D(A) \subset \mathcal{H}_n$ and $A\mathcal{H}_n \subset \mathcal{H}_n$.

Consider the unitary map which for each value of n diagonalizes $A(m)$; one has

$$A_n = U_n \, A \, U_n^{-1} \qquad (A_n f)(m) = E_n(m) f(m) \qquad f(m) \in L^2([0, 1], dm) \tag{12}$$

We prove that each A_n has purely continuous spectrum. Since $E(m)$ is analytic in neighborhood of [0, 1] and is not constant, for a theorem of Weierstrass its derivative $\frac{dE_n(m)}{dm}$ has at most a finite number of zeroes. Denote by these zeroes by $m_1, \ldots, m_{N-1}$ and set $m_0 = 0 \;\; m_N = 1$. One has

$$L^2[0, 1] = \oplus_1^N L^2(m_{j-1}, m_j) \tag{13}$$

The operator A_n leaves each summand invariant and acts there as indicated in (12). On each interval E_n is strictly monotone and differentiable and one can define a differentiable function α through $E_n(\alpha(\lambda)) = \lambda$ such that

$$d\alpha = \left(\frac{dE_n(m)}{dm} \right)^{-1} d\lambda, \quad m = \alpha(\lambda) \tag{14}$$

Define the unitary operator U on $L^2((m_{j-1}, m_j))$ by $(Uf)(\lambda) = \sqrt{\frac{d\alpha}{d\lambda}} f(\alpha(\lambda))$. Then

$$UA_nU^{-1}g(\lambda) = \lambda \; g(\lambda) \tag{15}$$

We have therefore constructed a spectral representation of A_n with Lebesgue spectral measure. The spectrum of each operator A_n is therefore absolutely continuous and so is the spectrum of A. ♡

4 One Particle in a Periodic Potential

We apply now Theorem 2 to the analysis of Schroedinger equation with periodic potentials in dimension d. We begin with the simplest case, $d = 1$ and use the decomposition $\int_{\mathcal{B}}^{\oplus} L^2(\mathcal{W})\, d\mu$.

Denote by $\hat{p}_\theta^2$ the self-adjoint extension of the positive symmetric operator $-\frac{d}{dx^2}$, which is defined on C^2 functions with support in $(0, 2\pi)$ with boundary conditions

$$\phi(2\pi) = e^{i\theta}\phi(0), \qquad \frac{d\phi}{dx}(2\pi) = e^{i\theta}\frac{d\phi}{dx}(0) \tag{16}$$

Theorem 3 *Let $V(x)$ a bounded measurable function on R with period 2π. Consider the operator on $L^2(0, 2\pi)$*

$$H_\theta = \hat{p}_\theta^2 + V(x) \tag{17}$$

and define

$$\mathcal{H} = \int_{0,2\pi}^{\oplus} \mathcal{K}_\theta \frac{d\theta}{2\pi}, \quad \mathcal{K}_\theta = L^2(0, 2\pi) \tag{18}$$

Let $U : L^2(R, dx) \to \mathcal{H}$ the unitary transformation defined on $\mathcal{S}$ by

$$(Uf)_\theta(x) = \sum_n e^{-i\theta n} f(x + 2\pi n) \tag{19}$$

Then one has

$$U\left(-\frac{d^2}{dx^2} + V\right)U^{-1} = \int_{[0,2\pi)} H_\theta \frac{d\theta}{2\pi} \tag{20}$$

◇

Proof Notice that, for $f \in \mathcal{S}$ and by the periodicity of V

$$(UVf)_\theta(x) = V(x)(Uf)_\theta(x) = \sum_n e^{-i\theta n} V(x + 2\pi n) f(x + 2\pi n) \tag{21}$$

and therefore on $\mathcal{S}$ one has $UVU^{-1} = \int_{[0,2\pi)} V \frac{d\theta}{2\pi}$.

On the other hand, taking Fourier transform and noting that for $f \in \mathcal{S}$ one has $\mathcal{F}[\hat{p}_\theta^2 f] = (\frac{\theta}{2\pi} + n)^2 \mathcal{F}f$

$$-U\frac{d^2}{dx^2}U^{-1} = \int_{[0,2\pi)} \hat{p}^2 n_\theta d\theta \tag{22}$$

Equation (20) follows because $\mathcal{S}$ is a core for $H + V$. ♡

As a consequence of Theorems 2 and 3 in order to study the spectral properties of $-\frac{d^2}{dx^2} + V(x)$ with V 2π-periodic it suffices to study $\hat{p}_\theta^2 + V(x)$ for $0 \le \theta < 2\pi$ in $L^2(0, 2\pi)$.

Lemma 1 *For each value of* θ*:*
(i) The operator $\hat{p}_\theta^2$ *has compact resolvent.*
(ii) $\hat{p}_\theta^2$ *is the generator of a positivity improving contraction semigroup.*
(iii) The resolvent of $\hat{p}_\theta^2$ *is an operator-valued function analytic in* θ *in a complex neighborhood of* $[0, 2\pi]$. ◇

Proof Items (i) and (ii) could be proved by general arguments. We give a constructive proof which provides also a proof of (iii).

Let $G_\theta = (\hat{p}_\theta^2 + I)^{-1}$. If $f \in C_0^\infty((0, 2\pi))$ both gf and $G_\theta f$ solve the equation $-u''(x) + u(x) = f(x)$ in $(0, 2\pi))$ and therefore their difference solves $-v_\theta'' + v_\theta = 0$.

It follows that there exist constants a and b such that $(G_\theta f)(x) - (Gf)(x) = ae^x + be^{-x}$.

The function $(G_\theta f)(x)$ must satisfy the boundary condition

$$(G_\theta f)(2\pi) = e^{i\theta}(G_\theta f)(0), \quad (G_\theta f)'(2\pi) = e^{i\theta}(G_\theta f)'(0) \tag{23}$$

and therefore

$$G_\theta(q, y) = \frac{1}{2}e^{-|x-y|} + \alpha(\theta)e^{x-y} + \bar{\alpha}(\theta)e^{y-x}, \quad \alpha(\theta) = \frac{1}{2(e^{2\pi - i\theta}) - 1} \tag{24}$$

Properties (i), (ii) follow from the explicit form of G_θ. Also (iii) is satisfied because $\theta \to G_\theta$ is analytic (as map from C to the Hilbert-Schmidt operators) for all $|Im\theta| < 2\pi$. ♡

We can now study the operators $H_\theta = \hat{p}_\theta^2 + V$

Theorem 4 ([2]) *Let* V *be piece-wise continuous and* 2π*-periodic. Then*
(i) H_θ *has purely point spectrum and is real-analytic in* θ.
(ii) H_θ *and* $H_{2\pi-\theta}$ *are (anti)-unitary equivalent under complex conjugation.*
(iii) For $\theta \in (0, \pi)$ *the eigenvalues* $E_n(\theta)$, $n = 1, 2, \ldots$ *of* H_θ *are simple.*
(iv) Each $E_n(\theta)$ *is real-analytic in* $(0, \pi)$ *and continuous in* $[0, \pi]$.
(v) For n *odd (resp. even)* $E_n(\theta)$ *is strictly increasing (resp. decreasing) in* θ *in the interval* $(0, \pi)$*. Moreover*

$$E_k(0) \le E_{k+1}(1) \le E_k(\pi) \le E_{k+1}(\pi) \quad k = 1, 2, \ldots \tag{25}$$

(vi) The eigenvectors $\phi_n(\theta)$ *can be chosen to be real-analytic in* θ *for* $\theta \in (0, \pi) \cup (\pi, 2\pi)$ *and continuous in* 0 *and* π *(with* $\phi_n(0) = \phi_n(2\pi)$*).* ◇

Proof (i) this follows from regular perturbation theory because the statement is true for $V = 0$.
(ii) this relation is verified for $V = 0$ and therefore holds if V is H_0-bounded.
(iii) If E is an eigenvalue of H_θ the equation $-u'' + Vu = Eu$ has a solution, but this can be true for at most one of the boundary conditions.
(iv) Consider $E_1(0)$. It is a simple eigenvalue because H_0 is the generator of a positivity preserving semigroup. Since H_θ is analytic in a neighborhood of 0 and $E_1(0)$ is simple there exists a neighborhood of 0 in which H_θ has a minimum eigenvalue $E_1(\theta)$ analytic and simple.

If the upper end of the analyticity interval does not coincide with π there is $\theta_0 < \pi$ such that $E_1(\theta) \to \infty$ when $\theta \to \theta_0$ (remark that H_θ is bounded below because $V \in L^\infty$). Therefore it is sufficient to prove that $E_1(\theta)$ is bounded in θ in $[0, \pi)$.

This is true because $E_1(\theta)$ is the lowest eigenvalue of H_θ.

This argument can be repeated for $E_n(\theta)$ $n > 0$. Notice that $E_n(0)$, $n > 1$ can be degenerate but $E_n(\epsilon)$ is simple for ϵ small and different from zero.
(v) We begin by proving $\forall\theta \quad E_1(0) \le E_1(\theta)$. Since e^{-tH_0} is positivity improving the eigenvector $\phi_1(0)$ can be chosen to be strictly positive and extends to a periodic function $\tilde{\phi}_0$ on R.

Consider its restriction to $(-2\pi n, 2\pi n)$ and denote by H_k the operator $-\frac{d^2}{dx^2} + V$ restricted to periodic functions in this interval. It is easy to prove that $E_1(0)$ is the lowest eigenvalue of H_k. It follows for all positive integers n and for every $\psi \in C_0^\infty(-2\pi n, 2\pi n)$

$$\left(\psi, \left[-\frac{d^2}{dx^2} + V\right]\psi\right) \ge E_1(\psi, \psi) \tag{26}$$

Since $\cup_n(C_0^\infty(-2\pi n, 2\pi n))$ is dense in $L^2(R)$ we conclude that for a.a. θ one has $E_1(\theta) \ge E_1(0)$; from the continuity of E_i the inequality holds for all values of θ.

Consider now the differential equation

$$-\frac{d^2u(x)}{dx^2} + V(x)u(x) = Eu(x) \tag{27}$$

Let $u_1^E(x)$ and $u_2^E(x)$ be the solutions with boundary conditions $u_1^E(0) = 1$, $(u_1^E)'(0) = 0$ and $u_2^E(0) = 0$, $(u_2^E(0))' = 1$ respectively.

Let $M(E)$ be the Hessian matrix corresponding to the two solutions and of their first derivatives in 2π and define

$$D(E) \equiv Tr\, M(E) = u_1^E(2\pi) + (u_2^E)'(2\pi) \tag{28}$$

$M(E)$ has determinant one (the Wronskian is constant); we denote by λ and λ^{-1} its eigenvalues.

If $v(x)$ is a solution of (27) then the matrix $M(E)$ provides a linear relation between the vector $v(0),\ v'(0)$ and the vector $v(2\pi), v'(2\pi)$. It follows that the equation $H_\theta\psi = E\psi$ admits solutions iff $e^{i\theta}$ is an eigenvalue of $M(E)$; therefore $D(E) = 2\cos\theta$ and

$$Arc\ \cos\left[\frac{1}{2}D(E_1(\theta))\right] = \theta \tag{29}$$

We know that $D(E_1(0)) = 2$. When θ increases from 0 to π the function $D(E)$ decreases monotonically from 2 to -2.

Therefore the first value of E for which $D(E) = -2$ is $E_1(\pi)$. The next value must be $E_2(\pi)$. In the interval $(E_2(\pi), E_2(0))$ the function $D(E)$ is increasing and takes the value 2 when $E = E_2(0)$.

There are therefore intervals of the real line (called *bands*) in which $D(E)$ increases from -2 to 2 followed by intervals in which it decreases from 2 to -2.

This intervals are $[E_{2k+1}(0), E_{2k+1}(\pi)]$ and $[E_{2k}(\pi), E_{2k}(0))]$. The band k can touch the band $k+1$ only if either $E_k(\pi) = E_{k+1}(\pi)$ or $E_k(0) = E_{k+1}(0)$ (therefore if the corresponding eigenvalue is degenerate). Notice that for $V = 0$ all eigenvalues are degenerate and the spectrum is the entire positive half-line.
(vi) It follows from regular perturbation theory of self-adjoint operators that the eigenvectors $\phi_n(\theta)$ can be chosen as functions of θ analytic $(0, \pi) \cup (\pi, 2\pi)$ and continuous in $[0,\ 2\pi]$. ♡

We summarize these results in the following theorem

Theorem 5 *Let* $H = -\frac{d^2}{d^2} + V(x)$ *on* $L^2(R)$ *where* $V(x)$ *is periodic and piece-wise continuous. Denote* $E_k(0)$ *the eigenvalues of the (self-adjoint) operator* H_p *on* $[0, 2\pi]$ *with periodic boundary conditions and* $E_k(\pi)$ *those of* $H_{a.p.}$ *with anti-periodic b.c.* $(\phi(2\pi) = -\phi(0))$.

Then
(i) $\sigma(H) = \cup_n([E_{2k+1}(0), E_{2k+1}(\pi)] \cup [E_{2k}(\pi), E_{2k}(0)])$
(ii) H does not have discrete spectrum.
(iii) H has absolutely continuous spectrum. ◇

Proof Item (i) is a consequence of Theorem 1 since $E_n(\theta)$ is continuous for all n. Item (ii) is a consequence of Theorem 2 since $E_n(\theta)$ is strictly monotone. Therefore for each E_0 the set of values of θ for which $E(\theta)$ is an eigenvalue consists of at most two points.
Item (iii) is a consequence of Theorem 3 and Lemma 1. ♡

We shall call *gap* any open interval that separates two disjoint parts of the spectrum.

Remark that the boundaries of a gap are given by eigenvalues corresponding to the periodic and anti-periodic solutions of the Schroedinger equation. The eigenvalues corresponding to other boundary conditions are internal points of the spectrum.

This property is *no longer true* [1] for dimension $d \geq 2$. In this case the eigenfunctions at the borders of the spectral bands (*edge states*) may not correspond to specific boundary conditions at the border of the Wigner-Satz cell.

This feature constitutes a problem in the extension of the analysis we have so far, based on a fibration of the Hilbert space with basis corresponding to $L^2(B, (L^2(K))$ and suggests the use of the fibration with base $L^2(K, (L^2(B))$.

The fibration has the property that each fiber corresponds to a given energy. The energy is given as a function of the quasi-momentum by a *dispersion relation* Recall that we are considering particles that obey the Fermi statistics, and therefore two particles cannot be in the same (not degenerate) energy state. Therefore the energy of the ground state of a system of finite size (and therefore discrete spectrum) is an increasing function of the number of states occupied.

This leads to the definition of *occupation number* and *Fermi surface* for finite systems and, upon taking limits, of density and Fermi surface for infinitely extended systems (as are the crystals we are considering).

5 The Mathieu Equation

Before studying this new fibration we give a concrete example of the analysis in dimension one, the *Mathieu equation* corresponding to $V(x) = \mu \cos x$.

Lemma 2 *In the case*

$$H = -\frac{d^2}{dx^2} + \mu \cos x, \quad \mu \neq 0 \tag{30}$$

every gap is open. ◇

Proof Let H_p (resp. H_a) the self-adjoint operators on $L^2(0, 2\pi)$ be defined as restriction of $H = -\frac{d^2}{dx^2} + \mu \cos x$ to functions that are periodic (resp. anti-periodic).

We must prove that H_p and H_a don't have multiple eigenvalues. Let us determine first the eigenvalues of H_p. Taking the Fourier transform one has

$$H_p \phi_n = n^2 \phi_n, \quad \phi_n = \frac{1}{\sqrt{2\pi}} e^{inx} \tag{31}$$

therefore the eigenvalues are given by

$$(n^2 - E)a_n + \frac{\mu}{2}(a_{n+1} + a_{n-1}) = 0 \quad \sum_n |a_n|^2 = 1 \tag{32}$$

We prove that the solution, if it exists, is unique. Suppose there are two distinct solutions corresponding to the same E. Let $\{a_n\}$, $\{b_n\}$ be the two solutions. Multiplying the equations by b_n and a_n respectively and subtracting we obtain

$$c_n \equiv b_n a_{n+1} - a_n b_{n+1} = c_{n-1} \tag{33}$$

Since both $\{a_n\}$ and $\{b_n\}$ are in l^2, $c_n = 0 \ \forall n$. It follows

$$a_{n+1}b_n = a_n b_{n+1} \tag{34}$$

and therefore $a_n = cb_n \ \forall n$.

We remark now that if two consecutive a_n are zero, then there is no non-zero solution of (34) (this is due to the specific form of the potential). Therefore at least one among a_n and a_{n+1} is not zero, and the same holds for b_n.

If E is doubly degenerate, since the potential is even, we can assume that one of the solution is even and the other is odd. If $\{b_n\}$ is odd, then $b_0 = 0$ and $b_1 \neq 0$.

On the other hand (32) for $n = 0$ gives $-Ea_0 + \mu a_1 = 0$. Since a_0 and a_1 cannot be both zero it follows $a_0\, b_1 \neq 0$. But $a_1\, b_0 = 0$ and this violates (33).

The contradiction we have obtained shows that the eigenvalue is not degenerate. We have proved that for the Mathieu potential all gaps are open. ♡

6 The Case $d \geq 2$. Fibration in Momentum Space

We have seen that for $d \geq 2$ it in convenient to consider a fibration in momentum space.

We begin to discuss this fibration in one dimension. Assume $V(x) \in C_0^\infty(R)$ so that $H = -\frac{d^2}{dx^2} + V(x)$ maps $\mathcal{S}$ into itself. Taking Fourier transform

$$\hat{H}fp = p^2 \hat{f}(p) + \frac{1}{\sqrt{2\pi}} \int \hat{V}(p-p')\hat{f}(p')dp' \tag{35}$$

If $V(x)$ is 2π-periodic one has

$$V(x) = \sum_{-\infty}^{\infty} \tilde{V}_n e^{inx} \tag{36}$$

where the sum is uniformly convergent. Correspondingly (35) reads

$$\hat{H}f(p) = p^2\hat{f}(p) + \sum_{-\infty}^{\infty} \hat{V}_n \hat{f}(p-n) \tag{37}$$

Theorem 6 *Let $\mathcal{H}' = l_2$ and define $\mathcal{H} = \int_{[-\frac{1}{2},\frac{1}{2}]}^{\oplus} \mathcal{H}' dq$.*

For $j \in (-\frac{1}{2}, \frac{1}{2}]$ let

$$(H_j g)(p) = (p+j)^2 g_j(p) + \sum_{-\infty}^{\infty} (\hat{V}_n g_{j-n})(p) \tag{38}$$

Setting $H \equiv -\frac{d^2}{dx^2} + V(x)$ *one has, denoting by con* $\mathcal{F}$ *the Fourier transform as unitary operator*

$$\mathcal{F}H\mathcal{F}^{-1} = \int_{(-\frac{1}{2},\frac{1}{2}]}^{\oplus} H_j dj, \qquad [(\mathcal{F}f)(q)]_j = \hat{f}(q-j) \tag{39}$$

◇

Remark that (38) defines the operator through its integral kernel. In this specific case, this kernel represents a differential operator on $L^2(\mathcal{W})$.

But if one restricts the operator to a spectral subspace one obtains in general a *pseudo-differential operator* (the projection is represented by an integral kernel in this representation). This is the reason why the mathematical theory of Schroedinger operators with periodic potentials makes extensive use of the theory of pseudo-differential operators

Before generalizing to the case $d \geq 2$ we give estimates which extend to periodic potentials Kato's estimates. Notice that the Rollnik class criteria are not applicable here because a periodic potential does not belong to L^p for any finite p.

Making use of periodicity it is sufficient to have *local estimates.*

Definition 3 A function V on R^n is *uniformly locally* in L^p iff there exists a positive constant M such that $\int_C |V(x)|^p d^n x \leq M$ for any unitary cube C. ◇

With this definition Kato's theory extends to perturbations uniformly locally in L^p.

Theorem 7 *Let* $p \geq 2$ *for* $d \leq 3$, $p > 2$ *for* $d = 4$ *and* $p \geq \frac{d}{2}$ *for* $d \geq 5$. *Then the multiplication for a function* V *which is uniformly locally in* L^p *is an operator on* $L^2(R^n)$ *which is Kato-bounded with respect to the Laplacian with bound zero.* ◇

Proof If $\frac{1}{p} + \frac{1}{q} = 1$ for any $\epsilon > 0$ there exists A_ϵ such that

$$\|f\|_q^2 \leq \epsilon \|\Delta f\|_2^2 + A_\epsilon \|f\|_2^2 \tag{40}$$

For each unit cube C define $\|f\|_{r,C} \equiv [\int_C \|f(x)\|^r d^n x]^{\frac{1}{r}}$. Let C_3 be the cube of side 3 with the same center.

We shall make use of a standard process of localization. Let η be a C^∞ function with support strictly contained in C_3 and taking value 1 on C.

From (40) we obtain

$$\begin{aligned} \|f\|_{q,C}^2 &\leq \|\eta f\|_q^2 \leq \epsilon \|\Delta(\eta f)\|_2^2 + A_\epsilon \|\eta f\|_2^2 \\ &\leq 3\epsilon \|\Delta f\|_{2,C_3}^2 + B\|\nabla f\|_{2,C_3}^2 + D\|f\|_{2,C_3}^2 \end{aligned} \tag{41}$$

We have used the identities (the constants B and D do not depend on C.)

$$\Delta(\eta f) = f\ \Delta\eta + \eta\ \Delta f + 2\nabla\eta.\nabla f, \qquad (a+b+c)^2 \le 3(a^2 - b^2 + c^2) \tag{42}$$

Choose now $\xi \in Z^n$ and let C_ξ be the unit cube centered in in ξ and $C_{\xi,3}$ the cube of side 3 centered in ξ. By assumption

$$\|V\| \equiv sup_\xi \|V\|_{p,C_\xi} < \infty \tag{43}$$

We have therefore, for $\frac{1}{p} + \frac{1}{q} = \frac{1}{2}$

$$\begin{aligned} \|V f\| &= \sum_\xi \|Vf\|^2_{2,C_\xi} \le \sum_\xi \|V\|^2_{p,C_\xi} \|f\|^2_{q,\xi} \\ &\le \|V\|^2 \sum_\xi (3\epsilon \|\Delta f\|^2_{2,C_{\xi,3}} + B\|\nabla f\|^2_{2,C_{\xi,3}} + D\|f\|^2_{2,C_{\xi,3}} \\ &\le \|V\| 3^n [4\epsilon \|\Delta f\|^2_2 + \left(D + \frac{B}{4\epsilon}\right) \|f\|^2_2] \end{aligned} \tag{44}$$

Notice that, a part from a set of zero measure, every x belongs to exactly 3^n cubes $C_{\xi,3}$ and we have made use also of Plancherel inequality

$$\|\nabla f\|^2_2 \le \delta \|\Delta f\|^2_2 + \frac{1}{4\delta} \|f\|^2_2 \tag{45}$$

which in turn follows from the numerical inequality $a \le \delta\ a^2 + \frac{1}{4\delta}$. ♡

7 Direct Integral Decomposition

We give now some details about the integral decomposition of a Schroedinger operator in R^d with periodic potential.

This theory is a particular case of the general theory of direct integral decomposition of Hilbert spaces and operators [1, 2].

If the periodicity lattice Γ has basis $\gamma_1, \ldots, \gamma_n \in R^d$, the *dual lattice* is defined by the *dual basis* $\gamma_1^*, \ldots, \gamma_n^* \in R^d$ with $(\gamma_i, \gamma_j^*) = 2\pi\delta_{i,j}$.

We shall call *Brillouin zone* the fundamental centered domain of Γ^*

$$\mathcal{B} = \left\{ \sum_{i=1}^n t_i\ \gamma_i^* \mid 0 \le t_i \le 1 \right\} \tag{46}$$

We can now generalize to $d > 1$ the analysis in momentum space that we have given for $d = 1$. With estimates similar to those for $d = 1$ and using Theorem 6 one proves that, denoting by $\mathcal{W}$ is the elementary cell

$$\mathcal{W} = \left\{ x \,:\, x = \sum_{i=1}^{n} t_i\, \gamma_i \,,\; 0 \le t_i \le 1 \right\} \tag{47}$$

if $V \in L^p(\mathcal{W})$ (where $p = 2$ if $d \le 3$, $p = 4$ if $d = 4$ and $p = \frac{d}{2}$ if $d \ge 5$), then $-\Delta + V$ is unitary equivalent to

$$\frac{1}{2\pi} \int_{[0,2\pi)^n}^{\oplus} H_\theta d^n\theta \quad H_\theta = H_\theta^0 + V \tag{48}$$

We have denoted by H_θ^0 the operator Δ on $L^2(\mathcal{W}, d^n x)$ with boundary conditions

$$\phi(x + a_j) = e^{i\theta_j} \phi(x), \qquad \frac{\partial \phi}{\partial x_j}(x + a_j) = e^{i\theta_j} \frac{\partial \phi}{\partial x_j}(x), \tag{49}$$

For every value of θ the potential V is Kato-infinitesimal with respect to H_θ^0.

As a consequence each H_θ has compact resolvent and a complete set of eigenfunctions $\phi_m(\theta, x)$ (that using (49) can be extended to R^n) and a corresponding set of eigenvalues $E_m(\theta)$.

It is possible to show that the functions $E_n(\theta)$ are measurable and that the corresponding eigenfunction can be chosen to be measurable.

The operator H is equivalent to $\int_{\mathcal{B}}^{\oplus} H_k dk$ where H_k is defined on $l_2(Z^n)$ by

$$(H_k g)_m = (H_k^0 g)_m + \sum_{l \in Z^m} \tilde{V}_l\, g_{m-l}, \quad (H_k^0 g)_m = \left(k + \sum m_j \alpha_j\right)^2 g_m \tag{50}$$

and has domain $\mathcal{D} = \{g \in l_2(Z^n), \;\sum k_j\, |\alpha_j|^2 < \infty\}$. In (50) $m \in Z^n$ and $\tilde{V}_m$ are the coefficients of V as a function on $\mathcal{B}$. Explicitly

$$\tilde{V}_m = (vol \mathcal{K})^{-1} \int_{\mathcal{K}} e^{-i \sum_{j=1}^{n} m_j\, (\alpha_j, x)} V(x) d^n x \tag{51}$$

with inverse relation $V(x) = \sum_{l \in Z^n} \tilde{V}_l e^{i \sum_{j=1}^{n} l_j\, (\alpha_j, x)}$. Notice that this sum is uniformly convergent since $V \in L^2_{loc}$.

Equation (50) can be used to extend to $k \in C^n$ the resolvent of H_k as an entire function. This will be useful to construct a basis of functions which decay exponentially (Wannier functions).

We prove now that $-\Delta + V$ has absolutely continuous spectrum. Since V is infinitesimal with respect to H_k^0 it is sufficient to give the proof for H_0. By Theorem 2 it is sufficient to prove that the eigenvalues and eigenfunctions of H_k^0 can be analytically continued.

Denote by $E_m(k)$ the eigenvalues, which from (50) are seen to be $E_m(k) = (k + \sum m_j\alpha_j)^2$. We proceed by induction on the number of degrees of freedom. Choose a basis $\alpha_k, k = 1, \ldots, n$ such that the first element be in the direction of the vector a_1 (the first element of the configuration lattice

$$k = s_1 a_1 + s_2\alpha_2 + \cdots s_n\alpha_n \tag{52}$$

From (50)

$$H_k^0 = \int_{s_\perp \in \mathcal{N}} ds_2 \ldots ds_N \int_{s_1 \in \mathcal{M}_{s_\perp}} ds_1 \left[H_k(s_1 a_1 + \cdots s_n k_n) + \left(k + \sum m_j\alpha_j\right)^2 \right] \tag{53}$$

where $s_\perp = \{s_2, \ldots, s_N\}$ and $\mathcal{N}, \ \mathcal{M}_\perp$ are chosen to cover all integration domain.

If we regard the eigenvalues of H_k as functions of $s_1, s_\perp$ thy are are continuous in all variables and analytic in s_1 in a neighborhood of $\mathcal{M}_{s_\perp}$. With this choice of basis (which depends on k) one has

$$E_m(s, s_\perp) = (1 + s_1)a_1^2 + \sum_{p\geq 2}(k.s_p)^2 \tag{54}$$

Moreover on can prove that if $\beta > \frac{n}{2}, \ \ \beta \geq n-1$ the series

$$f_\beta(y) = \sum_m |E_m(x + iy, s_\perp) + 1|^{-\beta} \tag{55}$$

converges uniformly in $s_\perp$ and that, if $\beta > n-1$, one has $lim_{y\to\pm\infty} f_\beta(y) = 0$. For each m this function admits a continuation $E_m(z, s_\perp)$ which is analytic in $z \in C$ and continuous in $s_\perp$.

Also the eigenvectors are analytic functions of z in a neighborhood of the real axis, continuous in $s_\perp$. From the explicit expression one sees that the function $f(s)$ is not constant for any value of s. This estimates prove analyticity off the real axis for the resolvent of H_k^0. We have proved

Theorem 8 *Let $\hat{V} \in l_\beta$, where $\beta < \frac{d-1}{d-2}$ if $d \geq 3$ and $\beta = 2$ if $d = 2$, the operator $-\Delta + V$ has absolutely continuous spectrum.* ◇

It is now useful to introduce for any $\phi \in \mathcal{S}$ the *Bloch-Floquet-Zak transformation*

$$(U\phi)(k, x) = \sum_{\gamma\in\Gamma} e^{-i(x+\gamma).k}\phi(x + \gamma), \quad x, k \in R^d \tag{56}$$

If $\phi(x) \in L^2(R^d)$ the series (56) converges $L^2(\mathcal{B}, L^2(\mathcal{W}))$. The choice of the exponential factor in (56) is convenient because it gives rise to simple properties under lattice translations. One has indeed

$$(U\phi)(k, x+\gamma) = (U\phi)(k,\gamma) \qquad (U\phi)(k, x+\gamma^*,) = e^{-ix.\gamma^*}(U\phi)(k,x) \tag{57}$$

Definition 4 The function $U\phi$ which is associated uniquely through (56) to the state described by $\phi(x)$ is called *Bloch function* associated to ϕ. ◇

For each $k \in R^d$, $(U\phi)(k, .)$ is Γ- periodic and therefore it can be regarded as L^2 function over $T^d \equiv R^d/\Gamma$ (the d-dimensional torus).

Remark that T^d can be realized as Bloch cell or as Brillouin zone, with opposite sides identified through the action of Γ^*. The vector $k \in R^d$ takes the name of *quasi-momentum* (notice the analogy with with Fourier transform).

The function $(U\phi)(k, x)$ can be written as

$$(U\phi)(k,x)) = e^{ikx} v_k(x) \tag{58}$$

where v_k is periodic in x for each value of k. Moreover if $\phi_\gamma(x) = \phi(x+\gamma), \quad \gamma \in \Gamma$ then

$$(U\phi_\gamma)(k,x) = e^{-ik.\gamma}(U\phi)(k,x) \tag{59}$$

For periodic potentials the Bloch functions and the Bloch-Floquet transform have a role similar to that of plane waves and Fourier transform for potentials vanishing at infinity, and one has analogues of the classical Plancherel and Paley-Wiener theorems.

Let L^2_a be the space of locally $L^2(R^d)$ that decay at infinity sufficiently fast

$$\phi \in L^2_a \Rightarrow sup_{\gamma\in\Gamma} e^{a\gamma} |\phi|_{L^2(\mathcal{W}+\gamma)} < \infty \tag{60}$$

We will say that a function ψ has *exponential decay* of type a if $\psi \in L^2_a$.

If $\mathcal{H}$ is a Hilbert space and $\Omega \subset C^d$ we will use the notation $\mathcal{A}(\Omega, \mathcal{H})$ for the space of $\mathcal{H}$-valued functions which are analytic in Ω (for the topology of uniform convergence on compacts). One has the following results

Theorem 9 ([1]) *(1) If $\phi \in L^2(R^d)$ the series (56) converges in $L^2(T^*, L^2(B))$ and the following identity holds (analog of Plancherel identity))*

$$|\phi|''_{L^2(R^d)} = \frac{1}{vol(\mathcal{B})}\int_{\mathcal{B}} |U(\phi)(k,.)|^2_{L^2(\mathcal{W})} dk = \frac{1}{vol(T^*)}\int_{T^*} |(U\phi)(z,.)|_{L^2(\mathcal{W})} dz \tag{61}$$

where dk is Lebesgue measure on $\mathcal{B}$ and dz is Haar measure on T^*.

(2) For each $0 < a < \infty$ the map $\phi \to U\phi$ is a topological isomorphism between $L^2_a(R^d)$ and $\mathcal{A}(\Omega_a, L^2(\mathcal{W}))$, where Ω_a is the strip $z \in C^d$, $|Imz| \le a$. This is the analogue of Paley-Wiener theorem.

(3) The following inversion formula holds

$$\phi(x) = \frac{1}{vol(T^*)} \int_{T^*} (U\phi)(k, x)dk \tag{62}$$

◇

If $H = -\Delta + V(x)$, V periodic, the Bloch-Floquet transform reduces H with respect to T^*.

Denote by $H(k)$ the reduction of the operator H to the function with fixed quasi-momentum $k \in T^*$ (a self-adjoint operator with compact resolvent if $V \in L^\infty$) and denote by $\lambda_1(k) \le \lambda_2(k) \le \ldots.$ its eigenvalues in increasing order. One proves without difficulties.
(a) The functions λ_k are continuous, Γ^*-periodic and piece-wise analytic.
(b) The spectrum of H is $\sigma(H) = \cup_m I_m$ where I_m is the collection of the $\lambda_m(k)$.

A detailed description of the *band functions* λ_k and of the corresponding *Bloch waves* $\psi_m(k, x)$ (solutions of $H\psi_m = \lambda_m\psi_m$) can be found in [3].

We have noticed that the Bloch waves for V periodic, are the analog of the generalized eigenfunctions for potentials that decrease at infinity. If $V = 0$ the latter are plane waves and the dual basis (Fourier transformed) are Dirac measures.

We will be mainly concerned with the case of the hamiltonian $H \equiv -\Delta + V_\Gamma(x)$ with V_Γ *real* and periodic of period Γ.

One has $[H, T_\gamma] = 0$ where T_γ is the unitary operator implementing translations by vectors in the Bravais lattice.

$$T_\gamma\phi(x) = \phi(x - \gamma) \tag{63}$$

The same analysis can be applied to the *periodic Pauli hamiltonian*

$$H_{Pauli} = \frac{1}{2}[(-i\nabla_x + A_\Gamma(x).\sigma)^2\phi(x) + V_\Gamma(x)\phi(x)] \tag{64}$$

where A_Γ $R^3 \to R^3$ is $\Gamma-$periodic and $\sigma \equiv \{\sigma_1, \sigma_2, \sigma_3\}$ are the Pauli matrices.

We shall consider here only the Schrödinger equation.

The lattice translation form an abelian group and therefore

$$T_\gamma\phi(x) = e^{ik.\gamma} \quad k \in T_d \quad k \in T_d^* \equiv R^d/\Gamma^* \tag{65}$$

where Γ^+ is the dual lattice of Γ. The quantum numbers $k \in T_d^*$ are called *Bloch momenta* and the quotient R^d/Γ^* is the *Brillouin zone* (or *Brillouin torus*).

Notice that the Bloch functions are written as

$$\phi(k, x) = e^{ikx}u(k, x) \tag{66}$$

where for each k the function $u(k.x)$ is periodic in x and an element of the Hilbert space $L^2(T^d)$.

Later we shall study the topological properties of the states (or rather of the wave functions representing the states). For this it is convenient to consider periodic functions, so that the topological properties (or rather homological properties) can be seen as obstructions to the continuation as a smooth function in the interior of a periodic cell a function that is periodic and smooth at the boundary.

We shall see that the phase of the function can be changed smoothly except in a point (e.g. the center of the cell). In dimension 2 the possible singularity is a *vorticity* and in cohomology it correspond to a non trivial element of the first Chern class.

We rewrite (66) making use the *Bloch-Floqet-Zak transformation*, a map between $L^2(R^d)$ and $L^2(\Gamma) \times L^2(\Gamma^*)$ defined on continuous functions by

$$(U_{BFZ})\psi(k,x) \equiv \left(\frac{1}{volB}\right)^{\frac{1}{2}} \sum_{y\in\Gamma} e^{-ik(x-y)}\psi(x-y)$$

where B denotes the fundamental cell for Γ^* i.e.

$$B = \left\{ k = \sum_{j=1}^{d} k_j b_j \right\} \qquad -\frac{1}{2} \leq k_j < \frac{1}{2}$$

Notice that this implies that $\phi \equiv U_{BFZ}\psi$ is Γ periodic in y and Γ^* pseudoperiodic in k i.e. $\phi(k+\lambda, x) = e^{-i\lambda.x}\psi(k,x)$. As a consequence of hte definitions one has, if $f(x)$ is Γ-periodic

$$(U_{BFZ}T_\gamma U_{BFZ}^{-1} = \int_B dk e^{ik\gamma} I \qquad U_{BFZ} f(x) U_{BFZ}^{-1} = \int_B dk f_\gamma(y)$$

$$U_{BFZ}\left(-i\frac{\partial}{\partial x}\right) U_{BFZ}^{-1} = \int_B dk \left(-i\frac{\partial}{\partial y_j} + k_j\right)$$

8 Wannier Functions

Analogous considerations lead for periodic potentials to the definition of *Wannier functions* [1–4].

Remark that, although Bloch waves are an important instrument in the analysis of some electronic properties of a crystal, in particular conduction, they are not a convenient tool for the analysis of other properties, especially those that refer to chemical bonds and other local correlations [2–5].

For comparison, recall that in the case of potentials vanishing at infinity in order to study *scattering* it is convenient to make use of the momentum representation (generalized plane waves) while to study *local properties* it is convenient to use position coordinates, i.e. measures localized in a point. In a similar way, to study local properties in a crystal it is convenient to make use of a complete set of functions which are localized as much as possible. These are the *Wannier functions*.

For example in the modern theory of polarization a fundamental role is played by the modification under the action of an external electric field of those Wannier functions which are *localized at the surface of a crystal.*

Since, as we shall see, the Wannier functions can be expressed as weighted integrals over k of Bloch functions $\phi_m(k, x)$, the possibility to be localized in a region of the size of a Wigner-Satz cell depends both on the weight and on the regularity in k of the Bloch functions.

Let $\phi_m(k, x) \in L^2(T^*, L^2(W))$ be a Bloch function relative to the function $\lambda_m(k)$. Notice that, even when the eigenvalue $\lambda_m(k)$ is simple the function $\phi_m(k, x)$ is defined for each value of x only modulo a phase factor that may depend on k. This freedom of choice (of gauge) will be useful in determining properties of the Wannier functions.

Definition 5 We say that the Wannier function $w_m(x)$ is associated to the wave function $\phi_m(k, x)$ if the following relation holds

$$w_m(x) = \frac{1}{vol(T^*)} \int_{T^*} \phi_m(k, x) dk, \qquad x \in R^d \tag{67}$$

◇

From the definition it follows that the Bloch function $\phi_m(k, x)$ is the Bloch-Floquet transform of w_m

$$\phi_m(k, x) = \sum_{\gamma \in \Gamma} w_m(x + \gamma) e^{-ik.\gamma} \tag{68}$$

Conversely

$$w_m(x + \gamma) = \frac{1}{vol(\mathcal{B})} \int_{\mathcal{B}} e^{ik.\gamma} \phi_m(k, x) dk \tag{69}$$

It is easy to verify that the Wannier functions belong to $L^2(R^d)$, that

$$\int_{R^d} |w_m(x)|^2 dx = \frac{1}{vol(\mathcal{B})} \int_{\mathcal{B}} |\phi_m(k, x)|^2 dk dx \tag{70}$$

and that the Wannier functions $w_m(x)$ and $w_m(x+\gamma)$ are orthogonal iff the functions $\phi_m(k, x)$ are chosen so that their $L^2(W)$ norm does not depend on k.

The most relevant property of the Wannier functions is their *localizability* [1, 5]. From Theorem 8 one sees that the dependence on k of $\phi_m(k, x)$ determines the local properties of the corresponding Wannier function. In particular
(a) If $\sum_{\gamma \in \Gamma} |w_m|_{L^2(\mathcal{W})+\gamma)} < \infty$ then $\phi_m(k, x)$ is a continuous function on T^* with values in $L^2(\mathcal{W})$.
(b) $|w_m|_{L^2(\mathcal{W}+\gamma)}$ decays when $\gamma \to \infty$ more rapidly that any power of $|\gamma|^{-1}$ iff $\phi_m(k, .)$ is C^∞ as a function on T^* with values in $L^2(\mathcal{W})$.

(c) $|w_m|_{L^2(\mathcal{W}+\gamma)}$ decays exponentially iff $\phi_m(k, .)$ is analytic as a function on T^* with values in $L^2(\mathcal{W})$.

When two bands cross the eigenvalue $\lambda_m(k)$ becomes degenerate for some value of k and the corresponding eigenfunctions are not in general continuous at the crossing [1, 4, 5]. In this case it more convenient to try to consider Wannier functions that are associated to a *band* i.e. to a set of Bloch eigenvalues that are isolated *as a set* from the rest of the spectrum.

A Wannier system $\{w_1, \ldots, w_m\}$ associated to a band is by definition a family of orthonormal functions which have the property that their translates by the generators of the Wigner–Seitz cells are mutually orthogonal

$$(w_{i,\gamma}, w_{i',\gamma'}) = \delta_{i,i'}\delta_{\gamma,\gamma'} \tag{71}$$

and the projection P_b onto the band can be written as

$$P_b = \sum_{i=1}^{m} \sum_{\gamma \in \Gamma} |w_{i,\gamma} >< w_{i,\gamma}| \tag{72}$$

Here m is the number of elements in the band.

A relevant question if $m \geq 2$ (the band contains m eigenvalues which cannot be disentangled) is whether one can always find a *Wannier system* which is composed of sharply localized functions, in particularly *exponentially localized.*

For $d = 1$ it is always possible to choose analytic Bloch functions and therefore exponentially decreasing Wannier functions [2]. For $d \geq 2$ and still $m = 1$ existence of exponentially localized Wannier functions was proved by Nenciu [7] (see also for an independent constructive proof [5]) under the assumption of time-reversal symmetry.

This assumption allow the construction of regular Bloch functions by joining smoothly a function constructed in the first half of the cell with the time-reversed (conjugated and reversed with respect to the middle point) defined on the second half of the cell.

A simplifying feature (also in the case m >1) [9] is provided by the fact that the existence of *continuous* Bloch functions implies the existence of *analytic* Bloch function (this is sometimes referred to as the *Oka principle*). Therefore exponential decay follows from a variant of the Paley-Wiener theorem.

For $m \geq 2$ Thouless [10] proved that there are topological obstructions to the existence of exponentially localized Wannier functions. It is not possible to choose exponentially localized Wannier functions if the first Chern class c_1 of the bundle given by the Bloch fibration does not vanish.

9 Chern Class

We recall briefly few elements of the theory of the Chern classes [8], for a brief introduction see [9].

Recall that a representative for the Chern classes of a hermitian bundle $\mathcal{V}$ of rank m over a smooth manifold $\mathcal{M}$ are the *characteristic polynomials* of the curvature form Ω of $\mathcal{V}$ which are defined as the coefficients in the formal series expansion as power series in t of

$$det\left(I - i\frac{t\Omega}{2\pi}\right) = \sum_k c_k(\mathcal{V})t^k, \qquad \Omega \equiv d\omega + \frac{1}{2}\omega \wedge \omega \tag{73}$$

where ω is the connection one-form of $\mathcal{M}$.

By construction this construction is invariant under addition of an exact differential form, i.e. the Chern classes are cohomology classes. This implies that that the Chern classes do not depend on the choice of a connection on $\mathcal{M}$.

If the bundle is trivial (diffeomorphic to $\mathcal{M} \times \mathcal{V}$) then $c_k = 0 \ \forall k > 0$ but the converse *is not true in general.* An important special case occurs when $\mathcal{V}$ is a vector bundle ($n = 1$). In this case the only non trivial Chern class is c_1. Since the existence of exponentially localized Wannier functions is equivalent to triviality of the Bloch bundle, the condition $c_1 = 0$ is necessary *but in general not sufficient* for their existence.

Time reversal invariance implies triviality [2, 4, 7]. This provides [5] triviality of the Bloch bundle in absence of magnetic fields for any $m \in \mathbf{N}$ and $d \leq 3$ (d is the dimension of space on which the lattice is defined). This covers the physical situation ($d = 3$) but not when a periodic external field is present. In this case there is an additional parameter so one led to study Bloch waves in four dimensions.

The limitation in the dimension comes from the classification theory of vector bundles [6]; since for $2j > d$ one has $c_j = 0$. The presence of a magnetic field alters the topology of the Bloch bundle and makes it non trivial in general. A very weak magnetic field does not change the triviality [7]: since c_n are integers, a small modification cannot change this numerical value. The results for strong magnetic fields are scarce.

To prove triviality one must find an analytic (or sufficiently differentiable) fibration of complex dimension one on R^d by solutions of $(H(k) - \lambda_m(k)I)u = 0$ in $L^2(\mathcal{W})$. If such section exists, the fiber bundle is trivial (isomorphic to the topological product $T^* \otimes \mathcal{W}$).

If the eigenvalue $\lambda_m(k)$ remains isolated and simple for every k then regular perturbation theory permits a local extension to a small neighborhood of T^* in C^d as an analytic function. Topological obstructions may occur to prevent to obtain an analytic fiber bundle Λ^a_m

$$\Lambda^a_m = \cup_z ker\ (H_k - \lambda_j(k), \quad z = e^{ik}, \quad |Imk| < a \tag{74}$$

The following proposition shows if these obstructions occur, they are present already at a differential level.

Proposition 1 ([5]) *By a theorem of H. Grauert [6] the fiber bundle Λ_m on T^* is topologically trivial iff it is analytically trivial (the transition functions can be chosen to be analytic).* ◇

The possible topological obstructions appear therefore at the continuous level. We have seen that this has a role the existence of Wannier functions that are localized exponentially well on Wigner cells and these may be used to describe *local* properties of the crystal.

Topological obstructions are frequent in analytic and differential geometry. For example one cannot have an oriented segment on a Moebius strip or a vector field without zeroes on a three-dimensional sphere.

In the case of the Bloch functions there is no topological obstruction to the existence of exponentially localized Wannier functions in the case the eigenvalue $\lambda_m(k)$ remains simple and does not intersect other eigenvalues as k varies.

One has indeed [7].

Theorem 10 (Nenciu) *Let $\lambda_m(k)$ be an analytic family of* simple *eigenvalues of $H(k)$ that does not intersect (as a family) other eigenvalues $H(k)$.*

For small values of Im k the fiber bundle Λ^a_k is analytically trivial. There exists therefore a complete orthonormal system of normalized Wannier functions $w_m(x)$ which decay exponentially and such that for all $\gamma \in \Gamma$ the functions $w_{m,\gamma} \equiv w_m(x - \gamma)$ and $w_m(x)$ are mutually orthogonal. ◇

The theorem, with the same proof, holds for more general self-adjoint strictly elliptic operators with *real* periodic coefficients (this excludes, e.g. the presence of a magnetic field.

Remark that, when the coefficients are real, if $\phi_\lambda(k,x)$ is an eigenfunction H to the (real) eigenvalue λ with quasi-momentum k then $\bar{\phi}_\lambda(k,x)$ is an eigenfunction of H to the same eigenvalue and with quasi-momentum $-k$ (this corresponds to invariance under time-reversal).

Consider next the case $m > 1$, i.e. there is a collection S of m bands that is separated from the rest of the spectrum but there exists no separated sub-band. The entire Hilbert space decomposes in the direct sum $\mathcal{H}_S \oplus \mathcal{H}_S^\perp$ where $\mathcal{H}_S$ is the union of the subspaces that correspond to bands in S.

A function $\psi \in \mathcal{H}_S$ corresponds, under the Bloch-Floquet transform, to a family of functions, parametrized by $k \in B$, which for each value of k belong to the spectral subspace $\mathcal{H}_{S,k}$ of the Floquet operator corresponding S. Correspondingly we define *generalized Wannier function* a function in R^d which can be represented as

$$w(x) = \frac{1}{vol(\Gamma^*)} \int_{\Gamma^*} \phi(k,x)dk \tag{75}$$

where $\phi(k,.) \in \mathcal{H}_S^k$.

The extension of the theorem of Nenciu to the case $m > 1$ presents new difficulties. It seems natural to consider instead of the Bloch functions (or their orthogonal projectors) the projection operators $P_S(k)$ on the subspace associate to the collection S

$$P_S(k) = \frac{1}{2\pi i} \int_{C_k} (\zeta I - H_k)^{-1} d\zeta \tag{76}$$

where for each value of k, C_k is a close path that encircles the eigenvalues which belong to S. In Eq. (76) we can now extend k to a small complex neighborhood in C^d. As in the case of separated bands, we can now construct the fiber bundle (of dimension m) on a small complex neighborhood Ω_a of R^d

$$\Lambda_S = \cup_{z \in \Omega_a} P_S(z) \tag{77}$$

We may ask whether there exists a family of N generalized Wannier functions w_j, $j = 1, \ldots, N$, with exponential decay which together with their translates by Γ form a complete orthonormal system in $\mathcal{H}_S$.

Theorem 11 ([1]) *Let the band be composed of N subspaces.*

Necessary and sufficient condition for the existence of a family of N generalized Wannier functions that together with their translates under Γ form a complete orthonormal in $\mathcal{H}_S$ is that the fiber bundle Λ_S be topologically trivial. ♢

It is easy to prove that the condition is sufficient because then it is easy to prove that triviality implies the existence of a family of m generalized Wannier functions $w_n(x)$ which, together with their translates, for a complete orthonormal system and satisfy

$$\sum_{\gamma \in \Gamma} |w_n|_{L^2(W+\gamma)} < \infty \tag{78}$$

In case the fiber bundle is not trivial, N such Wannier functions cannot exist. P. Kuchment [1] has shown that if N is the number of bands contained in S there exists $M > N$ such and M Wannier functions with exponential decay which, together with their translates under Γ are a complete (but not orthonormal) system of functions in $\mathcal{H}_S$ (their linear span is dense in $\mathcal{K}_S$.

This is due to a theorem of Whitney, according to which it is always possible to provide an analytic immersion of a manifold of dimension N in R^M, irrespectively of its topological degree, provided on chooses M sufficiently large.

References

1. P. Kuchment, *Floquet Theory for Partial Differential Operators* (Birkhauser, Basel, 1993)
2. M. Reed, B. Simon, *Methods of Mathematical Physics* (Academic Press, New York). Theorem XIII, 85

3. C. Wilcox, Theory of Bloch waves. J. Anal. Math. **33**, 146–167 (1978)
4. G. Nenciu, J. Math. Phys. **43**, 1273–1298 (2002)
5. W. Kohn, Theory of insulating state. Phys. Rev. **133**, A71 (1964)
6. H. Grauert, Math. Ann. **133**, 139–159 (1957)
7. G. Panati, Ann. Inst. Henry Poincaré **8**, 995–1011 (2007)
8. S.S. Chern, Characteristic classes of Hermitian manifolds. Ann. Math. **47**, 85–121 (1946)
9. J. Jost, *Riemannian Geometry and Geometric Analysis* (Springer, Berliln, 2005)
10. D. Vanderbilt, R.D. King-Smith, Phys. Rev. B **48**, 4442–4456 (1993)

Lecture 5: Connection with the Properties of a Crystal. Born–Oppenheimer Approximation. Edge States and Role of Topology

We return now to the problem of the connection of the theory of Bloch–Floquet with the properties of a finite, very large crystal. We take up again, form a slightly different viewpoint, the problem of mathematical formalization of problems in Solid State Physics.

Consider a model of crystal in the frame of the approximations we have made in Lecture 4. This model is a *one-electron model* because we regard the electrons as non-interacting with each other. In the limit of an infinite crystal the atomic lattice fills R^3 and therefore the number of electrons in the system is infinite.

It is convenient to introduce the *density of states* and keep into account that the electrons satisfy the Fermi–Dirac statistics. This has lead to the definition of the *Fermi surface*.

Let $\mathcal{B}$ be the Brillouin zone and for $k \in \mathcal{B}$ denote by $E_n(k)$ the eigenvalues of H_k in increasing order.

The *integrated density of states* ρ is defined by

$$\mu(E) \equiv \rho(-\infty, E] = (2\nu(\mathcal{B})^{-1} \sum_n \nu(\{k \in \mathcal{K}, \; : \; E_n(k) \leq E\}) \tag{1}$$

where ν is Lebesgue measure. Since $lim_{n\to\infty} E_n(k) = \infty$ uniformly in $\mathcal{B}$ one has $\rho((-\infty, E]) < \infty$ and ρ is absolutely continuous with respect to ν. We shall call *density of states* the Radon–Nikodym derivative $\frac{d\rho}{dE}$.

In nature the system to be described is a finite-size macroscopic crystal. Since one is interested in properties that depend little on the specific size, the role of the boundary is usually considered negligible. We shall see later the role that can assume the boundary.

As we have seen in the preceding lecture, if boundary properties are neglected it is convenient to study a model in which the crystal is represented as an infinite lattice.

G. Dell'Antonio, *Lectures on the Mathematics of Quantum Mechanics II: Selected Topics*, Atlantis Studies in Mathematical Physics: Theory and Applications 2, DOI 10.2991/978-94-6239-115-4_5

The mathematical treatment of this approximation requires control of the limiting process.

Let $\mathcal{W}$ be the Wigner–Seitz cell and let $\mathcal{W}_m, \ m \in Z$ be the cell of volume $m^3\nu(\mathcal{W})$ which is obtained by dilating by a factor m the linear dimension (we consider a three-dimensional solid). Let H_m be the operator $-\Delta_P + V$ in $L^2(\mathcal{W}_m)$ and let P_m be the spectral family of H_m. Define

$$\rho_m(-\infty, E]) = 2\frac{dimP_m(-\infty, E]}{m^3} \tag{2}$$

The factor 2 keeps track of the fact that the electron has spin $\frac{1}{2}$ but the hamiltonian has no spin-orbit coupling term, therefore all level are doubly degenerate.

The following theorem relates in special cases the density of states relative to the Hamiltonians H_k with a density of states for the infinite system. Notice the strong similarity with the procedures used in defining the *thermodynamic limit*

Theorem 1 $lim_{m\to\infty}\rho_m = \rho$

◇

Outline of the proof

The main point consists in proving that each function on $\mathcal{W}_m$ with periodic boundary condition when restricted to the elementary cells contained in $\mathcal{W}_m$ defines m^3 functions on the single cells with boundary conditions that prescribe a phase difference multiple of $e^{i\frac{2\pi}{m}}$.

If $\Gamma \equiv \sum_{i=1}^3 t_i\alpha_i, \ \ 0 \le t_i \le 1$ in this decomposition of the Hilbert space the hamiltonian H_m takes the form

$$H_m = \sum_{\beta_1,\beta_2,\beta_3=0}^{m-1} H_{\frac{\beta_1}{m}\alpha_1+\frac{\beta_2}{m}\alpha_2+\frac{\beta_3}{m}\alpha_3} \tag{3}$$

where we denoted by $H(k)$ the fibers of the operator H that we have constructed in the finite volume case. Therefore

$$\rho_m(-\infty, E] = \frac{2}{m^3}N^o\{n \ : \beta_i \in \{0, 1, \ldots m-1\} \ : \ E_n = \sum_j \frac{\beta_j, \alpha_j)}{m} \le E\} \tag{4}$$

Since the function $E(k)$ is continuous this expression converges to $\rho(\infty, E]$ when $m \to \infty$. ♡

In this *macroscopic formulation* when the crystal is in equilibrium a zero temperature the *Fermi level* $E(F)$ is the maximum value of the energy level $E(k)$ such that if $E > E(F)$ then $\rho(E) = 0$.

Correspondingly the *Fermi surface* is the collection of k in the Brillouin zone such that $E(k) = E(F)$.

In this description a crystal is regarded as an *insulating material* if the Fermi level is placed in between two occupied bands and is interpreted as a *conducting material* if the Fermi level lies inside a band (called *conducting band*).

This interpretation fits the experimental data, but at present there are suggestions (3) that its *microscopic* justification resides in the structure of the eigenfunctions at the Fermi level and of their *deformation* in presence of an external electric field (this deformation can be calculated to lowest order in perturbation theory).

The corresponding structure has been studied in some detail by W. Kohn [1]. More recent models and interpretations are discussed e.g. in [2–4]. The latter Authors attribute electric polarization and electric conduction to the different structure *of the ground state eigenfunctions* for the insulating and the conducting phases.

In the insulating case more states are available for the decomposition in Bloch waves, and this gives the polarization. The deformation produced by the electric field is in this case localized in a neighborhood of the boundary while in the conducting case it is extended to the bulk of the material. In the conducting case the modification gives rise to the flow of the electrical current.

Analytically this is due to the fact that the localization tensor (the mean value in the ground state of the operator $x_k x_h$) diverges in the infinite volume limit in the conducting case while it is bounded in the insulating case.

A similar analysis, based on the different structure of Wannier functions can be done for electric polarizability and attempts have been made to study of orbital magnetization in the insulating case (see e.g. [7]). This very interesting analysis has not been developed yet from a mathematical point of view.

1 Crystal in a Magnetic Field

Consider now the case in which the crystal is placed in an external magnetic field. Consider first the case in which the magnetic field M is constant. Under the assumption that the interaction among electrons be negligible, the motion of an electron in a crystal lattice $\Gamma \in R^3$ is given by the Hamiltonian, in units $\hbar = 2m = \frac{e}{2} = 1$

$$H_0 = (-i\nabla_x + M \times x)^2 + V(x) \quad x \in R^3 \tag{5}$$

where $V(x)$ is a real γ-periodic potential. We shall always assume that V be regular (e.g. of class C^∞).

Denote by $e_1,\ e_2\ e_3$ the generating base of the lattice Γ and by $\{e_i^*\}$ the dual basis (which generates the Brillouin cell Ω). Therefore $(e_i, e_j^*) = 2\pi\delta_{i,j}$. The operator H_0 is self-adjoint and commutes with the *magnetic translations* T_γ defined by

$$(T_f)(x) = e^{i(M\wedge x,\gamma)} f(x-\gamma) \tag{6}$$

We shall assume for each choice of $i,\ j$ $(M \cdot e_j \wedge e_i) \in 4\pi\ Z$ (the magnetic flux across every face of the lattice is a multiple of the identity). Under this assumption $\mathcal{G} \equiv \{T_\gamma, \quad \gamma \in \Gamma\}$ is an abelian group and we can reduce H_0 over the characters of $\mathcal{G}$ setting

$$\mathcal{D}_k \equiv \{\phi \in H^2_{loc}(R^3)\ ,\ T_\gamma\ \phi = e^{-i(k,\gamma)}, \quad \gamma \in \Gamma \quad k \in \mathcal{G}^*\} \tag{7}$$

We can now use the decomposition of $L^2(R^3)$ as direct integral over $\mathcal{G}^*$. It is easy to see that the operator $H_0(k)$ in this decomposition is self-adjoint and has compact resolvent.

Denote $E_1(k) \leq E_2(k) \leq \cdots$ its eigenvalues. The spectrum of H_0 is therefore

$$\cup_{k \in \mathcal{G}^*}\ \cup_{m=1}^{\infty} E_m(k) \tag{8}$$

Remark that for every m one has

$$\gamma^* \in \mathcal{G}^* \rightarrow E_m(k + \gamma^*) = E_m(k) \tag{9}$$

It follows from regular perturbation theory that $E_m(k)$ is a continuous function of k which can be continued to a function analytic in a neighborhood of those k for which

$$E_{m-1}(k) < E_m(k) < E_{m+1}(k) \tag{10}$$

The domain spanned by $E_m(k)$ when $k \in \mathcal{G}^*$ is the mth *magnetic band*. In what follows it will be convenient to consider on each fiber instead of $H_0(k)$ the operator

$$H_0'(k) = e^{-ikx}\ H_0(k)\ e^{ikx} = (-i\nabla_x + M \wedge x + k)^2 \tag{11}$$

(M is the constant magnetic field) with domain

$$\mathcal{D} = \{\phi \in H^2_{loc}(R^3), \quad T_\gamma\ \phi = \phi, \quad \gamma \in \Gamma\} \tag{12}$$

We shall regard $\mathcal{D}$ as a subspace of $L^2(R^3)$.

2 Slowly Varying Electric Field

Consider now the case in which to the crystal in the field B is applied also an electric field W varying slowly in space. The hamiltonian H_ϵ of the system is

$$H_\epsilon = (-i\nabla + \mu \times x + A(\epsilon\, x)^2 + V(x) + W(\epsilon\, x) \tag{13}$$

where W, A_1, A_2, A_3 are smooth functions. The standard method to treat slowly varying fields is to introduce a new independent variable y. At the end we shall put $y = \epsilon\, x$. Accordingly, introduce a new Hamitonian

$$\tilde{H}_\epsilon(x, y) = (-i\nabla_x + M \wedge x + A(y)^2 + V(x) - i\epsilon\nabla_y + W(y) \tag{14}$$

To a function $\phi(x, y)$ on $R^3 \times R^3$ we associate the function $w(x) \equiv \phi(x, \epsilon x)$. We shall use the *adiabatic method* based on the following identity

$$(\tilde{H}_\epsilon)\,\phi(x, \epsilon x) = (H_\epsilon w)(x) \tag{15}$$

This identity permits to solve the Schroedinger equation for H_ϵ uniformly in ϵ by solving the equation for $\tilde{H}_\epsilon$ uniformly in $y,\ \epsilon$.

We shall make the following assumptions:
(i) For each value of k The magnetic band we consider is is isolated and remains isolated after application of the electric field.
(ii) For each $\gamma^* \in \Gamma^*$ one has

$$\phi(x; k + \gamma^*) = e^{i(\gamma^*, x)}\phi(x, k) \tag{16}$$

(iii) The flux of the magnetic field M across any of the faces of the elementary cell is *an integer multiple of* 4π.

These hypotheses have the following consequences
• Under hypothesis (i) we can choose the eigenfunctions $\phi_m(x, k)$ associate to the eigenvalue $E_m(k)$ to be an analytic functions of k with values in $\mathcal{D}$.
• Under hypothesis (ii) the fiber bundle of complex dimension 1 on R^3/Γ given by the Bloch function $\phi(x, k)$ is trivial.

Remark that in general

$$\phi(x, k + \gamma^*) = e^{i[\gamma^* + x + \theta(\gamma^*, k)]}\phi(x, k) \tag{17}$$

where $\theta(\gamma^*, k)$ is a real valued function given by the structure of the fiber bundle. Since the gauge group is abelian

$$\theta\left(k, \sum_i m_i e_i^*\right) = \sum m_i \theta(k, e_i^*) \equiv c_2 \tag{18}$$

where the constant c_2 represents the second Chern class of the fiber bundle.

As we have seen in the discussion of the magnetic Weyl algebra in the first Volume of these lecture Notes, the presence of the magnetic field is equivalent to a modification of the symplectic two-form. The appearance of c_2 in (18) is therefore natural.

The role of assumption (iii) is to set $c_2 = 0$ (the bundle is trivial). If this term does not vanish the gradient with respect to k of the function $\phi(x, k)$ is not uniformly bounded in the elementary cell and the regularity assumptions we make in the multi-scale method are not satisfied.

We remark that condition (iii) can be replaced by
(iv) The magnetic flux across any face of the elementary cell is a rational multiple of 4π.

To see this, recall that the analysis we do refer to the limit in which the system covers the entire lattice, we can consider an elementary cell which is a multiple of the Wigner–Seitz cell by an arbitrary factor N. If assumption (iv) is satisfied, we can choose N in such a way that the magnetic flux across the faces of the new cell satisfies (iii).

Under assumptions (i), (ii), (iii) the multi-scale method can be used to provide an expansion of the hamiltonian in an asymptotic series in ϵ (see e.g. [9]).

Theorem 2 *For each positive integer N there exist operators* $P_N : L^2(R^3) \to L^2(\mathcal{W} \times R^3)$ *that are approximately isometric (i.e.* $P_N^* P_N = I + O(\epsilon^{N+1})$*) and can be written as* $P_N = F_0 + \epsilon F_1 + \cdots + \epsilon^N F_N$ *where* F_n *is bounded for every n., and exists an* effective Hamiltonian

$$H_{eff}^N = h_0 + \epsilon h_1 + \epsilon^2 h_2 + \cdots + \epsilon^N h_N \tag{19}$$

such that, for each $u_x \in S(R^3); \quad x \in \Omega$

$$\tilde{H}_\epsilon(P_N(x, y, \epsilon D_y, \epsilon)u_x - P_N(x, y, \epsilon D_y \epsilon)H_{eff}^N(y, D_y)u_x(y) = O(\epsilon^{N+1}) \tag{20}$$

Moreover if we set $\Pi_N = P_N P_N^*$ *the operator* Π_N *is an approximate projection* $\Pi_N^* = \Pi_N, \quad \Pi_N^2 = \Pi_N + O(\epsilon^{N+1})$ *and for* $\phi \in \mathcal{S}$,

$$\Pi_N \tilde{H}_\epsilon \phi = \tilde{H}_\epsilon \Pi_N \phi + O(\epsilon^{N+1}) \tag{21}$$

◇

Remark that the effective Hamiltonian *does not depend on* $x \in \mathcal{W}$ (though the operators P_N depend on x). The wave function of the electron is $\phi(x) = u_x(\epsilon x)$. To order zero

$$h_0 = E_m(k + A(y)) + W(y) \tag{22}$$

Equation (22) is called *Peierls substitution*. The term $E_m(k + A(y))$ which *substitutes* the kinetic energy is a pseudo-differental operator (it is not the Fourier transform of a polynomial).

Outline of the proof of Theorem 2

Define as before

$$\tilde{H}_\epsilon(k) \equiv e^{-ik.x}\tilde{H}_\epsilon e^{ik.x)} = (i\nabla_x - i\epsilon\nabla_y + M \times x + A(y) + k)^2 + V(x) + W(y) \tag{23}$$

and remark that

$$\tilde{H}_\epsilon P_N(x, y, \epsilon D_y, \epsilon)u = \left(\frac{1}{2\pi\epsilon}\right)^3 \int e^{i\frac{(k,(y-z))}{\epsilon}} \tilde{H}_\epsilon(k) P_N(x, y, k, \epsilon)u(z)dk \tag{24}$$

Expanding in powers of ϵ one obtains

$$\tilde{H}_\epsilon(k) = \tilde{H}_0(k)\tilde{+}\epsilon H_1(k) + \epsilon^2 \tilde{H}_2(k) + \cdots \tag{25}$$

$$\tilde{H}_0(k) = H_0(k + A(y)) + W(y) \qquad \tilde{H}_2(k) = -\Delta_y \tag{26}$$

$$\tilde{H}_1(k) = -2i[-i\nabla_x + M \wedge x + k + A(y)].\partial_y - i\partial_y.A(y) \tag{27}$$

The proof of (21) and (22) is now completed by iteration and this procedure provides also the explicit determination of the symbols $F_0,\ F_1, \ldots$ and of the Hamiltonians $h_0, h_1 \ldots$ In this process one uses the Fredholm alternative; the arbitrary term in this process is chosen so as to satisfy (22).

We only give explicitly the first step. Set

$$h_0(y, k) = E_m(k + A(y)) + W(y), \quad F_0(x, y, k) = \phi(x, k + A(y)) \tag{28}$$

Keeping into account the terms up to first order in ϵ we have

$$(\tilde{H}_0(k) - h_0)F_1 = -i\frac{\partial F_0}{\partial k}.\frac{\partial h_0}{\partial y} - (\tilde{H}_1(k) - h_1)F_0 \tag{29}$$

For the Fredholm alternative, to obtain solutions a necessary condition is that the right hand term of Eq. (27) be orthogonal to the kernel of $\tilde{H}_0(k) - h_0$.

This leads to a unique choice for $h_1(y, k)$ and gives $F_1(x, y, k)$ modulo addition of an element of the kernel. The result is

$$h_1(y, k) = (F_0(., y, k), i\frac{\partial F_0}{\partial k}\frac{\partial h_0}{\partial k}(y, k)) + \tilde{H}_1(k)F_0(., y, k)) \tag{30}$$

$$F_1(x, y, k) = (\tilde{H}_0(k) - h_0)^{-1}[-i\frac{\partial F_0}{\partial k}\frac{\partial h_0}{\partial k} + h_1 F_0 - \tilde{H}_1(k)F_0] + a_1(y, k)F_0 \tag{31}$$

where $a_1(y, k)$ is an arbitrary function.

This function is then fixed by the requirement $\Pi_N\ \Pi_N^* = I + O(\epsilon^{N+1})$. Here we give only the expression for h_1

$$h_1(y, k) = \frac{1}{2i}\frac{\partial}{\partial y}\left[\frac{\partial E_m(k + A(y))}{\partial k}\right] - (L.\nabla \times A(y) - i < \phi(., k + A(y)), \dot{\phi}(., k + A(y) > \tag{32}$$

where $E_1(k) \equiv < F_0(., y, k), \tilde{H}_1(k)F_0(., y, k) >$

$$L = Im\left[< M(y,k)\frac{\partial\phi}{\partial k_2}, \frac{\partial\phi}{\partial k_3} >< M(y,k)\frac{\partial\phi}{\partial k_3}, \frac{\partial\phi}{\partial k_1} >< M(y,k)\frac{\partial\phi}{\partial k_1}, \frac{\partial\phi}{\partial k_2} >\right] \tag{33}$$

$$M(y,k) = \tilde{H}_0(k) - h_0(y,k) \quad \dot{\phi}(x, k + A(y)) = \frac{\partial\phi(x, k + A(y)}{\partial y}.\dot{y} + \frac{\partial\phi(x, k + A(y)}{\partial k}\dot{k} \tag{34}$$

$$\dot{y} = \frac{\partial(E(m)(k + A(y)) + W(y)}{\partial k}, \quad \dot{k} = -\frac{\partial[E_m(k + A(y) + W(y)]}{\partial y} \tag{35}$$

The term $i < \phi(., k + A(y), \dot{\phi}(., k + A(y) >$ is precisely the term that gives rise to the geometrical Berry phase, which we have briefly treated in the first Volume of these lecture Notes.

♡

3 Heisenberg Representation

It is interesting to translate this result in the Heisenberg representation. For an observable $B(y, \epsilon D_y)$ the evolution is given by

$$i\epsilon\frac{dB}{ds} = [H_{eff}, B] \tag{36}$$

(the coefficient ϵ originates in the difference in time scale typical of the adiabatic approximation). The symbol $b(y, \xi)$ of the operator B in its dependence on time follows the trajectories of the classical system

$$\dot{y} = \frac{\partial H_{eff}}{\partial\xi}, \quad \dot{\xi} = \frac{\partial H_{eff}}{\partial y}, \tag{37}$$

with $y \in R^3, \quad \xi \in \Omega$. Remark that one can still modify the operators Π_N with the addition of terms with norm $O(\epsilon^{N+1})$ in such a way as to obtain a projection operator π_N (since the operators are bounded the formal series converges for ϵ small enough).

In the same way the formal series that defines H_N can be modifies so that for each value of k the group generated by $\hat{H}^N_{eff}$ leaves Π_N invariant; the corresponding subspace is therefore an *invariant subspace*.

As always in the theory of regular perturbations the projection operator is modified to first order but the energy is only modified to order two. Therefore h_1 has the form given in Eq. (32).

A detailed analysis of perturbation theory by small periodic electric and magnetic fields can be found in Nenciu [8]. The adiabatic method can also be used in case the small external electromagnetic varies slowly in time, e.g. is time-periodic with a very long period.

4 Pseudo-differential Point of View

For a presentation of the adiabatic Adiabatic method Multi-scale method and multi-scale methods, with particular reference to the Schroedinger equation in periodic potentials and in presence of weak external electromagnetic fields, also in the case of slow variation both in space and in time, a very useful reference is the book by S.Teufel [9].

The latter Author stresses the advantage of approaching adiabatic perturbation theory *through the Weyl formalism.* This procedure is useful in the study of the dynamics of the atoms in crystals but also in the study of a system composed of N nuclei of mass m_N with charge Z and of NZ electrons of mass m_e. In the latter case one chooses the ratio $\epsilon \equiv \frac{m_e}{m_N}$ as small parameter in a multi-scale approach.

Recall that to order zero in ϵ the nuclei are regarded as fixed centers of force which determine the dynamics of the electrons. This (very fast) dynamics gives rise to a mean potential of strength proportional to ϵ which acts on the nuclei. One then studies to first order in ϵ the dynamics of the nuclei in this mean field (Born–Oppenheimer approximation).

A multiscale approach can be used to study the motion of the electrons in the crystal acted upon by an electromagnetic field which varies slowly in space and time as compared to the linear size of the crystal cell (of order $\hbar$) and to the momentum of the electrons in units of $\hbar$; the parameter ϵ has the role of Planck's constant in the semiclassical limit.

If one neglects the interaction between electrons, the Hilbert space of the system decomposes as tensor product of a Hilbert space for the slow degrees of freedom and one for the fast (external) degrees of freedom.

$$\mathcal{H} = L^2(R^3) \otimes \mathcal{H}_{fast} \tag{38}$$

When the electron in a crystal are subject to an external electromagnetic field one can use the magnetic Weyl algebra (Volume I); in this case and the parameter ϵ characterizes the speed of variation of the external field. In this approach the hamiltonian $Op^w(H(z, \epsilon))$ generating the time evolution of the states is given by the Weyl quantization of a semiclassical symbol i.e.

$$Op^w H(z; \epsilon) \simeq \sum_{k=0}^{\infty} \epsilon^j Op^w H_j(z), \quad z \in C^3 \tag{39}$$

with values in self-adjoint operators on $\mathcal{H}_f$. The principal symbol $Op^w(H_0(z)$ describes the decoupled dynamics and therefore contains information on the structure of the energy band.

For this system it is generally assumed that in the spectrum there is an isolated band that remains isolated for small values of ϵ. The Hamiltonian in this smaller space takes a simpler form under a suitable change of variables that can be chosen in such a way that the resulting error can be made of order ϵ^N for any N.

A crucial point in this procedure is to be able to express the projection operator into a band, or a collection of bands, as a pseudo-differential operator. Expressing the projection as a power series in the parameter ϵ by Weyl quantization one obtains a bounded operator that is a *quasi-projection* i.e. it satisfies

$$Op^w(\pi^2) = Op^w(\pi) + 0(\epsilon^\infty), \quad Op^w(\pi^*) = Op^w(\pi), \quad [Op^w(H), Op^w(\pi)] = 0(\epsilon^\infty)$$

where the $O(\epsilon)^\infty$ terms are pseudo-differential operators of suitable class [9]. Notice that the estimate is generally not true in operator norm. Using the definition of the projection on bands as a Riemann integral of the resolvent on a well chosen path in the complex plane it possible then to construct a true projection operator Π that can be expressed as a sequence of pseudo-differential operators wit an error $0(\epsilon^\infty)$.

One can further simplify the problem by mapping the Hilbert space of the band into a reference Hilbert space. The resulting hamiltonian $Op^w(H_\epsilon)$ admits an expansion in powers of ϵ and to any order

$$Op^w(H_\epsilon) = \sum_n Op^w(H_n)\epsilon^n \tag{40}$$

and each $Op^w(H_n)$ is a pseudofferential operator. The *principal symbol* is the band eigenvalue $E(q, p)$. This gives the transcription of *Peierls substitution* in the Weyl formalism.

The next orders, in particular H_1 carry relevant information about the polarizability and the conductivity of the crystal. We remark that although the formulation of the Hamiltonin as pseudo-differential operator is optional, it simplifies the analysis of the operators H_n.

The operators H_n are essentially self-adjoint on a natural domain but in general unbounded. In order to give the estimates described above it is sometimes convenient to consider them as bounded operators between different Hilbert spaces (for example $\frac{d}{dx^2}$ is bounded if regarded as an application from $H^2(R)$ to $L^2(R)$).

We do not enter here in the details on how this generalization can be constructed. The difficult point is related mainly to the need, when solving by iteration the corresponding dynamics, to have a control the domains *uniformly in the parameter* ϵ. For example one makes use of a result analogous to the Calderon-Vaillantcourt theorem. If there exists a constant $b_n < \infty$ such that

$$a \in C^{2n+1}(R^{2d}, \mathcal{B}(\mathcal{K})) \qquad sup_x \|a_x\| = b_d \tag{41}$$

then $Op^w(a) \in \mathcal{B}(\mathcal{H})$ with the bound

$$\|Op^w(a)\|_{\mathcal{B}(\mathcal{H}} \leq b_d sup_{|\alpha|+|\beta|\leq 2n+1} sup_{q,p\in R^n} \|(\partial_q^\alpha \, \partial_p^\beta a)(q,p)\|_{\mathcal{B}(\mathcal{K})} = b_d \|a\|_{C_b^{2n+1}} \tag{42}$$

In the case of strong magnetic field it is convenient to make use of the *magnetic Weyl calculus* which we have briefly described in Volume I of the lecture Notes. This calculus is particularly useful if the magnetic field is constant plus a small perturbation so that the Landau gauge is a good approximation.

Denote by $Op^{w,M}$ the pseudo-differential operators associated to the magnetic Weyl system. They are defined using the unitary group associated to the canonical variables in the minimal coupling formalism for the vector potential A (for this reason they are particularly useful when the magnetic field is large and is approximatively constant). One has

$$Op^{w,A}(F) \int \Lambda^A Op^w(F), \qquad \Lambda^A(x,z) = e^{-i\int_{[x,z]}A} \tag{43}$$

where $[x, z]$ is the oriented segment for x to z in configuration space.

For the magnetic pseudo-differential operators one has the same results as for for the usual pseudo-differential operators; in particular a magnetic version of the Calderon-Vaillantcourt theorem holds and conditions to be bounded or in a specific Schatten class can be found. We don't develop here this very interesting line of research.

5 Topology Induced by a Magnetic Field

We shall give now a brief account of the way topology enters the description of the states of an electron in a periodic two-dimensional potential V defined in a plane Π in presence of a uniform magnetic field B perpendicular to Π. The stationary Schroedinger equation is

$$H\phi \equiv (\frac{1}{2m}(p - eA)^2 + U(x))\phi = E\phi \qquad p = -\frac{1}{\hbar}\nabla \tag{44}$$

where A is a vector potential such that $rotA = B$.

Consider for simplicity the case in which the two-dimenional lattice defined by the potential V is generated by two vectors $a,\ b \in R^2$ (Bravais lattice Λ) and consider a Bravais lattice vector $\lambda = na + mb, \quad n, m \in Z$. Define a magnetic translation operator [8].

$$T_\lambda(B) = T_\lambda e^{-i\frac{e}{2\hbar}(\lambda\wedge B).x} \quad x \in R^2 \tag{45}$$

where T_λ is the operator of translation by the Bravais lattice vector λ, i.e. $T_\lambda = e^{\frac{i}{\hbar}\lambda.\nabla}$.

Using the symmetric gauge ($A = B \wedge x$) one has $T_\lambda H = HT_\lambda$; therefore one can simultaneously diagonalize H and the operator of translation along any Bravais vector. One has

$$T_\lambda T_\sigma = e^{2\pi i\Phi} T_\sigma T_\lambda \tag{46}$$

where $\Phi = \frac{eB}{\hbar}ab$ is the magnetic flux across the unit cell.

When Φ is rational (say $\frac{p}{q}$ where p and q are relative prime integers with $p < q$) consider a new Bravais lattice Λ' with $R' = n(qa) + b$ with a new elementary cell, the *magnetic unit cell*.

One can now diagonalize simultaneously the magnetic translations $\hat{T}$ along the new lattice and the Hamiltonian. It is easy to see that the eigenvalues of $\hat{T}_{qa}$ and of $\hat{T}_b$ are respectively $e^{ik_1 qa}$ and $e^{ik_2 b}$ where k_i are Quasi-momentum *quasi-momenta* with range $0 \le k_1 \le \frac{2\pi}{qa}$ and with eigenfunctions which can be written (in Bloch form)

$$\psi_{k_1,k_2}(x,y) = e^{i(k_1 x + k_2 y)} u^{\alpha}_{k_1,k_2}(x,y) \tag{47}$$

Here α is a band index and the $u^{\alpha}_{k_1,k_2}(x,y)$ have the property

$$u^{\alpha}_{k_1,k_2}(x+qa,y) = e^{-i\frac{\pi p y}{b}} u^{\alpha}_{k_1,k_2}(x,y) \quad u^{\alpha}_{k_1,k_2}(x,y+b) = e^{i\pi\frac{px}{qa}} u^{\alpha}_{k_1,k_2}(x,y) \tag{48}$$

(the eigenvalues $E(k_1,k_2)$ vary continuously and the set of values that they take when k_1, k_2 vary in a magnetic Brillouin zone for a *magnetic sub-band*).

Since by a gauge transformation $A \to A + \nabla\phi$ one has $\psi \to e^{-i\frac{e\psi}{\hbar}}$ only the change of phase of the wave function after a complete contour of the magnetic unit cell is meaningful. This change of phase in $2\pi p$. Writing

$$u^{\alpha}_{k_1,k_2}(x,y) = |u^{\alpha}_{k_1,k_2}(x,y)| e^{i\theta_{k_1,k_2}(x,y)} \tag{49}$$

one has

$$p = \frac{1}{2\pi}\int dl \frac{d\theta_{k_1,k_2}(x,y)}{dl} \tag{50}$$

where the integral is over a clock-wise contour of the unit magnetic cell. The number p is *a topological property* of the Bloch wave

There is another topological property of the wave-functions in the magnetic (Brillouin) zone. It is related to the Hall conductance, but we shall not treat this connection here.

We have considered the Bloch wave $u_{k1,k_2}(x,y)$, but the waves are defined by states only modulo a phase. Therefore it is convenient to consider a principal $U(1)$-bundle over the magnetic zone which has the topology of a torus T^2.

A principal $U(1)$ bundle over T^2 is defined by the transition functions between overlapping patches that are topologically trivial (contractible). The two dimensional torus can be covered by four such patches, corresponding e.g. to neighborhoods $W_j,\ j = 1, \ldots 4$ of the four quadrants in the representation of the torus as a square (neglecting identifications at the boundary).

In each patch the Bloch functions can be chosen to be continuous (in fact C^∞). We assume that the Bloch functions do not vanish in the overlap regions (this can always be achieved, since the zeroes are isolated points). The principal $U(1)$ bundle is trivial (isomorphic to $W_j \times U(1)$) in each neighborhood.

Since the W_j are contractible, it is possible to choose a phase convention such that

$$e^{i\theta_j(k_1,k_2)} \equiv \frac{u_{k_1,k_2}(x,y)}{|u_{k_1,k_2}(x,y)|} \tag{51}$$

is smooth in each W_j (except possibly in the zeroes of $u_{k1,k_2}(x,y)$) . But *in general it is not possible* to have global continuity for θ_j , i.e. a global phase convention that holds *in all* W_j. We will have a transition function $U_{i,j}$ in the overlap $W_j \cap W_j$

$$U_{i,j} \equiv e^{i(\theta_j(k_1,k_2)-\theta_i(k_1,k_2))} \equiv e^{iF_{j,i}(k_1,k_2)} \tag{52}$$

The principal bundle in completely characterized by these transition functions. In order to connect with differential forms, recall that one can write the connection one-form ω (which gives the transition functions) as

$$\omega = g^{-1}Ag + g^{-1}dg = A + id\xi, \quad A \equiv a_\mu(k_1,k_2)dk_\mu \quad a(k_1,k_2) = \left(u_{k_1,k_2}, \frac{\partial}{\partial k_\mu}u_{k_1,k_2}\right) \tag{53}$$

with $g \equiv e^{i\xi} \in U(1)$.

It easy to prove that this choice gives a connection form. Indeed ω is invariant under the *gauge transformation*

$$u'_{k_1,k_2}(x,y) = e^{if(k_1,k_2)}u_{k_1,k_2}(x,y) \tag{54}$$

where $f(k_1,k_2)$ is an arbitrary smooth function. The curvature of this connection is

$$F = dA = \frac{\partial a_\mu}{\partial k_\nu}dk_\nu \wedge dk_\mu \tag{55}$$

By definition $\frac{i}{2\pi}F$ is the *first Chern form* and its integral over T^2 is called *first Chern number*

$$c_1 = \frac{i}{2\pi}\int_{T^2} F = \frac{i}{2\pi}\int_{T^2}\frac{\partial a_\mu}{\partial dk_\nu}dk_\nu \wedge dk_\mu \tag{56}$$

This number is always an integer and depends only on the topology of the principal bundle that we have constructed from the Bloch vectors in each patch. It represents the obstruction to the construction of Bloch vectors which are continuous (in fact C^∞) over T^2.

6 Algebraic-Geometric Formulation

We have so far considered the formulation of the geometrical aspects of phase in the Quantum Mechanics for Solid State Physics (theory of cristalline bodies) form the point of view of Schreodinger's Quantum Mechanics. This description, as remarked before, makes use of the *visual* features of the wave function and therefore describes the different phases as *geometrical objects*.

We have mentioned several times that the wave function (rather its modulo square) represents a probability density and *locally* the phase has no physical reality. We have however seen, when we have considered the *Berry phase*, that the modification to which is subjected the phase when the systems is periodic and depends on a cyclic parameter (maybe time) are expressible by means of observable quantities.

This *berryology* is at the base of most researches of *different phases of matter* (the meaning of *phase* is not the same as in the case of the wave functions). From the analytic point of view that we have followed so far these researches are aimed to analytic (and geometric) properties of the *Bloch bundle*.

For this purpose they employ methods of classical geometry, mainly connections and curvature, that rely on the *visual* aspects of the wave function. The *geometric* complexity of this *visual* bundle determine physical properties of the material considered, e.g. conductivity, polarizability (electric and magnetic).

This analysis, by its very structure, depends on the regularity of the crystal and regards the crystal as infinitely extended.

In case some (infinite) edges are present, it relies on the sharpness of the edges and their periodicity in the transversal direction so the the *edge currents* are defined within Bloch theory.

Slight deformations of this structure can be studied, relying on smooth perturbation theory, but major perturbations are outside the scope of this theory.

Since there are two formulations of Quantum Mechanics, one may wonder how the algebraic (Heisenberg) formulation is able to attack these problems. The resulting theory should have the same relation with the topological aspect of Bloch theory as modular theory has with K.M.S. theory Gibbs theory.

The algebraic approach to Solid State theory was initiated by J. Bellissard [10] and it has not been fully developed yet. It covers partially random structure, e.g. the relevant case of crystal with random defects.

The observables are described by a C^* algebra on which there is an action of a continuous group (or grupoid) taking the place of lattice translations.

The group acts ergodically and therefore there is an invariant regular measure. Other groups of transformations reflect other symmetries and properties of the system, such as invariance under space and time reflection, gauge invariance if the material is electrically charged or has an intrinsic magnetism.

One can consider also deformations of the algebraic structure (corresponding in case of a Weyl system to deformation of the Weyl structure) and the corresponding Piezoelectricity (electric effect due to deformation). Currents are defined relative to the continuous group.

The algebraic-geometrical structure that takes the place of the Chern number and of other topological quantum numbers (topological indices) is the *non-commutative index* [12, 13] and Kasparov classes and spectral triples in algebraic topology [14, 15]. As a consequence these systems have symmetries and invariants, typically Z_2 invariants, that are *protected* by these symmetries.

They are *protected* because one cannot pass from one value to another without violating the symmetry. In particular, in the models in which the particles are not interacting among themselves (but only with an external field) when there is a coupling between the spin and the angular momentum and the sample is two-dimensional an occupies a half-space, there is a $\{0, 1\}$ invariant which is interpreted as a current flowing along the edge of the sample in the up or down direction.

The two points of view, that of Schrödinger with topological invariants seen through the geometrical properties of the wave function and the algebraic (Heisenberg) in which the invariants are seen through the algebraic-geometrical properties of the representation of the observable, are connected through the Atiyah–Bott index theorem [16]. We will not expand here on the algebro-geometrical point of view, and refer to [14] for a clear exposition.

7 Determination of a Topological Index

In the final part of this lecture we treat concrete examples of determination of a topological index. For the first we follow [21] using a model hamiltonian suggested by Kane and Mele defined on a honeycomb lattice.

The substitution of the Scrhrödinger equation with a matrix equation on a lattice (in the present case a honeycomb lattice) is an instance of a strategy, frequently used, to substitute a P.D.E. problem with an O.D.E. one (tight-binding model). The matrix equation is an integrated form of the Schrödinger equation.

The matrix elements (hopping terms) should be considered as a result of two reductions: first the reduction of the system to the border of the cell (it is the topology of the wave function at the border that determines the properties of the system) and then substitution of the Schrödinger equation on the border with is *integrated version* the *hopping matrix elements* at each vertex. This is legitimate since the topological analysis should be model independent.

Our purpose is to relate the Chern number of the system to the Bott-Singer index of the projection onto the Fermi sea and to the magnetic flux operator (a non-commutative index according to [13, 17]).

Consider a tight-binding model of spin $\frac{1}{2}$ fermions on the two-dimensional square lattice Z^2. We denote by $\pm$ the spin indices. The wave function ϕ is an element of $l^2(Z^2, \pm)$ and the action of the Hamiltonian H is given by $(\alpha,\ \beta = \pm)$

$$(H\phi)_\alpha(n) = \sum_{n \in Z^2} \sum_{\beta \in \pm} t^{\alpha.\beta}_{n,m} \phi^m_\beta \tag{57}$$

Introduce an antiunitary *time-reversal* map Θ under which

$$\phi^\Theta = U_\Theta \bar{\phi} \tag{58}$$

where U_Θ is a unitary operator invariant under some *finite* (may be random) translation of the lattice. We assume

$$\Theta^2\phi = -\phi \tag{59}$$

Let A be an operator on $l^2(Z, \pm)$ *odd* under time reversal:

$$\Theta(A\phi) = -A\phi^\Theta \qquad \forall\phi \in l^2(Z, \pm) \tag{60}$$

Introduce another unitary operator U_a for $a \in (Z^2)^*$ where Z^* is the dual lattice

$$(U_a\phi)_{n,\alpha} = U_a(n)\phi(n,\alpha) \quad U_a(n) = \frac{n_1 + in_2 - (a_1 + ia_2)}{|n_1 + in_2 - (a_1 + ia_2)|} \tag{61}$$

U is simultaneous rotation of the wave function on lattice and of the dual lattice and therefore it does not chance the physical structure. We assume now that the Fermi level E_F lies in a spectral gap of the Hamiltonian H and let P_F be the projection on energies below E_F.

We restrict now transformation U_a to $P_F l^2(Z, \pm)$. We choose A to be

$$A = P_F - U_a P_f U_a^* \tag{62}$$

The operator A is the difference of two projections. One can verify that A^3 is of trace class and the relative index is

$$Ind(P_F, U_a P_F U_a^*) = dimKer(A-1) - \dim Ker(A+1) = TrA^3 \tag{63}$$

Recall that U_Θ is invariant under some *finite* (may be random) translation in the lattice. The index written above is therefore finite but depends on these finite translations.

We define the Z_2 index for the Hamiltonian H as [19]

$$Ind_2(P_F, U - aP_F U_a^*) = dimker(P_F - U_a P_F U_a^* - 1) \ \ mod2 \tag{64}$$

Consider a lattice Hamiltonian H which is odd under time reversal symmetry.

Lemma 1 *The Z_2 index so defined is robust under any perturbation of H (in particular under any modification of the choice of the* finite *translations described above), provided it has the same odd time-reversal symmetry as the unperturbed Hamiltonian.* ◇

The proof is a standard *supersymmetry* argument. Write

$$B = 1 - P_f - U_a P_F U_a^* \tag{65}$$

and then

$$AB + BA = 0, \qquad A^2 + B^2 = 1 \tag{66}$$

Note that the spectrum of A is discrete with finite multiplicity; we prove that the non-zero eigenvalues come in pairs related by the operator B. Let $A\phi_\lambda = \lambda\phi_\lambda$ $\lambda \in (0, 1]$ One has

$$AB\phi_\lambda = -BA\phi_\lambda = -\lambda\phi_\lambda \tag{67}$$

Moreover

$$B^2\phi_\lambda = (1 - A^2)\phi_\lambda = (1 - \lambda^2)\phi_\lambda \tag{68}$$

It follows that B is invertible on the subspace spanned by the eigenvalues in $(0, 1)$ and these eigenvalues come in pairs.

We remark now that the time-reversal transformation Θ shares with B this property. Let $A\phi = \lambda\phi$, $\lambda > 0$. From the definition of the operator A one has

$$\Theta(P_F - U_a P_F U_a^*)\phi = \lambda\phi \tag{69}$$

Choose now that the unitary operator U_Θ to satisfy

$$U_\Theta U_a U_\Theta^* = U_a \tag{70}$$

These relations can be written

$$(P_F - U_a^* P_F U_a)\phi^\Theta = \lambda\phi^\Theta \tag{71}$$

It follows

$$A(U_a\phi^\Theta) = -\lambda\phi^\Theta \tag{72}$$

Lemma 2 *Let ϕ be an eigenvector of A with eigenvalue $0 < \lambda < 1$. Then*

$$U_a(B\phi)^\Theta = B(U_a\phi^\Theta) \tag{73}$$

Proof
One has

$$\Theta(B\phi) = (1 - P_F - U_a^* P_F U_a)\phi^\Theta = U_a^* B U_a \Phi^\Theta \tag{74}$$

Therefore in the localization regime

$$U_a(B\phi)^\Theta = B(U_a\phi^\Theta) \tag{75}$$

We now prove that the eigenvectors ϕ and $U_a(B\phi)^\Theta$ are independent.

Lemma 3 *Let ϕ be an eigenvector of A with eigenvalue* $0 < \lambda < 1$ *Then* $(\phi, U_a(B\phi)^{\Theta}) = 0$

Proof Set $\psi = U_a(B\phi)^{\Theta}$.
One has

$$(\Theta\psi, \Theta\phi) = (\phi, U_a(B\phi)^{\Theta} \tag{76}$$

By the previous Lemma, and using $\Theta^2\phi = -\phi$

$$-(U_a^* B\phi, \Theta\phi) = (\phi, U_a(B\phi)^{\Theta} \tag{77}$$

From this one derives

$$(\phi, U_a(B\phi)^{\Theta} = 0 \tag{78}$$

♡

It is now possible to prove that the Z_2 index is invariant under perturbations ∂H of H that are odd under the same time-reversal transformation under which H is odd. Assume that the range of hopping of ∂H is finite and that $\|\partial H\| < \infty$ We assume that the Fermi level lies in the spectral gap if H. Let $H' = H + \partial H$. Let

$$P'_F = \frac{1}{2\pi i} \oint dz \frac{1}{z - H'} \tag{79}$$

Consider the operator

$$A' = P'_F - U_a P'_F U_a^* \tag{80}$$

We have

$$A' - A = (P'_F - P_f) - U_a(P'_F - P_F)U_a^* \tag{81}$$

and

$$P'_F - P_f = \frac{1}{2\pi i} \oint dz \frac{1}{z - H'} \partial H \frac{1}{z - H} \tag{82}$$

The operator is continuous with respect to the norm of ∂H By the min-max principle the non zero eigenvalues of A are continuous with respect to the norm of the perturbation ∂H. Notice that the proof we have presented is valid in the *localization regime*.

It can be proven that the same is true when the Fermi level lies in the regime in which invariance of the Hamiltonian under *finite* (may be random) translation holds. When one assumes only that the Fermi level lies *in the localization regime* the result still holds but one must prove localization separately. For this, one needs estimates on the resolvent $\frac{1}{z-H}$ $z \in C - R$.

8 Gauge Transformation, Relative Index and Quantum Pumps

The algebraic analysis of the last part of the lecture has a counterpart in the theory of quantum pumps, i.e. periodic structures that make one electron per cycle pass over the Fermi level. One may say the in one cycle an index varies by one unit.

The problem is again the determination of a *relative index* of two projections on infinite dimensional spaces, the projection operators on the Fermi level of an infinitely extended crystal. We review briefly this issue [20].

Recall again that if P and Q are orthogonal such that P_Q is compact, then by definition the *relative index* is defined as follows

$$Ind(P, Q) \equiv dim(Ker(P - Q - I)) - dim(Q - P - I) \tag{83}$$

It is easy to verify

$$Ind(P, Q) = -Ind(Q, P) = -Ind(P_\perp, Q_\perp) \tag{84}$$

and that the index is invariant under unitary transformations. Moreover of $(P-Q)^{2n+1}$ is trace class for some integer n then

$$Ind(P.Q) = Tr(P - Q)^{2n+1} \tag{85}$$

Indeed one verifies without difficulties that if $(P - Q)^{2n+1}$ trace class then

$$Tr(P - Q)^{2n+1} = TR(P - Q)^{2n+3m} \forall m > 0 \tag{86}$$

and (84) follows by taking $m \to \infty$. If there exist a unitary U such that $Q = UPU^*$ then

$$Ind(P, Q) = -Ind(PUP) \tag{87}$$

and for any three projection operators P, Q, R

$$Ind(P, Q) = Ind(P, R) + Ind(R, Q) \tag{88}$$

Recall that the unitary U exists always in case P and Q are infinite dimensional projections (as in the case if they project onto the states below a Fermi surface in an infinite-dimensional translation invariant system).

In [20] the unitary that relates the orthogonal projections P and Q is associated to the (singular) gauge transformation which is obtained by piercing a two-dimensional quantum system with a flux tube carrying an integral number of *flux quanta* (Bohm-Aharanov effect).

The unitary U is in this case a unitary multiplication of the wave function by a phase corresponding to the number of flux quanta carried by the flux tube.

This system is called *quantum pump* because the change in phase is related to the number of electrons passing in the tube while the system undergoes on cycle. We will show that in order to have $Ind(P.Q) \neq 0$ *time- reversal invariance* must be broken in the process.

To have a simple example, consider in R^2 the map

$$U_\alpha(z) = \frac{z^\alpha}{|z|^\alpha} \quad z \in R^2/[0,\infty) \qquad U_\alpha(z) = 1 \quad z \in [0,\infty) \tag{89}$$

In this case the projection P has an integral kernel $p(x, y)$ that satisfies

$$|p(x, y)| \leq \frac{C}{1 + dist(x, y)} \tag{90}$$

This assumption is used in the general case and it is precisely the assumption of this bound allows to restrict in the previous system to the case in which translation invariance of the Hamiltonian under *finite* (may be random) translation holds.

In the remaining part of this analysis we will assume that the following is true for the trace class operator K: the kernel $K(x, y)$ of K is jointly continuous away form a finite set of point so that $K(x, x) \in L^1$. is

Under this assumption $TrK = K(x, x)dx$. One can see that if $P - Q$ is trace-class, $Q = UPU^*$ in the previous example one has

$$Ind(P, Q) = Tr(P - Q) = 0 \tag{91}$$

Therefore to obtain a non trivial result one must have

$$dimP = dimQ = +\infty \tag{92}$$

In the Aharonov–Bohm example above, $(P - Q)^2$ is trace class, $Tr(P - Q)^3 \in \mathcal{Z}$ and

$$Ind(PUP) = \int_\Omega dxdydzp(x, y)p(y, z)p(z, x)(1 - \frac{u(x)}{u(y)})(1 - \frac{u(y)}{u(z)})(1 - \frac{u(z)}{u(x)}) \tag{93}$$

It can also be proved that the index is invariant under translations or deformations of the provided one keeps the flux constant. Finally we notices that $Ind(PUP) = 0$ if P is time reversal invariant. Indeed, $Ind(PUP)$ is real and even under conjugation. On the other hand it is odd under time-reversal.

To clarify the concept of charge transfer for the pair of projections P, Q in [21] one considers a canonical interpolation (time dependent hamiltonian)

$$H(t) = (-i\nabla - \phi(t)\nabla(argz) - A_0)^2 + V \qquad t \in [0, 1] \tag{94}$$

where $\phi(t)$ interpolates smoothly between zero and one. Here $\nabla(argz)$ is regarded as a vector field in the plane. $H(t)$ has a *time-dependent* domain and therefore *it is not* equivalent to H. In addition to the magnetic field there is an electric field, hence a charge experiences a Lorentz force and is pushed radially (Hall effect).

The force is quantized by the number of units of flux quanta (quantum Hall effect) [20].

References

1. W. Kohn, Theory Phys. Rev. 13 A (1964)
2. S. Vanderbilt, R. King-Smith, Phys. Rev. B **48**, 442–456 (1993)
3. R. Resta, Rev. Mod. Phys. **66**, 899–914 (1994)
4. R. Resta, S. Sorella Phys. Rev. Lett. **82**, 370–378 (1999)
5. I. Souza, T. Wilkems, R.M. Martin, Phys. Rev. B **62**, 1666–1680 (2000)
6. G. Nenciu, Comm. Math. Phys. **91**, 81–85 (1983)
7. D.J. Thouless, M. Kohomoto, B. Nightingale, M. de Nijs, Phys. Rev. Lett. **49**, 405–408 (1982)
8. C.L. Kane, E.J. Mele, Phys. Rev. Lett. **95/1**, 46802 (2005)
9. S. Teufel Adabatic, *Perturbation Theory Lect. Notes in Math*, 1821 (Springer, Berlin, 2003)
10. J. Bellissard, A. Van Est, H. Schulz-Baldes, J. Math. Phys. **35**, 5373–5451 (1994)
11. M.F. Atiyah, R. Bott, Shapiro Topol. **3**(Supplement 1), 3–38 (1964)
12. J. Kellendok. arXiv:1509.06271 (2015)
13. A. Connes, Noncommutative differential geometry. Publ. Math. IHES **62**, 257–360 (1986)
14. C. Bourne, A. Carey, A. Rennie, arXiv **1509**, 07210 (2015)
15. G.G. Kasparov, Inventiones Meth. **91**, 147–201 (1988)
16. M.F. Atiyah, I.M. Singer, Ann. Math. **93**, 139–149 (1971)
17. E. Prodan, B. Leung, J. Bellissard, J. Phys. A **46**, 5202 (2013)
18. C.L. Kane, E.J. Mele, Phys. Rev. Lett. **95**(1), M46802 (2005)
19. M. Aizenmann, G.M. Graf, J. Phys. A **31**, 6783–6806 (1998)
20. J. Avron, R. Seiler, B. Simon, Comm. Math. Phys. **159**, 399–422 (1994)
21. H. Katsura and T.Koma

Lecture 6: Lie–Trotter Formula, Wiener Process, Feynman–Kac Formula

We begin recalling the Lie–Trotter formula. Let A and B be $N \times N$ matrices. Lie's formula for product of exponentials asserts that

$$e^{A+B} = lim_{n\to\infty} \left(e^{\frac{A}{n}} e^{\frac{B}{n}} \right)^n \tag{1}$$

This formula can be easily verified expanding the exponentials in power series. A more elegant proof is obtained substituting A with tA and B with tB and noticing that the identity holds for $t = 0$ and the derivative with respect to t of the two sides coincide.

The formula is attributed to S. Lie, who discussed it in the context of Lie algebras; it had already used in implicit form by Euler in his treatment of the symmetric top. The formula extends, with the same proof, for A and B closed and bounded operators in a Hilbert space. We will see presently that it can be extended without much difficulty to the case when A and B are self-adjoint and the domain $D(A + B) = D(A) \cap D(B)$.

Trotter has given an extension to the case in which A, B and *the closure* of $A + B$ all are generators of C^0 semigroups. Here we consider two cases, in increasing order of difficulty.

Theorem 1 *Let A and B be self-adjoint operators on a Hilbert space $\mathcal{H}$ and suppose that $A + B$ is self-adjoint with dense domain $D(A + B) = D(A) \cap D(B)$.*

Then uniformly over compact sets,

(i)

$$e^{-it(A+B)} = s - lim_{n\to\infty} \left(e^{-i\frac{tA}{n}} e^{-i\frac{tB}{n}} \right)^n \quad t \in R \tag{2}$$

Moreover if A and B are bounded below then, uniformly over compact sets in R^+,

(ii)

$$e^{-t(A+B)} = s - lim_{n\to\infty} \left[e^{-\frac{tA}{n}} e^{-\frac{tB}{n}} \right]^n \quad t \in R^+ \tag{3}$$

◇

G. Dell'Antonio, *Lectures on the Mathematics of Quantum Mechanics II: Selected Topics*, Atlantis Studies in Mathematical Physics: Theory and Applications 2, DOI 10.2991/978-94-6239-115-4_6

Proof We give a proof of (i); the proof of (ii) follows the same lines keeping into account that for $t \in R^+$ the operators e^{-tA}, e^{-tB}, $e^{-t(A+B)}$ are bounded uniformly in t.

Since the operators e^{-itA} and e^{-itB} are bounded, it suffices to prove (2) on a dense set, which we choose to be $D(A) \cap D(B)$. A simple computation shows, for any $s > 0$ and $\phi \in \mathcal{H}$

$$\frac{1}{s}(e^{-isA}e^{-isB} - I)\phi = \frac{1}{s}(e^{-isA} - I)\phi + e^{-isA}\frac{1}{s}(e^{-isB} - I)\phi] \tag{4}$$

If $\phi \in D(A) \cap D(B)$ the right hand side converges when $s \to 0$ to $-i(A+B)\phi$. Moreover

$$lim_{s\to 0}\frac{1}{s}(e^{-i(A+B)s} - I)\phi = -i(A+B)\phi 2 \tag{5}$$

Therefore

$$\frac{1}{s}(e^{-isA}e^{-isB} - e^{-is(A+B)})\phi \to 0 \tag{6}$$

On the other hand one has

$$\begin{aligned}&[e^{-i\frac{tA}{n}}e^{-i\frac{tB}{n}}]^n\phi - e^{it(A+B)}\phi \\ &= \sum_k .[e^{-iAt/n}e^{-itB/n}]^k[e^{-iAt/n}e^{-itB/n} - e^{-it(A+B)/n}]e^{-it(n-k-1)(A+B)/n}\phi \end{aligned} \tag{7}$$

From this one derives

$$\begin{aligned}&|[e^{-itA/n}.e^{-itB/n}]^n\phi - e^{-it(A+B)}\phi|_2 \\ &\le |t| \max_{k=1,..(n-1)}.tn^{-1}|[e^{-itA/n}e^{-itB/n} - e^{-i\frac{t}{n}(A+B)}]\phi((n-k-1)s/n)|_2 \end{aligned} \tag{8}$$

where $\phi(r) \equiv e^{-ir(A+B)}\phi$ and we have denoted by $|\phi|_2$ the norm of ϕ as element of $\mathcal{H}$.

Each term of the series converges to zero due to (2) and $\phi(r)$ is continuous in r. For fixed t the set $\{\phi(r) \;:\; |r| < |t|\}$ is closed in the closed set $D(A+B)$. Since $\phi(r)$ is continuous and the convergence is uniform over compact sets in t by the Ascoli–Arzelá theorem and

$$lim_{s\to 0}sup_{|r|\le t}\, s^{-1}|(e^{-isA}e^{-isB} - e^{-is(A+B)})\phi(r)|_2 = 0 \tag{9}$$

♡

Remark that since the proof is given by compactness *there is no estimate* of the error one makes in truncating the series to order N. In the proof of Theorem 1 we have made *essential use* of the assumption that $D(A) \cap D(B)$ is *closed* (as domain of a self-adjoint operator).

In general if the operators are unbounded the set $D(A) \cap D(B)$ is only an *open* subset of $D(A + B)$. Therefore $\{\phi(r) | |r| < |t|\}$ is in general an *open* set and the compactness argument cannot be used.

Still the conclusions of Theorem 1 hold also if the operator $A + B$ is *essentially self-adjoint* in $D(A) \cap D(B)$ but the proof becomes less simple.

Theorem 2 *Let A and B self-adjoint operators. Let A + B be essentially self-adjoint on* $D(A) \cap D(B)$.

Then

(i)

$$e^{-it(A+B)} = s - lim(e^{-i\frac{tA}{n}} e^{-i\frac{tB}{n}})^n \quad t \in R, \tag{10}$$

uniformly over compact sets in R.

(ii) If moreover A and B are bounded below

$$e^{-t(A+B)} = s - lim(e^{-\frac{tA}{n}} e^{-\frac{tB}{n}})^n \quad t \in R^+ \tag{11}$$

uniformly over compact set in R^+.

◇

Proof Also in this case we will prove only (i). The proof is completed in several steps.

Step 1

Let $\{C_1, C_2, \ldots, C_n\}$ be a sequence of bounded operators with $Im\ C_n \equiv \frac{C_n - C_n^*}{2i} < 0$.

Let C be a self-adjoint operator such that $lim_{n\to\infty} C_n\phi = C\phi$ if ϕ belongs to a domain $\tilde{D}$ which is dense in $D(C)$ in the graph norm. Under these conditions

$$s - lim_{n\to\infty}(C_n - z)^{-1} = (C - z)^{-1} \tag{12}$$

for $Imz > 0$ ◇

Proof If $Imz > 0$ the operator $C_n - z$ has an inverse bounded uniformly in n; therefore it is sufficient to prove $lim_{n\to\infty}(C_n - z)^{-1}\phi = (C - z)^{-1}\phi$ if ϕ is in a dense subset of $\mathcal{H}$. We shall choose it to be $(C - z)D(C)$. Setting $\psi = (C - z)\phi$ one has

$$|(C_n - z)^{-1}\phi - (C - z)^{-1}\phi| = |(C_n - z)^{-1}(C - z)\psi - \psi|$$

$$= |(C_n - z)^{-1}(C - C_n)\psi| \leq (Imz)^{-1}|(C - C_n)\psi| \to_{n\to\infty} 0 \tag{13}$$

♡

Step 2
Under the hypothesis of step 1 one has, uniformly on the compacts in R^+

$$s - lim_{n\to\infty} e^{-itC_n} = e^{-itC} \tag{14}$$

◇

Proof Fix $\phi \in \mathcal{H}$. The subspace spanned by the action of bounded functions C_n and by C on ϕ is separable. Hence we can assume that $\mathcal{H}$ be separable. One has

$$\frac{d}{dt}|e^{-itC_n}\phi|^2 = \left(e^{-itC_n}\phi, \frac{t}{i}(C_n - C_n^*)e^{-itC_n}\phi\right) == |t|\,(e^{-itC_n}\phi, Im(C_n)e^{-itC_n}\phi) \le 0 \tag{15}$$

and therefore $|e^{-itC_n}| \le 1$ for $t \ge 0$. It is then sufficient to prove step 2 when $\phi \in \tilde{D}$.

We prove the thesis arguing by contradiction. Suppose that for some $\phi \in \tilde{D}$ the equality

$$lim_{n\to\infty} e^{-itC_n}\phi = e^{-itC}\phi \tag{16}$$

does not hold. There $\exists\{n'\}$, $t(n') > 0$ such that $|e^{-it(n')C_{n'}}\phi - e^{-it(n')C}\phi| \ge \delta > 0$. This implies $\exists l_{n'} \in \mathcal{H}$, $|l_{n'}| = 1$ such that

$$|(l'_{n'}, e^{-it(n')C_{n'}}\phi) - (l'_n, e^{-it(n')C}\phi)| \ge \delta \tag{17}$$

Since the unit ball in $\mathcal{H}$ is weakly compact there exist a sub-sequence, still named $\{n'\}$ which converges to l, $|l| \le 1$ and for n large enough

$$|(l_{n'}, e^{-it(n')C_{n'}}\phi) - (l, e^{-it(n')C}\phi)| \ge \delta \tag{18}$$

On the other hand the sequence $\{(l_{n'}, e^{-itC_{n'}}\phi)\}$ is equibounded in $t \ge 0$.

By the Ascoli–Arzelá lemma one can choose a sub-sequence such that

$$(l_{\bar{n}}, e^{-it(\bar{n})C}\phi) \to F(t) \tag{19}$$

uniformly on the compact sets in R^+, where $F(t)$ is a continuous function of t. Therefore $|F(t(\bar{n})) - (l, e^{-it(\bar{n})C}\phi)| \ge \delta$ Since the functions are continuous the relation is true in a neighborhood of $t(n')$.

Consider now the Laplace transform of $F(t)$. From step 1 and Lebesgue dominated convergence theorem

$$\begin{aligned}\int_0^\infty F(z)e^{itz}dt &= lim_{k\to\infty}\int_0^\infty (l_{n'}, e^{-it(n')C_{n'}}\phi)e^{izt}dz \\ &= (-i)lim_{n\to\infty}(l_{n'}, (C_{n'} - z)^{-1}\phi) = -i(l, (C - z)\phi) \quad Im z > 0\end{aligned} \tag{20}$$

Therefore the Laplace transforms of $F(t)$ and of $(l, e^{-itC}\phi)$ coincide, against the assumption made. ♡

Step 3

Let T be a contraction operator ($|T| \leq 1$). Then $t \to e^{t(T-1)}$ is a contraction semigroup. Moreover

$$|(e^{n(T-I)} - T^n)\phi| \leq \sqrt{n}|(T-I)\phi|, \quad n \geq 1 \quad \forall \phi \in H \tag{21}$$

◇

Proof Since T is bounded the function $e^{t(T-1)}$ is continuous operator. It is a contraction because

$$|e^{t(T-I)}| = e^{-t}\left|\sum_n \frac{t^n T^n}{n!}\right| \leq e^{-t}e^{t|T|} \leq 1 \tag{22}$$

Moreover $e^{n(T-1)} - T^n = e^n \sum_0^\infty \frac{nk}{k!}(T^k - T^n)$. Using the inequality

$$|(T^j - I)\phi| = \left|\sum T^k(T-I)\phi\right| \leq j|(T-I)\phi| \tag{23}$$

one has

$$|(e^{n(T-1)} - T^n)\phi| \leq e^{-n}\left[\sum_0^\infty \frac{n^k}{k!}|n-k|\right]|(T-I)\phi| \tag{24}$$

On the other hand

$$e^{-n}\sum \frac{nk}{k!}|n-k| \leq e^{-n}\left(\sum \frac{n^k}{k!}\right)^{\frac{1}{2}} = e^{-n/2}(n^2 e^k - (2n-1)ne^n + n^2 e^n)^{\frac{1}{2}} = \sqrt{n} \tag{25}$$

♡

With these steps we can complete the proof of Theorem 2. Let

$$F(t) = e^{-itA}e^{-itB}, t > 0 \quad C_n = i\left(\frac{t}{n}\right)^{-1}\left(F\left(\frac{t}{n}\right) - I\right), \qquad C = A + B \tag{26}$$

If $\phi \in D(A) \cap D(B)$ one has then

$$C_n\phi = i\left(\frac{t}{n}\right)^{-1}[e^{-i\frac{tA}{n}}e^{-i\frac{tB}{n}} - I]\phi = ie^{-i\frac{tA}{n}}[\left(\frac{t}{n}\right)^{-1}(e^{-i\frac{tB}{n}} - I)\phi + i\left(\frac{t}{n}\right)^{-1}e^{-i\frac{tA}{n}} - I)\phi \tag{27}$$

$$lim_{n\to\infty}(A+B)\phi = C\phi, \quad n \to \infty \tag{28}$$

From steps 1 and 2 one derives $s - lim_{n\to\infty}e^{n(F((\frac{t}{n})-1)} = s - lim_{n\to\infty}e^{-itC_n} = e^{-itC}$. From Step 3

$$\left|e^{n(F\frac{t}{n}-I)} - F\left(\frac{t}{n}\right)^n]\phi\right| \leq \sqrt{n}\left|\left(F\left(\frac{t}{n}\right) - I\right)\phi\right| = \frac{t}{\sqrt{n}}|C_n\phi| \tag{29}$$

Combining these result

$$|e^{-it(A+B)}\phi - e^{-it\frac{A}{n}}e^{-it\frac{B}{n}}\phi| = \left|e^{-itC}\phi - F\left(\frac{t}{n}\right)^n \phi\right| \leq |(e^{-itC} - e^{-itC_n})\phi| + \frac{t}{\sqrt{n}}|C_n\phi| \tag{30}$$

This expression tends to zero as $n \to \infty$. This concludes the proof of Theorem 2 ♡

Remark that since in Steps 1 and 2 we used compactness, we cannot estimate of the error made if we terminate the expansion at the nth order.

1 The Feynman Formula

We shall now use the Trotter–Kato formula to obtain *formally* the Feynman formula of integration over path space. This formula has *only formal meaning* because there is no regular measure supported on those paths for which the integrand is meaningful. We shall see later how to define a convenient measure space and a measure on it.

Consider first bounded continuous potentials $V(x)$ and set $H_0 \equiv -\frac{1}{2}\Delta$. $H_0 + V(x)$ is self-adjoint with domain $D(H_0)$. Taking into account the explicit form of the kernel of e^{-itH_0} i.e.

$$G_0(x-y;t) = (4i\pi t)^{-d/2}e^{-\frac{|x-y|^2}{4it}} \tag{31}$$

It follows from Theorem 2 that for each $\phi \in L^2(R^d)$

$$(e^{-itH}\phi)(x) = s.l._{N\to\infty}\left(\frac{N}{i\pi t}\right)^{\frac{Nd}{2}} \int e^{-iS_N(x,x_1,\ldots,x_N,t)}\phi_0(x_N)dx_1\ldots dx_N \tag{32}$$

where

$$S_N(x_1,\ldots,x_N) = \frac{t}{N}\sum_{i,j=1}^{N}\frac{|x_i-x_j|^2}{2\frac{t}{N}} + \sum_i V(x_i)\frac{t}{N} \tag{33}$$

In (32) the integral is understood in the following sense:

$$\int_{R^d} f(x)d^N x = lim_{R\to\infty}\int_{|x|\leq R} f(x)d^N x \tag{34}$$

and the limit is in the topology of $L^2(R^N)$. We would like to interpret the limit on the right hand side of (32) as integral over a space of paths. Let Γ^1 be the class of absolutely continuous functions of time with values in R^d.

Following a well established tradition we call such function *paths* and we call position of the path at time t the value of the function at the value t of the parameter. We study first the case $d = 1$.

We identify the variable x_k with the value that the coordinate takes at time $\frac{tk}{N}$ on the path $\gamma_{x,x';T} \in \Gamma^1$. For each path $\gamma_{x,x';T} \in \Gamma^1$ with $\gamma(T) = x$, $\gamma(0) = x'$ we have

$$\lim_{N\to\infty} S_N(x', x_1, \ldots, x_k, \ldots, x_{N-1}, x, t)dx_1 \ldots dx_k \ldots dx_{N-1} = S(\gamma_{x,x',t})$$

$$S(\gamma_{x,x';t}) = \int_0^t \left[\frac{1}{2}|\dot{x}(s)|^2 + V(x(s)\right]_{x(.)\in\gamma} ds \tag{35}$$

Remark that $S(\gamma_{x,x';t})$ is the integral of the classical Action along the trajectory $\gamma_{x,x';t}$.

If one takes *formally* the limit $N \to \infty$ in the right hand side of the equation, one writes the integral kernel $(e^{itH})(x, x')$ as *formal* integral over absolutely continuous trajectories γ in the interval $[0, t]$

$$(e^{-itH})(x, x') = \left(\lim_{N\to\infty} C_N\right) \int_{\gamma\in\Gamma_1, \gamma(0)=x',\ \gamma(t)=x} e^{-iS(\gamma_{x,x';t})} \Pi_t d\gamma_t \tag{36}$$

where C_N is a normalization constant and $\int \Pi_t d\gamma_t$ represents is (formally) the integration over a continuous product of Lebesgue's measures. But the right hand side is only formal: the constant $C_N = (\frac{i\pi t}{N})^{-\frac{N}{2}}$ in (36) diverges as $N \to \infty$ and the measure $d\gamma$ remains undefined (Lebesgue measure is not a probability measure and the classical construction of product measures does not apply).

Remark that the same procedure can be followed if one considers the Schroedinger equation in the domain $|x_i| < C \forall I = 1, \ldots, d$ (defining the Laplacian with suitable boundary conditions). In this case the limit measure exists (Lebesgue measure on $[-C, +C]$ can be made with a suitable normalization into a probability measure) but is can be seen, following a procedure similar to the one which we shall outline for Gauss's measure, that the set of absolutely continuous functions is contained in a set of measure zero.

We conclude that, while the limit in (32) certainly exists as integral kernel, its interpretation as integral over a class of trajectories is ill defined and, if not taken with a suitable care, may be the source of error. It should be remarked that for some class of potentials, e.g. if the potential is the Fourier transform of a measure, it is possible to give meaning to the limit to the right in (35) as limit of oscillating integrals and to interpret it in the framework of a stationary phase analysis in an infinite dimensional space [1].

The approach in [1] is not within the framework of measure theory and one cannot make use of standard tools, e.g. of Lebesgue comparison principle. Therefore it is difficult to compare results for different choices of V without making reference to the expression in terms of integral kernels. We shall not discuss further this very interesting and difficult problem.

2 Stationary Action; The Fujiwara's Approach

For completeness we reproduce here, with some further details, the remarks we have made in Volume I of these Lecture Notes.

If $t - s$ is sufficiently small (depending on x and y) the classical Action

$$S(t, s; x, y) = \int_s^t L\left(\tau, x(\tau), \frac{dx(\tau)}{d\tau}\right) d\tau$$

is *stationary* on the classical orbits (absolutely continuous functions solutions of Lagrange's equations with end points x and y) and is the generating function of the family of canonical transformations that define motion in phase space. One can expect, in the semiclassical limit, to be able to make use of the fact that the Action is stationary on the trajectories of the system associated to the Lagrangian L.

In this case it may be reasonable to approximate the full propagator by stationary point techniques with a careful estimate of the remainder terms rather than by the Trotter formula. Introducing Planck's constant $\hbar$ one considers in the approximation finite time intervals of order $\hbar^\alpha$ with $\alpha < 1$ and seeks an approximation to order $\hbar^{\frac{1}{2}}$.

One can prove in this way [2, 3] that if the potential $V(t, x)$ is sufficiently regular the propagator (fundamental solution) $U(t, s)$ satisfies for any function $\phi \in L^2(R^d)$

$$U(t, s)\phi(x) = exp\left(i \int_{R^d} lim_{\delta \to 0} I(\delta; t, s; x, y)\phi(y)dy\right) \tag{37}$$

where *the limit is understood in distributional sense.*

We have denoted by $\{t_j\}$ a partition of the interval $[s, t]$ in equal intervals of length $\delta = \frac{t-s}{N}$, $N = \hbar^\alpha$ and we have set

$$I(\delta; t, s; x, y) = \Pi_{j=2}^{N-1} \left[\frac{1}{\hbar} \frac{-i}{2\pi(t_j - t_{j-1}))}\right]^{\frac{d}{2}}$$

$$\int_{R^d} \dots \int_{R^d} \Pi_{j=2}^{N-1} a_\hbar(t_j, t_{j-1}; x_j, x_{j-1}) exp\left\{-\frac{i}{\hbar} S(t_j, t_{j-1}; x_j, x_{j-1})\right\} \Pi_1^{N-1} dx_j \tag{38}$$

The function $a_\hbar$ is defined by

$$a_\hbar(t_j, t_{j-1}; x_j, x_{j-1}) = exp\{-\frac{1}{2\hbar}\int_s^t (\tau - s)\Delta_x\omega(\tau, s; x(\tau), y)d\tau \tag{39}$$

where ω is defined by $S(t, s; x, y) = \frac{1}{2}\frac{|x-y|^2}{t-s} + (t-s)\omega(t, s; x, y)$ and S is the classical Action for the Hamiltonian $H_{class} = p^2 + V(q),\ q, p \in R^d$ evaluated on the classical trajectory that joins x to y in time $t - s$.

This formula is derived for small values of $t - s$ using in the Stationary Phase Theorem together with an estimate of the residual terms *without using a Trotter product formula*. Remark that on each interval the Action S is the integral of the Lagrangian over the classical trajectory but the trajectories we have used over consecutive intervals do not join smoothly because we have used Dirichlet boundary conditions at the extremal points.

Therefore we are considering trajectories which are continuous but not everywhere differentiable. The set of point where they are not differentiable becomes dense as $N \to \infty$ (i.e. $\hbar \to 0$). At the same time the limit "Lebesgue-like" measure does not exist. Still for N finite (i.e. $\hbar \neq 0$) this expression has the advantage, as compared to (32), that on each interval one considers the solution of the classical equation of motion with potential V rather than free motion as in (32).

For this reason, Fujiwara's approach has been successfully used in the study of the semiclassical approximation to Quantum Mechanics in particular in the scattering regime where in a suitable sense the evolution of the wave function in Quantum Mechanics has stricter links with evolution in Classical Mechanics (resembles more the free evolution) at large times.

3 Generalizations of Fresnel Integral

In [4] the Authors have introduced a version of oscillatory integrals that can be interpreted as Feynman integrals for a suitable class of potentials (those which are the sum of a positive quadratic term and a function which is the Fourier transform of a measure of bounded variation).

The integral introduced in [4] generalized Fresnel's integral $\int_R e^{\frac{i}{2}x^2}dx$. Fresnel's integral is an oscillatory integral that cannot be interpreted as a Lebesgue integral with respect to a regular complex measure (the total variation of the measure would be infinite). It is rather interpreted as improper Riemann integral, and the convergence is a result of the oscillatory behavior of the integrand, with the result

$$\int_R e^{\frac{i}{2}x^2}dx = \sqrt{2\pi} \tag{40}$$

In [4] a generalization of this procedure is given for an infinite dimensional separable Hilbert space providing, under suitable conditions, an *infinite-dimensional Fresnel integral* Let f be the Fourier transform of complex valued regular measure of bounded variation on a separable Hilbert space $\mathcal{H}$. The definition of the integral of f is given in [4] by duality

$$\int_{\mathcal{H}} f(x)e^{\frac{i}{2\hbar}\|x\|^2}dx \equiv \int_{\mathcal{H}} e^{\frac{i\hbar}{2}\|x\|^2}d\mu_f \tag{41}$$

The integral on the right is absolutely convergent and well defined as Lebesgue integral. It is proved in [4] with this procedure that, for potential which are the Fourier transform of a regular measure of bounded variation, the Feynman integral can be interpreted as infinite-dimensional Fresnel integral over the Hilbert space of trajectories (Cameron space) with scalar product

$$< \gamma_1, \gamma_2 >= \int_0^t (\dot{\gamma}_1(s), \dot{\gamma}_2(s))ds \tag{42}$$

where $\dot{\gamma}$ is the distributional derivative of the trajectory γ. In [4] there also an application of this formalism to the semiclassical limit.

We shall not discuss further this very interesting and difficult approach. For more details we refer to [4, 5].

4 Relation with Stochastic Processes

A study of scattering in the semiclassical limit can be done also through the study of the representation of $e^{-tH}\phi$ through an integral over the trajectories of Brownian motion. This requires a similar representation for the resolvent $\frac{1}{H-z}$, $Imz \neq 0$. This can be done (Gutzwiller trace formula) but the subject is outside of the scope of this lecture.

We will see that a formulation which introduces a bona-fide measure on a space of trajectories (and that under suitable conditions can be extended to the infinite dimensional case) can be obtained for the Trotter–Kato formula relative to semigroups.

This is due to the fact that the integral kernel of e^{-tH_0} is of *positive type*, (maps positive functions to positive functions) and can be interpreted as *transition function* for a stochastic process (Brownian motion). Recall that the solution $u(t,x)$, $x \in R^d$ of the heat equation

$$\frac{\partial u}{\partial t} = \Delta u, \qquad u_{t=0} = u_0 \tag{43}$$

is given by

$$u(t,x) = (2\pi t)^{-\frac{d}{2}} \int e^{-\frac{(x-y)^2}{2t}} u(0,y)dy \tag{44}$$

For positive initial data $u(t,x)$ is strictly positive for $t > 0$, and N. Wiener has shown that it can be represented as the mean value of the initial datum under a measure (Wiener measure) defined on continuous trajectories which start in y at time 0 and are in x a time t.

This measure characterizes *Brownian motion*, is a *stochastic process* that we shall describe presently. Changing in a suitable way the process one can equally well represent in a similar way the solutions of $\frac{\partial u}{\partial t} = \Delta u - Vu$ under some hypothesis on $V(x)$.

From this representation in term of a stochastic process one can derive regularity properties of the resolvent of $-\Delta + V$. We remark that there exists a generalization of the integral that makes it possible, for a large class of potentials, the construction of *generalized Feynman integrals*. This generalization is sometimes called *White Noise Process* and in a suitable sense the process which is obtained may be regarded as the (weak) derivative of Brownian motion.

One proves that the measure μ associated to white noise is a measure on $\mathcal{S}'$ that is introduced by duality from the characteristic function

$$\Phi(f) = e^{-\frac{1}{2}\|f\|_2^2}, \quad f \in \mathcal{S} \tag{45}$$

This means that μ is a Gaussian measure for which $\Phi(f) = \int_{\mathcal{S}'} e^{i(\omega,f)} d\mu(\omega)$. For comparison recall that in the case of Brownian motion the characteristic function is

$$\Phi_B(f) = e^{-\frac{1}{2}\|f\|_{-1}^2}, \quad f \in \mathcal{S} \tag{46}$$

where $\|f\|_{-1} = \int \|\hat{f}(p|)^2(1+|p|^2)^{-\frac{1}{2}} dp$. It follows that the space of functions that may be used to give a description of the White Noise Process is larger then the space of continuous function.

For example one may use the space $L^2(\mathcal{S}', \mu)$ for a suitable (Gaussian) measure μ. In this way one obtains *a version* of the White Noise Process as *weak derivative of Brownian motion* (recall that continuous functions can be regarded as differentiable functions in the distributional sense) and therefore also a realization of Brownian motion (different from the one introduced by Wiener).

A rigorous definition leads to the introduction of *Hida distributions* which become the natural candidates for describing generalized Feynman integrals.

It is in this way possible to study the possibility to write $e^{it(-\Delta+V)}$ as Feynman-like integral for a rather large class of potentials. We shall not discuss further this approach. Further details can be found in [6–8].

We shall come back later to the problem of the construction of measures on space of trajectories in R^∞ (or on the space of trajectories in the space of distributions if one considers Quantum Field Theory) associated to positivity preserving semigroups.

Before discussing the Feynman–Kac formula we digress to make a brief introduction to the theory of stochastic processes; we need some notions from this theory to provide a rigorous approach.

In Lecture 7 we provide the reader with some elements of probability theory, in particular some a-priori estimates that are frequently used. We will also describe there two alternative derivations of Brownian motion. The first is the original construction N. Wiener, the second is a construction of Brownian motion as limit of a random walk, in the spirit of the analysis of Brownian motion made by A. Einstein.

5 Random Variables. Independence

Recall that a *random variable* is a measurable function f on a regular measure space $(\Omega, \mathcal{M}, \mu)$ ($\mathcal{M}$ are the measurable sets and μ is the measure). We shall call *probability law* (or distribution) of the random variable f the distribution defined by

$$\mu_f(B) = \mu\{\omega \,:\, f(\omega) \in B)\} \tag{47}$$

for any Borel B set in R. We shall always *identify* two random variables which have the same probability law, independently from the probability space $(\Omega, \mathcal{M}, \mu)$ in which they are concretely realized.

A random variable is called *gaussian* if the (measurable) sets $\{\omega | f(\omega) \leq a\}$ are distributed according to a gaussian probability law, i.e. the *distribution density* of f belongs to the class of gaussian distributions

$$\mu\{f \leq C\} = \int_{-\infty}^{C} \left(\sqrt{b\pi}\right)^{-1/2} e^{-\frac{(x-a)^2}{b}} dx, \quad a \in R, \quad b > 0 \tag{48}$$

The *mean* (expectation) and the *variance* of f are then

$$E(f) = \left(\sqrt{b\pi}\right)^{-1/2} \int x e^{-\frac{(x-a)^2}{b}} dx = a \qquad Var f = E(f^2) - E(f)^2 = b. \tag{49}$$

Notice that the distribution density of a random gaussian variable, and therefore *the random variable itself* is *completely determined* by the two real parameters a and b.

Two measurable functions on a measure space $(\Omega, \mathcal{M}, \mu)$ represent *independent random variables* if for every pair of measurable sets I and J one has

$$\mu\{f(\omega) \in I, \;\; g(\omega) \in J\} = \mu\{f(\omega) \in I\}\mu\{g(\omega) \in J\} \tag{50}$$

In the same way, considering N-ples of measurable functions, one defines the independence of N random variables. Two gaussian random variables $f, \; g$ with zero mean (on can always reduce to this case by subtraction a constant function) are independent iff

$$E(fg) = 0 \tag{51}$$

6 Stochastic Processes, Markov Processes

We recall here briefly the definition of Stochastic Processes [9, 11].

Definition 1 (*stochastic process in R^d*) The family of random variables ξ_t, $t \geq 0$ is called *stochastic process* with values in R^d living in the time interval $[0, T]$ if there exists a measure space Ω with measurable sets $\mathcal{M}$ and measure μ such that

(a) for all $t \in [0, T]$ the function $\xi_t : \Omega \to R^d$ is μ-measurable (i.e. it is a random variable)
(b) $\forall \omega \in \Omega, \quad \forall t \in [0, T] \quad \xi_t(\omega) \in R^d$ (i.e. one can define the *evaluation map*)
(c) the map $(t, \omega) \to \xi_t(\omega)$ is jointly measurable in ω and t if $t \in [0, T]$ with the Borel sets as measurable sets.

Point (b) defines the *evaluation map* (giving the value of the random variable ξ at time t). Remark that a stochastic process can be defined on any topological space X, e.g. a space of distributions. This is important in treating systems with infinitely many degrees of freedom.

The natural σ-algebra of measurable sets are the Borel sets of X. One often requires the measure μ to be a *Radon measure* i.e. to be *locally finite* (for each $x \in X$ there exists a neighborhood U_x with $\mu(U_x) < \infty$) and *tight* i.e. for each Borel set B, $\mu(B) = sup\{\mu(K), \ K \subset B\}$, K compact. In particular the Gauss measure in R^d is a Radon measure.

The processes we shall analyze are *Markov processes* i.e. stochastic processes which *have no memory*. The precise definition is as follows.

Let the family ξ_t be defined for each $t \leq T$. Denote by $\mathcal{F}_{\leq t}$ the σ-algebra generated by the random variables $\xi_s, \ s \leq t$ and with $\mathcal{F}_{\geq t_1}, t_1 \geq t$, the σ-algebra generated by the random variables $\xi_s, \ t_1 \leq s \leq T$. We will call this structure a *filtration*.

Recall that, given a σ-algebra $\mathcal{F}$ of measurable functions in a probability space $(\Omega, \ \mathcal{M} \ \mu)$, a sub-$sigma$ algebra $\mathcal{G}$ and a function f on Ω which is measurable with respect to μ, the *conditioning* of f with respect to $\mathcal{G}$ (denoted $C_{\mathcal{G}}(f)$) is the *unique* function $f_1 \in \mathcal{G}$ such that for all bounded $g \in \mathcal{G}$

$$\int_\Omega f_1 \ g \ d\mu = \int_\Omega f \ g \ d\mu \tag{52}$$

Definition 2 (*Markov processes*) The process $\{\xi_t\}, \quad t \in [0, T]$) is a Markov process iff for any pair $t, \tau < t$ the following relation holds

$$\mathcal{F}_{\leq\tau}(\xi_t) = C_{\mathcal{F}_\tau}(\xi_t), \quad C_{\mathcal{F}_t}(\xi_t) = \xi(t) \tag{53}$$

In other words, the dependence of ξ_t from $\mathcal{F}_{\leq\tau}$ can be expressed as dependence only from the σ-algebra generated by ξ_τ (the future depends on the past only via the present).

If the family $\{\xi_t\}$ is associated to an evolution in a Banach space has the Markov property the expectations have a semigroup property, i.e. for any measurable integrable real function f one has $E(f(\xi_t) = e^{-tL}E(f(\xi(0))$ where L is a positivity preserving operator on the space $L^1(\Omega, d\mu)$. Remark that the evolution described by a Hamiltonian system has the Markov property.

A stochastic process is fully described by the joint distributions of all *finite collections* of the random variables in the process. Different *realizations* differ only by the choice of the space Ω and of the measurable sets. A specific choice may be dictated by the convenience of enlarging the set of measurable functions to include also *weak limits* of measurable functions of the ξ_t.

The possibility of this extension depends in general from the specific probability space chosen in the realization. For example in the case of Brownian motion, the existence as measurable function of

$$lim_{t\to s}\frac{\xi_t - \xi_s}{|t-s|^p}, \quad p < \frac{1}{2} \tag{54}$$

holds only in a representation in which the Hölder-continuous functions of order p are a set of full measure.

7 Construction of Markov Processes

We shall now introduce a general procedure to construct Markov processes; this links them to *positivity preserving semigroups*. For the moment our interest lies in the connection between stochastic processes and Schroedinger operators. We begin from a particular case, Brownian motion. Denote by

$$K_t^0(q, q') = (4\pi t)^{-\frac{d}{2}} e^{-\frac{|q-q'|^2}{4t}} \quad q,\ q' \in R^d \tag{55}$$

the integral kernel of the operator $e^{t\Delta}$. The solution of the heat equation

$$\frac{\partial u}{\partial t} = \frac{1}{2}\sum_1^3 \frac{\partial_k^2 u}{\partial q_k^2} \tag{56}$$

is

$$u_t(q) = \int K_t^0(q, q')u_0(q')dq' \tag{57}$$

It is easy to verify that K_t has the following properties

(a)

$$K_t^0(q, q') > 0 \quad \forall q,\ q' \tag{58}$$

(b)

$$\int K_t^0(q,q')dq' = 1 \quad \forall t \tag{59}$$

(c)

$$K_{t+s}^0(q,q') = \int K_t^0(q,r)K_s^0(r,q')dr \tag{60}$$

Property (c) reflects the fact that the equation is autonomous and therefore the solutions define a semigroup.

We shall now define a measure *on continuous functions (paths)* $x(t)\ \ t \in [0,T]$ such that $x(0) = q,\ \ x(T) = q',\ \ q, q' \in R^d$. We shall denote by $W_{q,q',T}$ this measure and call it *Wiener measure conditioned to* $(q, q', [0,T])$. From this measure we will construct Wiener measure on continuous paths in $[0,T]$ with $x(0) = 0$ by translation and integration over the final point of the trajectory

Notice that since the points ω of the measure space are R^d-valued continuous functions of time, we can define the *evaluation map* that for each value of t assigns to the point ω the value of the corresponding function at time t. The total mass of $W_{q,q',T}$ is $K_T^0(q,q')$.

By definition a *generating family* of measurable sets are the cylinder sets of continuous functions defined by

$$\{x(s) : x(0) = q_0,\ x(T) = q,\ x(t_k) \in I_k, k = 1,\ldots,N\} \equiv M(\{t_k\}, I_k) \tag{61}$$

where t_k are arbitrary in (0,T) with $t_k < t_{k+1}$ and I_k are measurable sets in R^3. The term *cylindrical* is used to stress that the indicator function of $M(\{t_k\}, I_k)$ belongs to the σ-algebra of measurable functions of the $\xi_{t_1},\ldots,\xi_{t_N}$.

This σ-algebra depends only from a subset of the coordinates and therefore *has the structure of a cylinder*. The measure of the set $M(\{t_k\}, I_k)$ is by definition

$$\mu_{W_{q,q',T}}(M(\{t_k\}, I_k)) = \int_{I_1} dq_1 \ldots \int_{I_N} dq_N K_{t_1}^0(q_0,q_1)K_{t_2-t_1}^0(q_1,q_2)\ldots K_{t-t_N}^0(q_N,q) \tag{62}$$

Theorem 3 (Wiener [9]) *The measure we have defined is countably additive on the collection of cylindrical sets and has a unique extension to a completely additive measure on the Borel sets of the space of continuous functions* $q(s)$, $[0 \le s \le T]$ *for which* $q(0) = q_0,\ \ q(T) = q$. ◇

The proof of Wiener theorem has been given by Kolmogorov as a special case of a general theorem. We will give in Lecture 7 the proof of Kolmogorov theorem. Wiener's own proof is more constructive; we shall sketch it in Lecture 7.

There we will also sketch the construction of Brownian motion given by Einstein as limit of a random walk. Uniqueness in distribution follows from uniqueness on cylindrical sets.

Theorem 4 (Kolmogorov [9, 10]) *Let I be an infinite (may be not denumerable) collection of indices, and for each $\alpha \in I$ let X_α be a separable* locally compact *metric space. Let F be a* finite *subset of I and define*

$$X_F \equiv \otimes_{\alpha \in F} X_\alpha \tag{63}$$

with the product topology. Denote by B_F the Borel sets of X_F and denote by $\mathcal{F}$ the collection of finite subsets of I.

For $F,\ G \in \mathcal{F}$ and $F \subset G$, consider the natural projection of X_G on X_F, denoted with π_F^G. Then $(\pi_F^G)^{-1}$ maps Borel sets in F to cylindrical Borel sets in G and provides a conditional probability.

Suppose that on each X_F there exists a completely additive measure of mass one (probability measure), denoted by μ_F, that satisfies the following compatibility property

$$\mu_F(A) = \mu_G\left((\pi_F^G)^{-1}(A)\right) \tag{64}$$

Under this hypothesis there exist a finite measure space $\{X, \mathcal{B}, \mu\}$, with completely additive finite measure μ_X and a natural projection π_F of X on X_F such that $\mu_F = \mu(\pi_F^{-1})$. ◇

In the specific case of Wiener measure, I is the interval $[0, T]$. We remark that the space X with the properties we have described *is not unique*: different choices of the maps π_F^{-1} lead to different spaces. In the case of Brownian motion, Wiener has shown that is possible to choose X as the space of continuous functions in $[0, T]$ with prescribed value at $t = 0$ and $t = T$. Another choice may lead to a Sobolev space.

8 Measurability

According to Kolmogorov theorem, a given subset of X which is not in X_F need not be measurable. For example in the previous case the measurability of the set of functions orthogonal to a fixed continuous function is not guaranteed.

In order to be sure that a pre-assigned set be measurable one must choose π_F^{-1} properly. If $X_\alpha = R$ and $I = [0, T]$ one can e.g. make use of compactness a and convergence results to prove the following criterion:

Wiener's criterion [4]

Let

$$\Omega \equiv \{C([0, T], R^d,\ x(0) = x_0,\ \ x(T) = x\} \tag{65}$$

The set Ω has $\mu-$measure one if for every *denumerable* collection of points $\mathcal{N} \in [0, T]$ the set of those functions whose evaluation in $\mathcal{N}$ is uniformly continuous has measure one (but this set may depend on the choice of $\mathcal{N}$). ◇

In particular a *sufficient* condition is given by the following theorem [9]

Theorem 5 *If a stochastic process $\xi(s)$ with values in R satisfies for some α $\beta > 0$ and $0 < C < \infty$*

$$E(|\xi_s - \xi_t|^\beta) \leq C|t-s|^{1+\alpha} \tag{66}$$

for all $0 \leq s \leq t \leq 1$, then there is measure on $C[0,1]$ with the same finite-dimensional distributions for ξ_s. ◇

Proof The proof consists in constructing successive approximations of evaluation processes at fixed times and then prove *almost surely* uniform convergence. Notice that *almost sure convergence* means that the set of trajectories on which one does not have uniform converges is a set of measure zero, while *convergence in measure* means that the set of points for which one does not have convergence has measure which tends to zero; *but this set may depend on n* and the union over n of these sets may have finite measure.

At all times $\xi_t(\omega) \equiv x(t) \in R^d$ is defined for each ω since the process is defined on R^d; the question is whether there is a realization of the process for which $x(t)$ can be chosen to be continuous in t with probability one. At step n for each ω let $x_n(t)$ be equal to $x(t)$ for $t = \frac{j}{2^n}$. At the other times define $x(t)$ by linear interpolation

$$x_n(t) = 2^n\left(t - \frac{j}{2^n}\right)x\left(\frac{j+1}{2^n}\right) + 2^n\left(\frac{j+1}{2^n} - t\right)x\left(\frac{j}{2^n}\right) \tag{67}$$

for $t \in [\frac{j}{2^n}, \frac{j+1}{2^n})$. We can estimate the difference

$$\begin{aligned} sup_{0\leq t\leq 1}|x_{n+1}(t) - x_n(t)| &= sup_{1\leq j\leq 2^n}\, sup_{\frac{j-1}{2^n}\leq t\leq \frac{1}{2^n}}|x_{n+1}(t) - x_(t)| \\ &= sup_{1\leq j\leq 2^n}\left|x_{n+1}\left(\frac{2j-1}{2^{n+1}}\right) - x_n\left(\frac{2j-1}{2^{n+1}}\right)\right| \\ &\leq sup_{1\leq j\leq 2^n}\max[x\left(\frac{j-1}{2^n}\right) - x\left(\frac{2j-1}{2^{n+1}}\right)|, \\ &\quad \left|x\left(\frac{2j-1}{2^{n+1}}\right) - x_n\left(\frac{j}{2^n}\right)\right| \end{aligned} \tag{68}$$

Therefore for any positive γ

$$\begin{aligned} P[sup_{0\leq t\leq 1}|x_{n+1}(t) - x_n(t)| \geq 2^{-\gamma n}] &\leq 2^{n+1} sup_j P[|x\left(\frac{j}{2^{j+1}}\right) - x\left(\frac{j+1}{2^{j+1}}\right)| \\ &\leq C2^{n+1}2^{-(n+1)(1+\alpha)}2^{n\gamma(1+\beta)}] \end{aligned} \tag{69}$$

In the last inequality we made use of the assumption on $E(|\xi_s - \xi_t|^\beta)$. Choosing γ such that $1 + (1 + \beta)\gamma < 1 + \alpha$ one obtains

$$\sum_n P[sup_{0 \le t \le 1}|x_{n+1}(t) - x_n(t)| \ge 2^{-n\gamma}] < \infty \tag{70}$$

We now make use of the Borel–Cantelli lemma (see next lecture) to conclude that with probability one the limit

$$lim_{n \to \infty} x_n(t) \equiv x^*(t) \tag{71}$$

exists uniformly.

By the Ascoli–Arzelá compactness lemma $x^*(t)$ is a continuous function of t. By construction $x(t) = x^*(t)$ with probability one at dyadic points. Since both processes ξ and ξ^* are continuous in probability it follows that they have the same finite dimensional distributions and in fact $P[\xi(t) = \xi^*(t)] = 1$ for all $0 \le t \le 1$. ♡

Using Hölder norms instead of the sup norm one can prove that there is a realization of the process ξ^* supported on functions that satisfy a Hölder condition with exponent δ if $\delta < \frac{\alpha}{\beta}$. In this way one can prove that γ may be any positive number smaller then $\frac{1}{4}$.

Considering higher moments one can obtain realizations in spaces of functions that satisfy higher order Hölder conditions. For example in the case of Brownian motion one has

$$E[|\xi(t) - \xi(s)|^{2n}] = c_n([\xi(t) - \xi(s)^2])^n = c_n|t - s|^n \tag{72}$$

and by the procedure outlined above one can obtain any Hölder exponent smaller that $\frac{n-1}{2n}$. It follows that Brownian motion can be realized in spaces of functions that are Hölder continuous of exponent γ for any $\gamma < \frac{1}{2}$. It is worth remarking that $\gamma = \frac{1}{2}$ *cannot be reached.*

Suppose that there is a positive constant A such that for ant s, t

$$P[x(.) \ : \ |x(t) - x(s)| \le A|t - s|^{\frac{1}{2}}] = 0 \tag{73}$$

But one has

$$A \ge sup_{0 \le s \le t} \frac{|x(t) - x(s)|}{\sqrt{|t - s|}} \ge sup_j \sqrt{\left| x\left(\frac{j+1}{n}\right) - x\left(\frac{j}{n}\right) \right|} \tag{74}$$

The constant A must therefore be larger that the maximum of the absolute value of N independent gaussian variables. Since N is arbitrarily large and a gaussian variable is unbounded, A must be infinite.

9 Wiener Measure

In the following we will consider only the realization of Brownian motion on the space of continuous functions. We remark explicitly that for the construction of the process we could have use the positivity preserving contraction semigroup associated to any operator $\Delta - V$ with V Kato-small with respect to Δ.

In doing so we would construct a process in which ξ_t are not gaussian random variables, and it would be more difficult to find the joint distributions. The only potentials that lead to gaussian random variables are zero and the harmonic potential which we shall use presently.

It is of interest for us the find a measure on the continuous paths in the time interval $[0, T]$ with the only condition $x(0) = 0$ and no conditions on $x(T)$. We do this by distributing the location of the end point $x(T)$ according to a uniform distribution. Since Lebesgue measure is a limit form of gaussian measures what we obtain is still a gaussian measure this time on the continuous path in the interval $[0, T]$ starting at zero.

To compute expectation and variance of this new measure one has to do a further integration over the endpoint $x(T)$. One verifies by explicit computation that for the new gaussian measure

$$\forall t \quad E(\xi_t) = 0 \qquad E(\xi_t^2) = (2\pi t)^{-3/2} \int q^2 e^{-\frac{q^2}{2t}} dq = t \tag{75}$$

$$E(\xi_t \xi_s) = (2\pi t)^{-\frac{3}{2}} \int q' e^{-\frac{(q-q')^2}{2(t-s)}} q e^{-\frac{q^2}{2s}} dq' dq = s \quad s \leq t \tag{76}$$

$$E(\xi_t, \xi_s) = \min(t, s) \qquad E((\xi_t - \xi_s)^2) = t - s \tag{77}$$

$$E((\xi_t - \xi_\tau)(\xi_\sigma - \xi_s)) = 0, \qquad s < \sigma < \tau < t \tag{78}$$

$$E((\xi_t - \xi_\tau))^2 (\xi_\sigma - \xi_s)^2) = (t - \tau)(\sigma - s) \qquad s < \sigma < \tau < t \tag{79}$$

From the last equation it follows that the random variables $(\xi_t - \xi_\tau)$ and $(\xi_\sigma - \xi_s)$ are *independent gaussian random variables* if the segments (a, b) and (c, d) *are disjoint*. Therefore Wiener process has *independent increments* over disjoint intervals.

For comparison, notice that the Wiener process conditioned by fixing the starting and end points (*Brownian bridge*) *does not have* independent increments. Notice the following: Let $A_k(x)$ $k = 1, \ldots, N$ measurable functions. Denote by $E_{x_0,x;T}$ the expectation with respect to Wiener measure conditioned to $q(0) = x_0$, $q(T) = x$. Then if $t_{k+1} \geq t_k$

$$E_{x_0,x;T} \left(\Pi_1^N A_k(\xi_{t_k}) \right) = (e^{-t_1 H_0} A_1 e^{-(t_2 - t_1) H_0} A_2 \ldots . e^{t_n - t_{n-1}) H_0} A_N e^{-(T - t_n) H_0})(x, x_0) \tag{80}$$

where A_k is the operator that acts as multiplication by $A_k(x)$ and H_0 is the generator of the heat semigroup. It will be convenient in what follows to consider measures on paths defined in the interval $[-T, T]$ with $x(-T) = q$ and $x(T) = q'$. In analogy with what we have done so far one has, denoting $W_{q,q',[-T,T]}$ Wiener measure conditioned by $x(-T) = q$ and $x(T) = q'$

$$E_{W_{q,q',[-T,T]}}(\Pi_k A(\xi_{t_k}))$$
$$= (e^{-(t_1-T)H_0} A_1 e^{-(t_2-t_1)H_0} A_2 ... e^{-(t_{N_1}-t_N)H_0} A_N e^{-(T-t_N)H_0})(q, q') \quad (81)$$

If we choose $A_k = 1, \quad \forall k$ we obtain for every integer N

$$E_{W_{(q,q',[-T,T])}}(\iota) = \left(\frac{N}{2\pi}\right)^{\frac{3}{2}} \int ... \int e^{-\frac{1}{2}\sum\left(\frac{\Delta q_i}{\Delta t_i}\right)^2 \Delta t_i} \Pi_i dq_i, \quad \iota(q) = 1 \quad \forall q \quad (82)$$

Remark that if the function $x(t)$ were absolutely continuous, the last sum would converge to $\int_0^T e^{-\frac{1}{2}\dot{x}(t)^2} dt$. But we have seen that Wiener measure gives *weight zero* to the set of absolutely continuous trajectories.

We have considered up to now mainly processes in $[0, T]$ with value in R^1. The same considerations and formulae are valid for processes which take value in R^d for arbitrary *finite* value of d. Since we have made extensive use of compactness arguments, the case $d = \infty$ is not covered by the simple analysis presented here.

10 The Feynman–Kac Formula I: Bounded Continuous Potentials

According to the Trotter–Kato theorem, if $A = H_0$ and $B = V$

$$e^{-2T(H_0+V)} = s - lim \left(e^{-\frac{2TH_0}{n}} e^{-\frac{2TV}{n}}\right)^n \quad t \in R \quad (83)$$

The convergence is understood in the weak sense, as integral kernel of an operator, and the limit is the nucleus of the operator $e^{-2T(H_0+V)}$. Therefore, if we choose $A_t = V(\xi_t)$ where V (the potential) is suitably regular potential, we have proved

$$(e^{-2T(H_0+V)})(q, q') = \lim_{N\to\infty} \int dW_{q,q';[-T,T]} e^{-\sum_{k=1}^{2N} \frac{t}{N} V(q(-T+\frac{kT}{2N})} \quad q, q' \in R^d \quad (84)$$

(we have chosen to divide the interval [−T,T] in $2(N-1)$ disjoint intervals of equal length). The limit is understood in distributional sense.

Choose now a realization of Brownian motion *in which the measure is supported by continuous functions* $\omega(.)$ with value in R^d and such that

$$\xi(t)(\omega) = \omega(t) \quad (85)$$

If $V(x)$ is Riemann integrable the exponent in (84) converges to $\int_{-T}^{+T} V(\omega(t))dt$ *for each path* ω point-wise as a function of q, q', If V is bounded below, the integrand in (84) is bounded above by a constant C. Therefore the dominated convergence theorem of Lebesgue applies (Wiener measure is finite and completely additive). The right hand side of (84) converges therefore to

$$\int e^{\int_{-T}^{+T} V(\xi(s))ds} dW_{q,q';[-T,T]} \tag{86}$$

Since $V(x)$ is bounded below the sequence of integral kernels is uniformly bounded and therefore it converges in $L^1_{loc}(R^d \times R^d)$ and in distributional sense. Since the limit is unique we have proved that, in the case of potential which are bounded below and integrable

$$(e^{-2TH}\phi)(x) = \int dq'\phi(q') \left(\int_\Omega \int e^{-\int_{-T}^{+T} V(\xi(s))ds} dW_{q',x;[-T,T]} \right) \quad x \in R^d \tag{87}$$

This equation is known with the name *Feynman–Kac formula*. It has been obtained formally by R. Feynman in the case of the one -parameter group e^{-itH} and proved rigorously, by the use of Wiener measure, by V. Kac for the semigroup e^{-tH}, $H = -\frac{1}{2}\Delta + V$ under suitable assumptions on V, in particular if V is small with respect to the Laplacian.

11 The Feynman–Kac Formula II: More General Potentials

We shall now prove the Feynman–Kac formula under less restrictive assumptions on V.

Theorem 6 (Feynman–Kac formula, general case) *Let*

$$V = V_+ - V_-, \quad V_+ \geq 0 \;\; V_+ \in L^2_{loc}(R^d), \quad V_- \in S_d \tag{88}$$

where S_d stands for Stummel class.

Let $H = H_0 + V$, $H_0 = -\Delta$. For every $\phi \in L^2(R^d)$, for every $x \in R^d$ and every $t \in R^+$

$$(e^{-2tH}\phi)(x) = \int dy \int_\Omega e^{\int_{-t}^{+t} V(w(s))ds} \phi(y) dW_{x,y;[-t,t]} \tag{89}$$

(the second integral is over the paths are located in y at time $-t$). ◇

Proof We have already seen that the formula holds if $V \in L^\infty(R^d)$. Recall that $V \in S_d$ if

$$d = 3 \;\; : \;\; sup_{x \in R^3} \int_{|x-y|<1} |V(y)|^2 < \infty \tag{90}$$

$$d = 4 \;\; lim_{\alpha\to 0} sup_{x\in R^4} \int_{|x-y|\le\alpha} log|x-y|^{-1} V(y)^2 dy \le \infty \tag{91}$$

$$d \ge 5 \;\; lim_{\alpha\to 0} sup_{x\in R^4} \int_{|x-y|\le\alpha} |x-y|^{4-d} V(y)^2 dy \le \infty \tag{92}$$

Let $V_n \equiv \max(V, -n)$. Then $V_n \in L^\infty$ and $V_n(x) \to V(x)\ \forall x$. By monotone convergence

$$\int_0^t V_m(\omega(s))ds \to \int_0^t V(\omega(s))ds \tag{93}$$

and therefore, again by monotone convergence, for each $\phi \in L^1(R^d)$

$$\int \phi(y)dy \int_\Omega e^{\int_{-t}^{+t} V_n(w(s))ds} dW_{x,y;[-T,T]} \to \int dy \int_\Omega \int e^{\int_{-t}^{+t} V(w(s))ds} \phi(y) dW_{x,y;[-T,T]} \tag{94}$$

Assume now that V satisfies the assumptions of the Theorem. Defining $V_n(x) = \min(V(x), n)$ one has $lim_{n\to\infty} V_n(x) = V(x)$. Recall that C_0^∞ is a core for H and therefore

$$e^{-t(H_0+V_n)}\phi \to_s e^{-tH}\phi \quad \phi \in L^1 \tag{95}$$

The Feynman–Kac formula holds for $H_0 + V_n$; passing to the limit $m \to \infty$ one proves it for $\phi \in L^1 \cap L^2$ using the dominated convergence theorem. One makes use next of the regularity of $W_{x,y,(-t,t)}$ to extend the result to $\phi \in L^2$. ♡

Remark that, strictly speaking, we have not proved that under our assumptions $V(\omega(t))$ is measurable with respect to $dW_{x,y;[-T,T]}$. But it is certainly measurable if $V(x)$ is continuous since the integrand is limit of regular functions on Ω.

Since the measure $dW_{x,y;[-T,T]}$ is regular and the integral is equi-bounded with respect to N we can make use of Lebesgue criterion, substituting on a set of measure zero $V(\omega(t))$ with a measurable function $\tilde{V}$ without modifying the integral. After this rewriting, the integral $\int \tilde{V}(\omega(t))dt$ is rigorously defined and is measurable with respect to $dW_{x,y;[-T,T]}$. Notice that from the Feynman–Kac formula one sees that for every t the operator e^{-tH} *is positivity preserving*. This property plays an important role in the study of Markov processes.

We have associated to the Laplacian in R^d the Wiener process on the interval $[0, T]$. From the construction it is apparent that we can associate a stochastic process with continuous trajectories to any positivity preserving contraction markovian semigroup with a suitably regular generator, i.e. for which the procedure we followed for the Laplacian can be repeated.

We will return in Lecture 14 to the problem of the properties that operators and quadratic forms must have to define a stochastic process. In the Lecture 8 we will use the hamiltonian of the harmonic oscillator to construct a stochastic process (the Ornstein–Uhlenbeck process).

As remarked above, to construct a Feynman–Kac formula we can use any Schrödinger hamiltonian associated to a self-adjoint operator given by the Laplacian plus a potential of a suitable class, but only for the Laplacian and the harmonic oscillator one has simple expression for the kernel of the associated semigroup. One may also use as generator the Laplacian in $[0, K]$ with Neumann boundary conditions, denoted $\Delta^N_{[0,K]}$.

This would give a process with continuous trajectories with values in the interval $[0.K]$ but the kernel of $e^{-t\Delta^N_{[0,K]}}$ has a complicated expression which makes it inconvenient for explicit estimates.

References

1. S. Albeverio et al., Rigorous Feynman path integrals, *Stochastic Analysis and Mathematical Physics* (World Scientific, River Edge, 2004), pp. 1–60
2. C. Morette, Phys. Rev. **81**, 848–852 (1951)
3. D. Fujiwara, Duke Math. J. **47**, 559–600 (1980)
4. S. Albeverio, R. Hoegh-Krohn, Invent. Math. **40**, 59–106 (1977)
5. S. Mazzucchi, *Mathematical Feynman Integrals and their Applications* (World Scientific, Singapore, 2009)
6. T. Hida, H. Kuo, J. Pothoff, L. Streit, *White Noise, An Infinite Dimensional Calculus* (Kluwer Academic, London, 1993)
7. Yu. Berezansky, Yu. Kondriatev, *Spectral Methods in Infinite-Dimensional Analysis* (Kluwer, Dordrecht, 1995)
8. H.H. Kuo, *White Noise Distribution Theory* (CRC, New York, 1996)
9. J. Doob, *Probability Theory* (Wiley, New Jersey, 1953)
10. B. Gnedenko, *The Theory of Probability* (Mir, Moscow, 1976)
11. P. Billingsley, *Convergence of Probability Measures* (Wiley, New York, 1968)

Lecture 7: Elements of Probability Theory. Construction of Brownian Motion. Diffusions

We return briefly in this lecture to the realization of the Wiener process; we study here its realization from the point of view of semigroup theory, using transition functions. The same approach will be used in the next lecture to study the Ornstein–Uhlenbeck process.

We begin with some more elements of Probability Theory, giving in particular some useful a-priori estimates.

Definition 1 (*measure spaces*) A *measure space* is a triple $\{\Omega, \mathcal{F}, P\}$ where Ω is a set, $\mathcal{F}$ is a $\sigma-$ algebra of subsets C_i (the measurable subsets) and P is a probability measure on $\mathcal{F}$ i.e. a function on $\mathcal{F}$ with the following properties
(1) $\forall C \in \mathcal{F} \quad P(C) \geq 0$
(2) $P(\Omega) = 1$
(3) $C_i \in \mathcal{F} i = 1, 2, \ldots, \quad C_i \cap C_j = \emptyset, \Rightarrow P(\cup_{i=1}^{\infty} C_i) = \sum_{i=1}^{\infty} P(C_i)$
The positive number $P(C)$ is the *probability* of C. ◇

Definition 2 (*σ-algebras*) A collection $\mathcal{F}$ of subsets of Ω is a *σ-algebra* if $C_1, C_2, \ldots, C_k \ldots \in \mathcal{F}$ implies $\cup_i C_i \in \mathcal{F}$ and $\Omega - C_i \in \mathcal{F}$. It is easy to see that if $C_1, C_2, \ldots, C_k \ldots \in \mathcal{F}$ then $\cap_i C_i \in \mathcal{F}$.

Equivalent conditions on a σ-algebra are as follows
(a) If $C_i \in \mathcal{F}, \quad C_i \subset C_{i+1}$ then $P(\cup_i C_i) = lim_{i \to \infty} P(C_i)$
(b) If $C_i \in \mathcal{F}, \quad C_{i+1} \subset C_i$ then $P(\cap_i C_i) = lim_{i \to \infty} P(C_i)$
Notice that without σ-additivity one has only $P(\cup_i C_i) \leq \sum_i P(C_i)$. ◇

Definition 3 Let $\mathcal{A}$ be a family of subsets of Ω. The σ-algebra *generated* by $\mathcal{A}$ is the smallest $\sigma-$algebra of subsets of Ω which contains $\mathcal{A}$; it is denoted by $\mathcal{F}(\mathcal{A})$. ◇

Often in the applications Ω is a metric space. We will consider only this case. We denote by ω a generic point and we choose as $\sigma-$algebra the Borel algebra (the $\sigma-$algebra generated by the open sets in Ω).

G. Dell'Antonio, *Lectures on the Mathematics of Quantum Mechanics II: Selected Topics*, Atlantis Studies in Mathematical Physics: Theory and Applications 2, DOI 10.2991/978-94-6239-115-4_7

Definition 4 (*probability distribution*) Let ξ be a *random variable*, i.e. a real valued function $\xi(\omega)$ which is P-measurable. If there exists a positive measurable function $\rho(t)$ such that for every interval $[a, b]$

$$P(\{a \leq f(\omega) \leq b\}) = \int_a^b \rho(t)dt \tag{1}$$

we say that the random variable ξ has a *probability distribution* with *density* $\rho(t)$. More generally one can define a probability distribution in case the exists a positive Borel measure μ such that for each Borel set B and each continuous function f one has $P(f(\omega) \in B) = \mu(B)$. ◇

Definition 5 (*Expectation. Variance*) The *mathematical expectation* (*mean value*) $E_P(\xi)$ of the random variable ξ is

$$E_P(\xi) = \int \xi(\omega)dP(\omega) \tag{2}$$

where $\xi(\omega)$ is the evaluation map, a measurable function.

The *variance* Var is defined as

$$Var(\xi) \equiv E(\xi - E(\xi))^2 = E(\xi^2) - E^2(\xi) \tag{3}$$

◇

It is easy to see that if $a \leq \xi(\omega) \leq b$ then $Var(\xi) \leq (\frac{b-a}{2})^2$.

Definition 6 (*independence*) Two random variables ξ_1 and ξ_2 defined on the same probability space are said to be *independent* if

$$P[\xi_1(\omega) \in \mathcal{B}_1, \xi_2(\omega) \in \mathcal{B}_2] = P(\xi_1(\omega) \in \mathcal{B}_1).P(\xi_2(\omega) \in \mathcal{B}_2) \tag{4}$$

In the same way one defines the independence of a finite collection of random variables. In the case of an infinite collection, independence holds if it holds for any finite subset. ◇

1 Inequalities

The following inequalities hold [2]

Tchebychev inequality I

If $\xi \geq 0$ and $E(\xi) < \infty$ then for each $t > 0$

$$P\{\omega : \xi(\omega) \geq t)\} \leq \frac{E(\xi)}{t} \tag{5}$$

◇

Proof

$$P\{\omega\ :\ \xi(\omega) \geq t\} \leq \int_{\omega:\xi(\omega)\geq t} \frac{\xi(\omega)}{t} dP(\omega) \leq \frac{1}{t} E(\xi) \tag{6}$$

♡

Tchebychev inequality II
If $Var(\xi) < \infty$ then

$$[P(\omega\ :\ \xi(\omega) - E(\xi)] \geq t\} \leq \frac{Var(\xi)}{t^2} \tag{7}$$

◇

Proof From Tchebychev inequality I applied to the random variable $\eta \equiv (\xi - E(\xi))^2$ one has

$$\{\omega\ :\ |\xi - E(\xi)| \geq t\} = \{\omega\ :\ \eta(\omega) \geq t^2\} \tag{8}$$

Therefore $P(\omega||\xi(\omega) - E(\xi)| \geq t\} \leq \frac{E(\eta)}{t^2} = \frac{Var\xi}{t^2}$. ♡

An important result is described by the two limit theorems of De Moivre–Laplace that we will state without proof. Consider the binomial distribution with probabilities $p,\ q$ i.e. on N objects

$$P_k^N = \frac{N}{k} p^k (1-p)^{N-k} \tag{9}$$

We seek the asymptotic distribution in k for large values of N.

De Moivre–Laplace *local* limit theorem [1]
Let $Np + a\sqrt{N} \leq k \leq Np + b\sqrt{N}$. Then

$$P_k = \frac{1}{2\pi Np(1-p)} e^{-\frac{(k-Np)^2}{2N(p(1-p))^2}} (1 + R_N(k)) \tag{10}$$

where the remaining term $R_N(k)$ converges to zero $N \to \infty$ *uniformly in k in bounded intervals*

$$lim_{N\to\infty} \max_{Np+a\sqrt{N}\leq k\leq Np+b\sqrt{N}} |R_N(k)| = 0 \tag{11}$$

◇

De Moivre–Laplace *integral* limit theorem [1]
Let $a < b$ be real numbers. Then

$$lim_{N\to\infty} \sum_{Np+a\sqrt{Np(1-p)}\leq k\leq Np+b\sqrt{Np(1-p)}} = \frac{1}{2\pi} \int_a^b e^{-\frac{x^2}{2}} dx \tag{12}$$

◇

2 Independent Random Variables

We now give some inequalities which refer to a sequences of independent random variables.

Kolmogorov inequality

Let $\xi_1, \xi_2, \ldots, \xi_n$ be a sequence of independent random variables. Suppose that $E(\xi_i) = 0, \quad Var(\xi_i) < \infty \quad i = 1, \ldots, n$. Then

$$P(\{\omega \,:\, \max_{1,\ldots,n} |\xi_1 + \cdots + \xi_k| \ge c\}) \le \frac{1}{c^2} \sum_{k=1}^{n} Var\ (\xi_k) \tag{13}$$

◇

Proof Denote by A_k the set of points ω for which

$$\max\{|\xi_1|,\ |\xi_1 + \xi_2|,\ \ldots |\xi_1 + \cdots \xi_{k-1}|\} < c, \quad |\xi_1 + \cdots \xi_k| \ge c \quad 1 \le k \le n \tag{14}$$

Denote by η_k its indicator function, which is by construction measurable with respect to $\xi_j, \quad 1 \le j \le n$. The sets A_k are pairwise disjoint and

$$P(\max\{|\xi_1|,\ |\xi_1+\xi_2|, \ldots |\xi_1+\cdots \xi_{k-1}|\} < c) = P(A_1 \cup A_2 \cup \cdots \cup A_n) = \sum_{k=1}^{n} P(A_k) \tag{15}$$

From $E(\xi_k) = 0 \;\; \forall k$ and $\sum_{k=1}^{n} \Xi_k \le 1$ it follows

$$Var\ (\xi_1 + \cdots + \xi_n) = E((\xi_1 + \cdots \xi_n)^2) \ge \sum_{k=1}^{n} E(\eta_{A_k} (\xi_1 + \cdots \xi_n)^2) \tag{16}$$

Consider now the identity

$$\begin{aligned} E(\eta_{A_k}(\xi_1 + \cdots + \xi_n)^2) = {} & E(\eta_{A_k}(\xi_1 + \cdots + \xi_k)^2) \\ & + 2E(\eta_{A_k}(\xi_1 + \cdots + \xi_k)\eta_{A_k}(\xi_{k+1} + \cdots + \xi_n)) \\ & + E(\eta_{A_k}(\xi_{k+1} + \cdots + \xi_n)^2) \end{aligned} \tag{17}$$

By definition the measure of A_k is not smaller than $c^2 P(A_k)$. The second term is zero since it is the expectation of the product of two independent mean zero random variables. The third term is positive. Therefore

$$Var\ (\xi_1 + \cdots + \xi_n) \ge c^2 \sum_{k=1}^{n} P(A_k) \tag{18}$$

Since ξ_k are independent variables the left hand side is $\sum_{k=1}^{n} Var\ (\xi_k)$. ♡

Kolmogorov zero-one law
Let $(\Omega, \mathcal{F}, \mu)$ be a probability space and let $\xi_1, \xi_2, \ldots$ a collection of independent random variables equally distributed (*permutable*).

Suppose that a set A is measurable with respect to ξ_n for all values of the index n.

Then either $\mu(A) = 0$ or $\mu(A) = 1$ where $\mu(A)$is the measure of A (the integral of its indicator function). $\diamond$

Proof By definition of product measure there exists an integer N sufficiently large and a set A_ϵ measurable with respect to the collection $\xi_1, \ldots, \xi_N$ (a cylinder set) such that $|\mu(A) - \mu(A_\epsilon)| < \epsilon$. By the substitution $\xi_i \to \xi_{i+N}$ we construct another measurable set A' with the properties that $\mu(A) = \mu(A')$ and that A' and A_ϵ are mutually independent.

Therefore, denoting by $\Xi(A)$ the indicator function of the set A and with $P(A)$ its expectation

$$P(A' \cap A_\epsilon) = P(A')P(A_\epsilon) = P(A)P(A_\epsilon) \tag{19}$$

But $lim_{\epsilon \to 0} P(A' \cap A_\epsilon) = P(A')$ and therefore $P(A)^2 = P(A)$ i.e. either $P(A) = 0$ or $P(A) = 1$. $\heartsuit$

3 Criteria of Convergence

We turn now to convergence criteria for sequences of random variables.

Definition 7 Let ξ_i be a sequence of (real valued) random variables. We say that the sequence converges to the random variable ξ
(i) *in probability* (in measure) if

$$\forall \epsilon > 0 \quad lim_{n \to \infty} P(|\xi_n - \xi > \epsilon) = 0 \tag{20}$$

(ii) *almost surely* (a.s.) if *for almost all* ω (i.e. except for a set of zero measure)

$$lim_{n \to \infty} \xi_(\omega) = \xi(\omega) \tag{21}$$

$\diamond$

Notice that a.s. convergence implies convergence in probability but the converse is not true. Let $\{\xi_n\}$ be a sequence of random variables with finite mean. Denote by $\zeta_n = \frac{1}{n}(\xi_1 + \cdots + \xi_n)$ its arithmetic mean.

We will use Kolmogorov zero-one law to prove the very useful *Borel–Cantelli lemma* which states, roughly speaking, that if $\xi_1, \xi_2, \ldots$ s a sequence of independent equally distributed random variables in a probability space Ω, then the sets of ω's such that the series $\sum_n \xi_n(\omega)$ converges have measure either zero or one. Similarly, under the same assumptions, the measure of a set of ω such that

$$lim_{n\to\infty}\frac{1}{n}\sum_{k=1}^{n}\xi_k(\omega)=0 \tag{22}$$

is either zero or one.

Borel–Cantelli lemma I [2, 3]
Let A_n be a sequence of measurable sets (a sequence of *events*) in a probability space $\{\Omega, \mathcal{F}, P\}$ and assume $\sum_n P(A_n) < \infty$.

Let $\eta(A)$ be the indicator function of the set A of those ω's for which there is an infinite sequence $\{n_i(\omega)\}$ such that $\omega \in A_i \; i = 1, 2, \ldots$. Then $P(A) = 0$ (i.e. A occurs with zero probability). ♢

Proof We can write A as $A = \cup_{k=1}^{\infty} \cup_{n=k}^{\infty} A_n$. Then

$$P(A) \leq P(\cup_{n=k}^{\infty} A_n) \leq \sum_{n=k}^{\infty} P(A_n) \to 0, \quad k \to \infty \tag{23}$$

Since $\sum_1^{\infty} P(A_k) < \infty$ one has $lim_{n\to\infty} \sum_n^{\infty} P(A_k) = 0$. Therefore $P(A) = 0$. ♡

Borel–Cantelli lemma II [2, 3]
Let $\{A_n\}$ be a sequence of mutually independent events in a probability space $\{\Omega, \mathcal{F}, P\}$ and suppose $\sum_n P(A_n) = \infty$.

Let A be the collection of points ω for which there exists an infinite sequence $\{n_i(\omega)\}$ such that $\omega \in A_k \; k \in n_i(\omega)$. Then the measure of A is one. ♢

Proof Write A as $A^c = \cup_{k=1}^{\infty} \cap_{n=k}^{\infty} A_n^c$ and therefore for each value of n one has

$$P(A^c) \leq \sum_{k=1}^{\infty} P(\cap_{k=n}^{\infty} A_n^c) \tag{24}$$

We have denoted by B^c the complement of B in Ω. Since the A_n are mutually independent also the A_n^c are mutually independent

$$P(\cap_{n=k} A_n^c) = \Pi_{n=k}^{\infty}(1 - P(A_n)) = 0 \tag{25}$$

From $\sum_{n=1}^{\infty} P(A_n) = \infty$ it follows $lim_{n\to\infty} \Pi_{n=k}^{\infty}(1 - P(A_n)) = 0$ and therefore $P(A^c) = 0$. ♡

4 Laws of Large Numbers; Kolmogorov Theorems

We shall now briefly mention one of the theorems in probability theory which is more frequently used in applications, the *laws of large numbers* (Kolmogorov theorems)

Definition 8 We say that the sequence of random variables ξ_n
(1) satisfies the *weak* law of large numbers if $\zeta_n - E(\zeta_n)$ converges to zero *in probability* as $n \to \infty$ i.e. for every $\epsilon > 0$ one has $lim_{n\to\infty} P(|\zeta_n - E(\zeta_n)| = 0$.
(2) satisfies the *strong* law of large numbers if $\zeta_n - E(\zeta_n)$ converges to zero *almost surely* (i.e. for almost all ω one has $lim_{n\to\infty}(\zeta_n - E(\zeta_n)) = 0$). Remark that in the weak form of the law the sets considered *may depend on n*. ◇

Kolmogorov theorem I
A sequence of mutually independent random variables $\{\xi_n\}$ with $\sum_{n=1}^{\infty} \frac{1}{n^2} Var(\xi_i) < \infty$ satisfies the strong law of large numbers. ◇

Kolmogorov theorem II
A sequence $\{\xi_n\}$ of mutually independent and identically distributed random variables such that $E(\xi_n)^2 < \infty$ satisfies the strong law of large numbers. ◇

Remark that both laws of large numbers imply that for a sequence of random variables which satisfy the assumptions of Kolmogorov, for N large enough the random variable *arithmetic mean* $m_N = \frac{1}{N}\sum_{n=1}^{N} \xi_n$ differs little from its expectation.

Therefore *asymptotically* the mean does not depend on ω, i.e. it *tends to be not random*. This property can be expressed in the following way: in a long chain of random equally distributed variables there appear almost surely regular sequences (which are not random). The statement that a gas occupies almost surely the entire available space can be considered as an empirical version of the strong law of large numbers.

We shall give a proof of Kolmogorov theorem I. For the proof of Kolmogorov theorem II one must show that if one assumes that $\sum_{i=1}^{\infty} \frac{1}{i^2} Var\xi_i < \infty$, then the ξ_k are equally distributed with finite mean of the squares.

For this one uses the properties of product measures and Kolmogorov inequality that we recall here

$$P(\{\max_{1\le k\le n}|(\xi_1 + \cdots \xi_k) - (E(\xi_1) + \cdots E(\xi_k))| \ge t\} \le \frac{1}{t^2}\sum_{i=1}^{n} Var(\xi_i) \quad (26)$$

Proof of Kolmogorov Theorem I
Replacing the random variables ξ_k with $\xi_k - E(\xi_k)$ we can assume $E(\xi_k) = 0\ \forall k$. We must show that $\zeta_N \equiv \frac{1}{N}\sum_1^N \xi_i$ converges to zero a.s. when $N \to \infty$.

Choose $\epsilon > 0$ and consider the event (measurable subset) $\mathcal{B}(\epsilon)$ of the points $\omega \in \Omega$ such that there exists $N = N(\omega)$ such that for all $n \ge N(\omega)$ one has $|\zeta_n(\omega)| < \epsilon$. By definition

$$\mathcal{B}(\epsilon) = \cup_{N=1}^{\infty} \cap_{n>N(\omega)} \{\omega|\ |\zeta_n(\omega)| < \epsilon\} \quad (27)$$

Define $\mathcal{B}_m(\epsilon) \equiv \{\omega\ :\ \max_{2^{m-1}\le n\le 2^m}|\zeta_n| \ge \epsilon\}$. From Kolmogorov inequality

$$P(\mathcal{B}_m(\epsilon)) = P\left(\max_{2^{m-1}\le n\le 2^m}\left|\sum_{i=1}^{n}\xi_i\right| \ge \epsilon\, n\right) \le \max_{2^{m-1}\le n\le 2^m}(P\left(\left|\sum_{i=1}^{n}\xi_i\right| \ge \epsilon 2^{m-1}\right)$$
$$\le \frac{1}{\epsilon^2}\, 2^{2m-2}\sum_{i=1}^{n}\max_{1\le n\le 2^m} P\left(\left|\sum_{i=1}^{n}\xi_i\right| \ge \epsilon 2^{m-1}\right)\sum Var(\xi_i) \tag{28}$$

Therefore

$$\sum_{m=1}^{\infty} P(\mathcal{B}_m(\epsilon)) \le \frac{1}{\epsilon^2}\sum_{i=1}^{\infty} Var(\xi_i) \sum_{n\ge m_i,\; 2^{m_i-1}\le i\le 2_i^m} \frac{1}{2^{2n-2}} \le \frac{16}{\epsilon^2}\sum_{i=1}^{\infty}\frac{Var(\xi_i)}{i^2} \tag{29}$$

and this sum is finite by assumption. It follows from the Borel–Cantelli lemma that for a.a. ω there exists an integer $M(\omega)$ such that for $m \ge M$

$$\max_{2^{m-1}\le n\le 2^m}|\zeta_n| < \epsilon \tag{30}$$

Therefore $P(\mathcal{B}(\epsilon)) = 1$ for each $\epsilon > 0$. In particular $P(\cap_k \mathcal{B}(\frac{1}{k})) = 1$. If $\omega \in \cap_k \mathcal{B}(\frac{1}{k})$ there exists $N(\omega, k)$ such that for every $n \ge N(\omega, k)$ one has $|\zeta_n| < \frac{1}{k}$. It follows that for almost all ω, $lim_{n\to\infty}\zeta_n = 0$. ♡

5 Central Limit Theorem

Using the law of large numbers one can derive the important *Central Limit Theorem*. In its most commonly used version this theorem is about the sum of independent identically distributed random variables. This theorem plays an important role in Statistical Mechanics and provides a link between Statistical Mechanics and Thermodynamics.

According to the strong law of large numbers the difference between the arithmetic mean of N independent identically distributed random variables and the arithmetic mean of their expectation values $E(\xi_k)$ converges to zero as $N \to \infty$. It is natural to enquire about the rate of convergence.

From Tchebychev inequality one derives that the order of magnitude of the error is $\sqrt{N}$. Therefore it is of interest to study the convergence of the sequence

$$\zeta_N \equiv \frac{1}{\sqrt{N}}\sum_{k=1}^{N}\xi_n \qquad E(\xi_k) = 0. \tag{31}$$

The *Central Limit Theorem* states that the random variables ζ_N do not in general converge strongly but, under suitable assumptions, *their distributions have a limit that does not depend on the details of the distribution of the* ξ_i.

Let is recall the definition of *characteristic function* of a random variable.

Definition 9 (*characteristic function*) The *characteristic function* ϕ_ξ of the random variable ξ is by definition

$$\phi_\xi(\lambda) \equiv E(e^{i\lambda\xi}) \quad \lambda \in R \tag{32}$$

◇

It is easy to see that the characteristic function determines *the distribution* of the random variable ξ and that convergence of a sequence of characteristic functions is equivalent to convergence *in distribution* (not in probability) of the corresponding sequence of random variables.

The use of the characteristic function simplifies the study of the sum of independent random variables. Let $\zeta_N = \sum_{k=1}^N \xi_k$. It is easy to see that $\phi_{\zeta_N}(\lambda) = \Pi_{k=1}^N \phi_{\xi_k}(\lambda)$.

We can now state the Central Limit Theorem.

Central Limit Theorem

Let $\{\xi_1, \ldots, \xi_n \ldots\}$ *be a sequence of independent identically distributed random variables and let their common distribution* $f(x)$ *have finite second moment. Denote by m the (common) expectation and with v the common variance* $v = m_2 - m^2$*. Then for* $N \to \infty$ the distribution of their average

$$\eta_N \equiv \frac{1}{\sqrt{Nv}} \sum_{n=1}^N (\xi_n - m) \tag{33}$$

converges weakly to a gaussian normal distribution with density $\frac{1}{\sqrt{2\pi}} e^{-\frac{x^2}{2}}$). ◇

Proof The characteristic function of the gaussian distribution is

$$\phi(\lambda) = E\left(\frac{1}{\sqrt{2\pi}} e^{i\lambda x - \frac{x^2}{2}}\right) = e^{-\frac{\lambda^2}{2}} \tag{34}$$

while the characteristic function of the random variables η_N is

$$\phi_{\eta_N} = \phi\left(\frac{\lambda}{\sqrt{Nv}}\right) e^{-iN\frac{\lambda m}{\sqrt{Nv}}} \tag{35}$$

where ϕ is the common characteristic function of the ξ_k. It is sufficient therefore to prove that for each value of λ

$$lim_{N\to\infty} \phi_{\eta_N}(\lambda) = e^{-\frac{\lambda^2}{2}} \tag{36}$$

Because $m_2 < \infty$ the function $\phi(\lambda)$ is twice differentiable with continuous second derivative. Therefore for λ small

$$\phi(\lambda) = 1 + im\lambda - \frac{\lambda^2}{2} m_2 + o(\lambda^2) \tag{37}$$

It follows for each value of λ

$$\phi_{\eta_N} = (1 - \frac{\lambda^2}{2N} + o\left(\frac{\lambda^2}{N}\right)^N \to_{N\to\infty} e^{-\frac{\lambda^2}{2}} \tag{38}$$

(remark that the linear terms vanish by symmetry). ♡

There are generalizations of the Central Limit Theorem, e.g. to the case in which the random variables are not identically distributed or are only approximately independent or if one considers other averages. instead of the mean. In particular it can be shown that if the random variables are identically distributed with distribution function $p(x)$ such that $p(x) = p(-x)$ and $p(x) \simeq \frac{c}{|x|^{\alpha+1}}$ for $\alpha \in (0, 2)$, then *the distribution* of the random variable $\eta_N(\alpha) = N^{-\frac{1}{\alpha}}(\xi_1 + \cdots \xi_N)$ converges when $N \to \infty$ to a limit distribution with characteristic function $Ce^{-b|\lambda|^\alpha}, \quad b > 0$.

6 Construction of Probability Spaces

We end this description of results about collections of identically distributed random variables presenting theorems about the construction of probability spaces in which one can realize collections of random variables (given through their characteristic functions) preserving their joint distributions. These constructions are analogous to the construction of product measures. It should be stressed that the construction *is not unique*.

We begin with a theorem of Kolmogorov on the existence of a measure space in which can be realized a collection (not necessarily denumerable) of random variables preserving joint distributions.

Theorem 1 (Kolmogorov) *Let I a set. Let $\mathcal{F}$ be the collection of the finite subsets of I and assume that for each $F \in \mathcal{F}$ there exists a completely additive measure μ_F of total mass one on the Borel sets $\mathcal{B}(R^{N(F)})$ (we have denoted by $N(F)$ the number of elements in F). Assume that this collection of measures satisfies the compatibility requirements for the inclusion of the subsets.*

Then there exists a (not unique) probability space (X, M, μ) and functions $\{f_\alpha, \quad \alpha \in I\}$ such that μ_I be the joint probability of $\{f_\alpha \quad \alpha \in I\}$. Moreover if $\mathcal{F}$ is the smallest σ-algebra that contains all measurables f_α, the measure μ is unique modulo homeomorphisms. ◇

Proof Let $\dot{R} \equiv R \cup \infty$ be the one-point compactification of R and set $X \equiv (\dot{R})^I$. Let C_{fin} the set of function which depend only on a finite number ξ^I of α. If $f \in C_{fin}$ define $l(f) = \int f(x^I) d\mu_I(x^I)$ By construction X is compact in the product topology.

By the Stone–Weierstrass theorem C_{fin} in dense in $C(X)$. Indeed the polynomials in C_{fin} coincide with those in $C(X)$. Therefore the functional l extends to $C(X)$. By the Riesz–Markov representation theorem, there exists a Baire measure μ on X such that $l(f) = \int f(x)d\mu(x)$.

Let f_α be equal to ξ_α if $|x_\alpha| < \infty$, 0 otherwise. Then, if the set J is finite, $d\mu_I$ is the joint probability of $f_\alpha, \quad \alpha \in J$. This proves existence.

To prove uniqueness it is sufficient to prove that C_{fin} is dense in $L^2(X, d\mu)$. Let $\mathcal{H}$ be the closure of C_{fin} in $L^2(X, d\mu)$. For any Borel set $A \subset X$ the indicator function $\eta(A)$ can be approximated in $L^2(X, d\mu)$ by linear combinations of $\eta(A_n), \quad A_n \subset B_{fin}$ (the cylindrical Borel sets with finite dimensional basis). Therefore the collection of A_n is closed for finite intersections.

Since the complement of a cylinder set is itself cylindrical it follows that the collection of A_n is also closed under complementation and denumerable union. Therefore

$$\{A \ : \ \eta(A) \in \mathcal{H}\} \tag{39}$$

is a σ-algebra. But by assumption $\mathcal{F}$ is the smallest σ-algebra that contains all Borel sets. Hence

$$\{A \ : \ \eta(A) \in \mathcal{H}\} = \mathcal{H} \tag{40}$$

and therefore $\mathcal{H} = L^2(X, d\mu)$. ♡

Remark that one can use R^I as a model because $\mu\{x \ : \ \exists\alpha, \ \ |x_\alpha| < \infty\} = 1 \ \ \forall\alpha$ and, for every finite J

$$\mu\{x \ : \ |x_\alpha| = \infty \qquad \forall\alpha \in J\} = 0 \tag{41}$$

From the σ-additivity of the measure one derives then $\mu(\dot{R}^I - R^I) = 0$.

7 Construction of Brownian Motion (Wiener Measure)

We give now two constructions of Brownian motion. One is the original construction of Wiener as measure on continuous functions [2, 4]. The other is the construction, due to Einstein, of Wiener measure as limit of measures on *random walks* on a lattice. We also give a modification of Brownian motion which is obtained through a modification of its paths.

Wiener construction

Using Kolmogorov estimates to bound the measure of the part of the measure space in which a given random variable exceeds a prefixed value, and elementary probabilistic estimates, in particular on product measure, it is possible to prove that if $c_0, c_1, c_2, \ldots$ are independent gaussian variables the series

$$X_N(t) \equiv c_0 t + \sum_{n=1}^{N} \sum_{k=2^{n-1}}^{2n} c_n \frac{\sqrt{2} sen \pi k t}{\pi k} \tag{42}$$

converges *in distribution* when $N \to \infty$ uniformly over compact sets with probability one. This means that, a part a set of measure zero, one has -uniform convergence in $L^1(R)$ of the distribution of the sequence

$$X_N(t,\omega) \equiv c_0 t + \sum_{n=1}^{N} \sum_{k=2^{n-1}}^{2n} c_n(\omega) \frac{\sqrt{2}\, sen \pi k t}{\pi k} \tag{43}$$

The limit function is continuous and is zero for $t = 0$ because each term is zero. We have thus defined for each value of $T \in R^+$ a correspondence Φ_T between a set of full measure Y of points ω in the probability space (Ω, M, μ) and continuous functions vanishing at the origin.

We define now a probability measure μ' on continuous functions X_T on $[0, T]$) vanishing at the origin by setting

$$\mu'(\Phi_T^{-1}(Y)) = \mu(Y), \qquad \mu'(X_T - \Phi_T^{-1}(Y)) = 0 \tag{44}$$

This is *Wiener measure*. Wiener has proved that Y is dense in X_T in the C^0 topology.

For each value of t the $X_n(t)$ are independent gaussian variables, (being sum of independent gaussian random variables) and therefore also their limit in distribution is a gaussian

$$\xi_t(\omega) = lim_{N\to\infty} X_N(t) \tag{45}$$

Due to the correspondence between a set of measure one in Ω and a dense subset of continuous functions, the random variable ξ_t can be seen as an element of the dual of continuous functions. It follows from the definitions that ξ_{t_0} assigns to the function $x(t)$ the number $x(t_0)$.

From the definition one verifies $E(\xi_t) = 0$. Using the trigonometric relations and the independence of $c_k(\omega)$ and performing the limit that defines ξ_t one has

$$E(\xi_t \xi_s) = \int \xi_t(x(.))\xi_s(x(.))d\mu' = \min(t, s) \tag{46}$$

Since the ξ_t are random gaussian variables this determines completely their distribution and we see that the random variables ξ_s and ξ_t *are not mutually independent*. Remark that we have assumed that the random variables we consider take value in R.

An identical construction can be made under the assumption that the random variables c_k take value in R^d and that the components are gaussian independent random variables with mean zero and variance one. One obtains in this way the Wiener process in R^d.

We have used the fact that the class of continuous functions is closed under uniform convergence (the convergence we have proved is in the uniform topology outside a set of measure zero). This follows because closed sets in R^d are compact.

This *is not true* for an infinite dimensional Banach space X. Still we shall see, in Lecture 15, through the theory of Dirichlet forms, processes that play the role of Wiener processes in infinite dimensional Banach spaces.

Remark that using Kolmogorov inequality one proves that the set of ω for which the limit is an absolutely continuous function has measure zero. The representation we have given of Wiener process is particularly convenient to determine the regularity of the trajectories making use of theorems about Fourier transforms. Further analyses of this problem are e.g. in [4].

8 Brownian Motion as Limit of Random Walks

We now construct the Wiener process as limit of random walks on a lattice. Our exposition follows closely the construction given by Einstein. We will consider only the case of one space dimension and we will study the motion of a heavy particle which moves due to elastic collisions with very many light particles which move independently from each other.

This is the model introduced by Einstein to give a mathematical treatment the phenomenon described by R. Brown in 1927 [5] of the erratic movement of pollen particles suspended in water. Einstein [6] described the motion of pollen as due to the (random) collisions with the molecules of water. Einstein's theory was verified experimentally by J. Perrin [7] who used it to give a (precise) estimate of Avogadro's number. Perrin's experiments constituted at that time the best evidence for the existence of atoms and molecules.

Consider the motion in one space dimension. The light particles come at random form the right or the left; in each unit of time the heavy particle is hit by a light particles and moves to left or to the right of one unit of space. Since the direction of the light particle is random, if at time 0 the heavy particle is at the origin at (*microscopic*) time n it will be in position given by $\sum_{i=1}^{n} \xi_i$ where ξ_i are independent random variables with common distribution $P(\xi_i = \pm 1) = \frac{1}{2}$.

On a *macroscopic scale* of space and time there are ϵ^{-2} collisions in each unit of time, and the absolute displacement in each collision is ϵ. Therefore after the macroscopic time t the heavy particle will be in (macroscopic) position $X_\epsilon(t) = \epsilon \sum_{i=1}^{\epsilon^{-2}t} \xi_i$. We will construct Brownian motion as limit *in distribution* of the random variable $X_\epsilon(t)$. More generally, consider a probability space $\{R, \mathcal{B}(R), \mu\}$ such that

$$\int_R x^4 d\mu(x) < \infty \qquad \int x d\mu(x) = 0 \qquad \int x^2 d\mu(x) = 1 \tag{47}$$

and let ξ a random variable.

Consider now a product space and for $\epsilon > 0$ define by linear interpolation for each realization of ξ a continuous path $t \to \psi_\epsilon(\xi; t)$ through

$$\psi_\epsilon(\xi; t) = \epsilon \sum_{i=1}^{[\epsilon^{-2}t]} \xi_i + \epsilon(\epsilon^{-2}t - [\epsilon^{-2}t])\xi_{[\epsilon^{-2}t]+1} \tag{48}$$

where $[y]$ is the integer part of y. Define $P_\epsilon(A) = P(\psi_\epsilon^{-1}(A)$ for any cylindrical set of paths.

Theorem 2 *When $\epsilon \to 0$ the sequence P_ϵ converges weakly to Wiener measure.* ◇

Proof We give the proof in three steps
(i) At each time $0 \le t \le T$ the distribution converges to the distribution of Brownian motion.
(ii) The finite dimensional distributions converge to those of Brownian motion
(iii) The family P_ϵ is tight
Step (i)
This is a consequence of the central limit theorem. Introduce the characteristic function $\phi_\epsilon(\lambda)$, which is the Fourier transform of the distribution of $\xi_\epsilon(t)$ under P_ϵ. It is easy to prove that convergence in distribution is equivalent to the convergence of the characteristic function. and that the characteristic function of the sum of independent random variable is the product of the characteristic functions

$$\phi_\epsilon(\lambda) = \phi_\epsilon(\lambda\epsilon)^{\frac{t}{\epsilon^2}} = \left[1 - \frac{1}{2}\lambda^2\epsilon^2 + +o(\epsilon^2)\right]^{\frac{t}{\epsilon^2}} \to e^{-\frac{1}{2}\lambda^2 t} \tag{49}$$

Step (ii)
From step (i) one sees that adding the terms

$$\epsilon(\epsilon^{-2}t - [\epsilon^{-2}t])\xi_{[\epsilon^{-2}t]} \tag{50}$$

one obtains a continuous path. This term goes to zero uniformly as $\epsilon \to 0$. The proof of step (ii) follows then because P is a product measure.
Step (iii)
To prove relative compactness we make use of *Prohorov criterion* (see e.g. [3]; this book is a basic reference for weak convergence and compactness criteria).

9 Relative Compactness

Let $\mathcal{S}$ be a metric space and $\mathcal{B}(\mathcal{S})$ its Borel sets. Denote by $\mathcal{C}(\mathcal{S})$ the continuous function on $\mathcal{S}$.

Recall that a family of probability measures Π_α on $(\mathcal{S}, d)$ is *relatively compact* iff for any bounded sequence P_n it is possible to extract a weakly convergent subsequence (i.e. there exists a probability measure P such that $lim_{n\to\infty} \int f dP_n = \int f dP$ for every bounded $f \in C(S)$).

We shall denote weak convergence by $P_n \to_w P$. In case $S = R$ we can characterize weak convergence by means of the characteristic function $\phi(\lambda) \equiv \int e^{i\lambda x}\mu(dx)$ Weak convergence is equivalent to point-wise convergence of the characteristic function.

A collection Π_α of probability measures is *tight* iff for each $\epsilon > 0$ there exists a compact set $\mathcal{K}$ such that $P(\mathcal{K}) > 1 - \epsilon$ for each $P \in \Pi$. We now use Prohorov criterion

Prohorov criterion [3]

If the collection Π_α is tight, then it is relatively compact. If $\mathcal{S}$ is complete and separable, the condition is also necessary. ◇

This criterion is particularly useful if S is the set C of continuous functions on R^N, $N < \infty$. In this case the compact sets are characterized by the Ascoli–Arzelá theorem. Let the *continuity modulus* if $x(t) \in C$ be $\omega_x(\delta) \equiv sup_{|t-s|<\delta}|x(t) - x(s)$. The Ascoli–Arzelá theorem states that a set $A \in C$ has compact closure iff

$$sup_{x\in A}|x(0)| < \infty, \qquad lim_{\delta\to 0} sup_{x\in A}\omega_x(\delta) = 0 \tag{51}$$

It follows from the definition that if A has compact closure, then its elements are equi-bounded and equi-continuous. It is then easy to see that in this case the sequence P_n is tight iff

(i) for each $\eta > 0$ there exists $a > 0$ such that $P_n(x : |x(0)| > a) \leq \delta \quad \forall n \geq 1$.

(ii) For every i $\eta > 0$, $\epsilon > 0$ there exist $\delta \in (0, 1)$ and $n_0 \in N$ such that

$$P_n(x : \omega_x(\delta) \geq \epsilon) \leq \eta \qquad \forall n \geq n_0 \tag{52}$$

Returning now to the construction of Wiener measure, notice that if s and t, $0 \leq s < t \leq T$ are such that $\epsilon^{\cdot 2}t$ and $\epsilon^{-2}s$ are integers one has

$$\begin{aligned}\int dP_\epsilon|x(t) - x(s)|^4 &= E\left(\epsilon \sum_{i=\epsilon^{-2}s+1}^{\epsilon^{-2}t} \xi_i\right)^4 \\ &= \epsilon^4 \sum_{i=\epsilon^{-2}s+1}^{\epsilon^{-2}t} E(\xi_i^4) + 6\epsilon^4 \sum_{\epsilon^{-2}s+1\leq i\leq j\leq\epsilon^{-2}t} E(\xi_i^2\xi_j^2) \\ &\leq C(\epsilon^2(t-s) + (t-s)^2) \leq 2C(t-s)^2\end{aligned} \tag{53}$$

By interpolation this inequality is valid for $0 \leq s \leq t \leq T$. Remark now that if $\exists\, \alpha,\ \beta\ C < \infty$ such that

$$E(|x(t) - x(r)|^\beta) \leq C|t-s|^{1+\alpha} \tag{54}$$

then $\exists\, c_1,\ c_2 < \infty$ such that

$$P\left[sup_{0\leq s\leq t\leq T}\frac{|x(t)-x(s)|^{\beta}}{|t-s|^{1+\alpha}}\geq c_1\lambda\right]\leq c_2\frac{1}{\lambda} \tag{55}$$

This inequality, a version of Tchebychev inequality called also Garcia's inequality, can be found in [2]. With the choice $\alpha = 1$ and $\beta = 4$ it follows

$$P_\epsilon(x,\ \frac{\omega_x}{\delta})\geq\eta)\leq P_\epsilon\left(\delta^{\beta}sup_{|t-s|\leq\delta}\frac{|x(t)-x(s)|}{|t-s|^{\beta}}\geq\eta\right)\leq c_2\left(c_1\frac{\delta^{\beta}}{\eta}\right)^4 \tag{56}$$

and then $lim_{\delta\to 0}sup_{\epsilon>0}P_\epsilon(x:\ \omega_x(\delta)\geq\eta)=0$. This implies relative compactness.

10 Modification of Wiener Paths. Martingales

We consider now the process obtained modifying the Laplacian by a *drift* $b(x)$ (which need not be a gradient). We want to interpret the modification of Brownian motion as *modification of the Brownian trajectories.* This will give us a version of the modified process which *has continuous trajectories defined in any finite interval of time.* The generator of the semigroup is now

$$L=\frac{1}{2}\frac{d^2}{dx^2}+b(x)\frac{d}{dx} \tag{57}$$

We assume that the vector field $b(x)$ is Lipshitz continuous. Consider the modification of the Brownian trajectories under the following rule, for each $t \geq 0$

$$\beta\to\xi(t)\equiv\Phi_b(t),\quad \xi(t)=x+\beta(t)+\int_0^t b(\xi(s))ds\quad x(t)\in R^d \tag{58}$$

where $\xi(t)$ is the trajectory associated to the path $\beta(t)$ of Brownian motion (we make use of the evaluation map for all trajectories in the support of Brownian motion).

Consider the dual action of this modification on the measures on continuous trajectories in any finite interval of time, and call $\mu_b(t)$ the resulting measure. Since we have assumed that $b(x)$ is Lipshitz continuous one can use the Picard iteration scheme to prove that the map $\mu_0 \to \mu_b(t)$ is well defined. We recall some definitions

Definition 10 (*martingale* [2–4]) Given a probability space $\{\Omega, \mathcal{F}, P\}$ and a filtration $\mathcal{F}_t \in \mathcal{F}$ (a family of sub-sigma fields such that $\mathcal{F}_s \subset \mathcal{F}_t$ for $s < t$) a family $M_t(\omega)$ of random variables is called a *martingale* if
(1) For almost all ω, $M_t(\omega)$ has left and right limits and is continuous to the right.
(2) For each $t \geq 0$ $M_t(\omega)$ is measurable and integrable.

(3) For $0 \leq s \leq t \quad E(M_t, \mathcal{F}_s)_t = M_s$ *almost surely*, where $E(X, \mathcal{F}_s)$ denotes conditional expectation of X with respect to the σ algebra $\mathcal{F}_s$ (a subalgebra of $\mathcal{F}_{\leq t}$. ◇

The role of this definition of martingale can be seen from the following theorem.

Theorem 3 (Girsanov's formula) *Denote by P_x^0 the measure that Brownian motion defines on the space Ω of continuous trajectories starting from x at time 0. Let $b(x)$ be a Lipshitz continuous vector field and by P_x^b the measure of the stochastic process with drift $b(x)$. Then P_x^b is absolutely continuous with respect to P_x^0.*

The Radon–Nikodym derivative is given by

$$R_t^b(\omega) = e^{\int_0^t b(\xi_s(\omega))ds - \frac{1}{2}\int_0^t b^2(\xi_s(\omega))ds} \qquad b^2 = \sum_{k=1}^d b_k^2 \tag{59}$$

The process defined by P_x^b is a Markov process because R_t is a martingale with respect to $\{\Omega, \mathcal{F}_t, P_x\}$. As usual we have denoted by $\xi_s(\omega)$ is the evaluation map. ◇

Proof We shall give the proof only in the case when the vector field is bounded. The proof in the general case will follow by approximation and a limit procedure.

Define a new measure $\hat{Q}_x$ by

$$\frac{d\hat{Q}_x}{dP_x} = R_t^b \tag{60}$$

where R_t^b is given by Eq. 59. We prove first that R_t^b is a martingale.

By inspection, this is true when b is a piecewise constant function b_n. Denote by $R_t^{b_n}$ the corresponding martingale. One can verify that $(R_t^b)^2 \leq R_{2t}^b e^{tC^2}$ where $C(\omega)$ is chosen such that $C(\omega) \geq |b_s(\omega)|$ for $0 \leq s \leq t$.

A bounded progressively measurable function b can be approximated by piecewise constant functions b_n which are uniformly bounded. Therefore when $b_n \to b$ the martingales $R_t^{b_n}$ are uniformly bounded in $L^2(P_0)$ and the limit R_t^b exists and is again a martingale.

Since the distributions are consistent for different times, it follows that

$$R_t(\theta, \omega) = e^{\int_0^t (\theta - b(x(s))dx(s) - \frac{1}{2}\int_0^t (b(x(s)) - \theta)^2 ds} \tag{61}$$

is a martingale for every θ. This implies that

$$\begin{aligned} S_t(\theta) &= e^{\int_0^t \theta dx(s) - \frac{t\theta^2}{2} - \frac{1}{2}\int_0^t \theta b(x(s))ds} \\ &= e^{\theta(x(t) - x - \int_0^t b(x(s))ds) - \frac{t\theta^2}{2}} \end{aligned} \tag{62}$$

is a martingale with respect to $\hat{Q}_x$. Therefore

$$y(t) \equiv x(t) - x - \int_0^t b(x(s))ds \tag{63}$$

is distributed as the Brownian motion. Since $\Phi_x(y(.)) = x(.)$ one has $\hat{Q}_x = Q_x$. ♡

We call attention to the second term in the exponential in Girsanov's formula, which has its origin in the fact that Brownian motion is a process with increments in time which are independent for *disjoint intervals of time*. For closed intervals which have a point in common the increment is not equal to the sum of increments in the two parts (Wiener measure is not a product measure).

The difference is encoded in the quadratic term $b^2(x(s))$. Recall that Wiener's paths are nowhere differentiable and therefore this term is *not related to jumps in the derivative*

So far we have studied Brownian motion, a Markov process that has the Laplacian as generator. We have also studied modifications obtained by adding a potential and/or a drift. The same analysis can be done for Markov processes which have a generator of the form

$$H_{A,b} = -\sum \frac{\partial}{\partial x_k} a_{k,h}(x) \frac{\partial}{\partial x_h} + b_k(x) \frac{\partial}{\partial x_k} \qquad a_{h,k} = A_{k,h} \quad A > 0 \quad k, h = 1 \dots d \tag{64}$$

provided the coefficients $a_{h,k}$ and $b_(x)$ are sufficiently regular. All these processes are defined for $0 \leq t \leq T$ (for some $T > 0$), have continuous trajectories in R^d, are recurrent for $d < 3$ and have a measure equivalent to Wiener measure on path in $0 \leq t \leq T$.

They are a special class of *diffusion processes* (or for short *diffusions*), stochastic processes that behave *locally* as a Brownian motion. We restrict ourselves here to the one-dimensional case. The idea is to realize a process with *increments* which satisfy

$$[x(t+\epsilon) - x(t)|_{\mathcal{F}(t)} = \epsilon b(t, x(t)) + o(\epsilon), \quad E[x(t+\epsilon) - x(t)]^2_{\mathcal{F}(t)} = o(\epsilon) \tag{65}$$

where the expectation E is with respect to the (Markov) σ-field $\mathcal{F}(t)$ generated by $x(t)$

Let Ω be Wiener space on $[0, T]$ i.e. the space of continuous trajectories $\omega(t)$, $t \in [0, T]$). Let $\xi(t)$ be a function on this space on this space, continuously progressively measurable. We are looking for a continuous progressively measurable function $x(t, \omega)$ that satisfies for almost all Brownian paths,

$$x(t+\epsilon, \omega) - x(t, \omega) = \sqrt{a(t, x(t)}[\beta(t+\epsilon) - \beta(t)] + \epsilon b(t, x(t)) + o(\epsilon) \tag{66}$$

Notice that now both the drift and the covariance depend on the path A shorthand (*formal*) notation commonly used is

$$dx(t) = \sqrt{a(t, x(t)} d\beta(t) + b(t, x(t))dt \tag{67}$$

where $\beta(t, x(t))$ is Brownian motion.

11 Ito Integral

More precisely we are looking for a progressively measurable (i.e. measurable with respect the the σ-algebra generated in time by Brownian motion) function $x(t, \omega),\ t \geq s$ that satisfies for each $\omega \in \Omega$ the integral relation

$$x(t, \omega) = \beta(\omega) + \int_0^t \sqrt{a(\tau, x(\tau, \omega))} d\beta(\tau) + \int_0^t b(\tau, x(\tau, \omega))d\tau \qquad 0 \leq t \leq T \tag{68}$$

The first integral, *Ito stochastic integral*, is defined for a special class of function $\mathcal{F}$ (that we describe now) by convergence in measure of the corresponding Riemann integral for approximating functions that are piecewise constant on (almost all) paths. The class $\mathcal{F}$ is made of functions f mapping $[0, T] \times \Omega \to R$ which satisfy
(i) $\forall t > 0$ the function f is jointly measurable with respect to the σ-field of the Brownian motion
(ii) $\forall t > 0 \ \ E \int_0^t |f(s, \omega)| ds < \infty$ where E is expectation with respect to Wiener measure μ_0.

One has then [1, 2, 4].

Theorem 4 *For $F \in \mathcal{F}$ one defines the* stochastic Ito integral

$$X_f(t) = \int_0^t f(s, \omega) dx(s) \tag{69}$$

with the following properties that characterize it completely
(1) the map $f \to X_f(t)$ is linear
(2) $X_f(t)$ is progressively measurable, continuous as a function of t and a martingale with respect to $(\Omega, \mathcal{F}_t, P_0$ where P_0 is Wiener measure on paths stating at the origin.
(3) $X_f^2 - \int_0^t f(s, \omega)^2 ds$ is a martingale with respect to $\Omega, \mathcal{F}_t, P_0$. ◇

Recall that a process $M(t)$ is a martingale with respect to $(\Omega, \mathcal{F}_t, P_0)$ if the following are true
(1) for almost all ω, $M(t, \omega)$ has right and left limits and is continuous from the right
(2) for each $t \geq 0,\ M(t)$ is measurable and integrable
(3) for $0 \leq s \leq t$ one has $E[M(t)_{\mathcal{F}}(s)] = M(s)$ almost everywhere.

Sketch of the proof of Theorem 7.7
One start as usual with simple functions (piece-wise constant for every Ω) Let $0 = t_0(\omega) < t_1(\omega) \cdots < t_n(\omega) < \infty$ be n times, let $t_n(\omega)$ be measurable and let the

function $f(\omega, t)$ be constant in these intervals. For these function one can define the integral as Riemann sums.

$$X_f(t) = \sum_{i=1}^{k-1} f_{i-1}(\omega)[x(t_i) - x(t_{i-1})] + f_{k-1}(\omega)[x(t) - x(t_{k-1})] \tag{70}$$

It is easy to check that properties (1) to (2) are satisfied. To verify (3) we have to prove

$$E\left[X_f^2(t) - X^2(s) - \int_s^t f(\tau, \omega)d\tau|_{\mathcal{F}_s}\right] = 0 \tag{71}$$

where E is the expectation with the probability measure of the Brownian motion starting at the origin. This can be verified by using the properties

$$E[X_f(s)[X_f(t) - X_f(s)]]|_{\mathcal{F}(s)}] = 0 \tag{72}$$

$$E[f_{j-1}(\omega)(x(t) - x(s)^2|]_{\mathcal{F}_s} = E[f_{j-1}(\omega)(t - s)]|_{\mathcal{F}_s} \tag{73}$$

which follow from the properties of Brownian motion. As a consequence of Doob's inequality for Brownian motion [1, 2] one has also

$$E[sup_{0\leq s\leq t}(X_f(s)|^2] \leq 4E\int_0^t f^2(s, \omega)ds] \tag{74}$$

These estimates lead to the definition of the *Ito integral* for a large class $\mathcal{F}$ of functions. The class $\mathcal{F}$ is made of function $f(s, \omega) \in \mathcal{F}$ which can be approximated in measure with a sequence of simple functions such that

$$E\left[\int_0^t |f_n(s, \omega) - f(s, \omega)|^2 ds\right] \to 0 \tag{75}$$

Then the limit X_f of X_{f_n} exists in the sense that

$$lim_{n\to\infty} E[sup_{0\leq s\leq t}|X_{f_n}(s) - X_f(s)|^2] = 0 \tag{76}$$

It is easy to see that the approximation can be done if one shows [1, 2, 4] that every bounded progressively measurable (i.e. measurable with respect to $\mathcal{F}_s$ for $0 \leq s \leq t$) can be approximated by bounded progressively measurable *almost surely continuous functions*. ♡

As a consequence of the definition of Ito integral one has the *Ito formula*, an important result in the theory of stochastic integrals. Let $f(t, x)$ be a bounded continuous function of t and x with a bounded continuous derivatives in t and two bounded continuous derivatives in x. Then

$$f(t,x(t)) = f(0,x(0)) + \int_0^t f_s(s,x(s))ds + \frac{1}{2} f_{xx}(s,x(s))ds + \int_0^t f_x(s,x(s))dx(s) \quad 0 \le t \le T \tag{77}$$

One can verify that if $g(s,x) = f_s(s,x) + \frac{1}{2} f_{xx}(s,x)$ is a bounded continuous function , then

$$f(t,x(t)) - f(0,x(0)) - \int g(s,x(s)ds \tag{78}$$

is a martingale.

In particular if $a(x,\omega) \equiv 1$ and for almost all paths $b(t,\omega) = b(\omega(t)$ this procedure leads to the construction of the process which corresponds to the Brownian motion with a drift $b(x)$ (the process has generator $\frac{1}{2}\frac{d^2}{dx^2} + b(x)\frac{d}{dx}$) If $b(x,\omega)$ for almost all ω is locally Lipschitz but *depends on the path* we still have a Markov process with paths given by

$$x(t),\omega) = x + b(t,\omega) + \int_0^t b(x(s),\omega)ds \tag{79}$$

In this case Wiener measure on paths induces a distribution $\mathcal{Q}_x$ on the paths of this process on $[0,T]$ starting in x at time zero. This measure is absolutely continuous with respect to Wiener measure. The Radon–Nikodym derivative on the σ-field $\mathcal{F}_t$ is given by

$$RD_t(\omega) = exp\left[\int_0^t b(x(s))dx(s) - \frac{1}{2}\int_0^t b^2(s,x(s))ds\right] \tag{80}$$

References

1. J. Doob, *Probability Theory* (Wiley, New Jersey, 1953)
2. B. Gnedenko, *The Theory of Probability* (Mir, Moscow, 1976)
3. P. Billingsley, *Convergence of Probability Measures* (Wiley, New York, 1968)
4. K. Ito, H. Mc Kean, *Diffusion Processes and Their Sample Paths* (Springer, Heidelberg, 1965)
5. R. Brown, Philos. Mag. **4**, 161–173 (1828)
6. A. Einstein, zur Theorie der Brownschen bewegung. Ann. del Physik **19**, 180–188 (1908)
7. J. Perrin Annales, de Chemie et de Physique Mouvement Brownien et realité moleculaire. **8**, 5–114 (1909)

Lecture 8: Ornstein–Uhlenbeck Process. Markov Structure. Semigroup Property. Paths Over Function Spaces

The structures we have analyzed so far describe random processes in the time interval $[0, T]$ with T arbitrary but finite. One can equivalently consider processes in the time interval $[-T, T]$.

We have associated to the Laplacian in R^d the Wiener process on the interval $[0, T]$. It is easy to see that the procedure followed in the case of the Laplacian Δ can be repeated for the generator $\mathcal{L}$ of any positivity preserving contraction markovian semigroup. Therefore we can associate to $\mathcal{L}$ a stochastic process on $[-T, T]$.

We will return in Lecture 11 to the properties that operators and quadratic forms must have to define a stochastic process. We want now to construct a process for which the paths are defined for all times. Moreover we want that the process has an invariant measure and that the group of time-translations acts as measure-preserving transformations.

We can do this if the generator $\mathcal{L}$ has a unique (positive) ground state. For example we can take $\mathcal{L} = H_V = -\Delta + V(x)$ where $V(x),\ x \in R^d$ is a potential of a suitable class and H a unique ground state with eigenfunction $\phi_0(x)$ which is strictly positive.

Or one can choose $\mathcal{L} = \Delta^N_{[0,1]}$, the Laplacian in $[0, 1]$ with Neumann boundary conditions. The invariant measure would in this case have constant density in $[0, 1]$.

These choices would give a process with trajectories in $-\infty \leq t \leq +\infty$ but the kernel of $e^{-t\mathcal{L}}$ has a complicated expression which makes it inconvenient for explicit estimates. We choose the Hamiltonian of the harmonic oscillator

$$H_0 = -\frac{\Delta}{2} + \frac{x^2}{2} - \frac{d}{2} \qquad x \in R^d \tag{1}$$

1 Mehler Kernel

For the Hamiltonian of the harmonic oscillator the kernel of the associated semigroup has a simple form. Moreover it possible in a simple way to extend the construction to the infinite-dimensional case, and construct processes with paths in Sobolev spaces.

G. Dell'Antonio, *Lectures on the Mathematics of Quantum Mechanics II: Selected Topics*, Atlantis Studies in Mathematical Physics: Theory and Applications 2, DOI 10.2991/978-94-6239-115-4_8

The integral kernel of the corresponding semigroup is known explicitly (Mehler kernel)

$$K_t^0(x,y) = (1-e^{-2t})^{-1/2} exp\left\{-y^2 + \frac{(e^{-t}y-x)^2}{1-e^{-2t}}\right\} \tag{2}$$

We briefly indicate a derivation of Mehler formula for $d = 1$. Using the creation and destruction operators

$$a^* = \frac{1}{\sqrt{2}}\left(x - \frac{d}{dx}\right), \quad a = \frac{1}{\sqrt{2}}\left(x + \frac{d}{dx}\right) \tag{3}$$

one has

$$H_0(a^*)^n \Omega_0 = n(a^*)^n \Omega_0 \qquad \Omega_0 = \frac{1}{\sqrt{\pi}} e^{-\frac{x^2}{2}} \tag{4}$$

Therefore

$$e^{-tH_0} e^{i\nu \frac{a^*}{\sqrt{2}}} \Omega_0 = e^{i\nu \frac{e^{-t}}{\sqrt{2}} a^*} \Omega_0 \tag{5}$$

(both sides are analytic in t because a^* $(H_0 + 1)^{-\frac{1}{2}}$ is a bounded operator). From the commutation relations between a and a^*

$$\int e^{-tH_0}(x,y) e^{i\nu y} e^{-\frac{y^2}{2}} = e^{-\nu \frac{e^{-t}}{\sqrt{2}} e^{i\nu x}} e^{-\frac{x^2}{2}} \tag{6}$$

From this one derives Mehler's formula. A similar construction can be done for generators of the form

$$H_{A,b} = \sum_{k,h=1}^{d} \frac{\partial}{\partial x_k} A_{k,h} \frac{\partial}{\partial x_h} + b_k \frac{\partial}{\partial x_k} \qquad x \in R^d \tag{7}$$

where A is a positive definite matrix. The Mehler kernel can be derived with the same procedure as above by using a suitable representation of the canonical commutation relations.

Formally the same is true in infinite dimensions but one must pay attention to the fact that not all the representations are equivalent and the choice of A and b fixes the representation in which the process is defined. Notice that

$$lim_{t\to\infty} K_t^0(x,y) = \frac{1}{\pi} e^{-\frac{x^2}{2} - \frac{y^2}{2}} = P_0(x,y) \tag{8}$$

This is the kernel of the projection operator on the ground state of the harmonic oscillator. The ground state is unique, because H_0 is strictly positive on the complement of the ground state.

2 Ornstein–Uhlenbeck Measure

Proceeding as in the case of the heat kernel, one can verify that K_t^0 defines for any finite T a process in $[-T, T]$ and fixed $q,\ q' \in R^d$, called *Ornstein–Uhlenbeck bridge*. The paths start in q at time $-T$ and end in q' at time T.

There is a corresponding measure on these paths (the *Ornstein–Uhlenbeck bridge measure*) $W^{OU}_{q,q';T}$. This measure is supported on continuous paths for $t \in [-T, T]$ conditioned to be in q' at time $-T$ and in q at time T.

We want now to describe processes that are defined for all times. Time translation will act as measure-preserving group of transformations and there will be an invariant measure (state). We start with the Ornstein–Uhlenbeck bridge.

Let $\mathcal{M}_T$ be the space of continuous functions in the interval $[-T, T]$ with values in R^d. Define a measure $\Phi_0(T)$ on $\mathcal{M}_T$ as product measure of $d\mu^{OU}_{y,y';[-T,T]}$ times the measure on $R^d \times R^d$ having for each factor as density the (positive) eigenfunction Ω_0 of the ground state of the harmonic oscillator

$$d\Phi_0(T) = d\mu^{OU}_{y,y';[-T,T]}\Omega_0(y)dy\Omega_0(y')dy' \tag{9}$$

Denote by A_k the operator which act as multiplication by $A_k(x),\ k = 1, \dots, n$

$$(\Omega_0, A_1 e^{-(t_2-t_1)H_0} A_2 \dots e^{-(t_n-t_{n-1}H_0} A_n \Omega_0) = \int \Pi_k A_k(\xi(t_i)) d\Phi_0(t). \tag{10}$$

From the invariance of Ω_0 under e^{isH_0} (which implies $e^{-tH_0}\Omega = \Omega$) it follows that for $S > T$ the measure $\Phi_0(T)$ can be regarded as the *conditioning* of $\Phi_0(S)$ to the paths in $[-T, T]$.

Moreover for $|S| \leq T$ the random variables $\xi_T(s)$ have a the same distribution as ξ_S for $S > T$ and can be realized in the same probability space. Therefore we are justified in identifying them.

This compatibility property allows, by Kolmogorov theorem, the construction (in several ways) a common probability space. But since Kolmogorov theorem is very general, in principle in a representation the measure is carried by the continuous product of R_t, $t \in R$ and a priori it is not obvious that the process can be realized in a smaller function space, e.g. the continuous functions of t with values in R^d (the space of continuous trajectories in R^d.

It is therefore convenient to describe the limit Ornstein–Uhlenbeck measure μ_0 as a measure on a *set of measurable functions* on R and of their expectations instead of a measure on measurable sets of paths.

The measurable sets are then recovered using characteristic functions. Notice that the same procedure was followed by Wiener and the conclusion that there is a realization in the space of continuous function was derived from the smoothness properties of the covariance.

Therefore the process is *indexed* now by a set of functions of time. For any continuous function $f(t)$ with values in R^d and support in $[-T, T]$ define the function on path space

$$\phi(f) = \int f(t, \xi_t)dt \tag{11}$$

which is measurable with respect to tho measure $dW^{OU}_{y,y';[-T,T]}$. It is also measurable with respect to the measure $dW^{OU}_{y,y';[-S,S]}$ with the same expectation.

Therefore we *have defined a measure on continuous functions* $f(t)$ with values in R^d and ξ_t is the evaluation map for the Ornstein–Uhlenbeck process defined in bounded time interval that contain the support of the function f.

Remark that since both the ground state of the harmonic oscillator and the measure of the Ornstein–Uhlenbeck bridges are gaussian, also the limit measure μ_0 is gaussian (as limit *in measure* of gaussian measures).

The invariant measure of the Ornstein–Uhlenbeck process is gaussian and therefore completely determined by its mean and its covariance. Formally the limit measure can be written

$$d\phi_0 = \int_{R^d \times R^d} \Omega_0(q)\Omega_0(q')dU_{q,q'} \tag{12}$$

and defined on continuous paths in any *bounded* interval. For any collection f_i of bounded function of $q \in R^d$ and for any polynomial P and for any value of T one has

$$\int \Pi f_i d\omega = (\Omega_0, f_1 e^{-(t_1-t_2)} f_2 \ldots f_{n-1} e^{-(t_{n-1}-t_n)} f_n \Omega) \tag{13}$$

for $|t_i| < T, \forall i$. This measure is constructed by a weak limit procedure for $T \to \infty$.

Therefore the support of the limit measure can in principle be any measurable subset of the continuous product of R^d. It is therefore advisable, as we did, to consider first the measure on measurable functions and then, if needed, enquire about a space of paths on which the measure can be realized. This space will be not unique. We shall denote by Ω any of the measure space we can choose and by ω its "points".

The most natural functions in Euclidian space are the coordinates. By the explicit form of the Mehler kernel one derives

$$\int q_k(t)d\omega = 0 \quad \forall t \tag{14}$$

$$\int q_k(t)q_h(\tau)d\omega = e^{-|t-\tau|}(\Omega, q^2\Omega) = C_{k,h}e^{-|t-\tau|} \tag{15}$$

where $C_{k,h}$ is a covariance matrix (the measure is a Gaussian measure). Denote by $\phi(f)$ the random variable associate.d to the function f. One has

$$E(\phi(f)) = 0 \quad E(\phi(f)\phi(g) = (f, g)_{-1} \qquad (f, g)_1 \equiv \int (\hat{f})^*(k)\hat{g}(k)(k^2+1)^{-1}dk \tag{16}$$

$$E(\phi(f)^{2n}) = (2n-1)!!f_1^{2n}, \qquad E(\phi(f_i)^{2n+1}) = 0, \qquad E(e^{i\phi(f)}) = e^{-\frac{1}{2}(f,f)_{-1}} \tag{17}$$

Notice that the process is defined for all times and has a positive invariant measure. As a stochastic process it has a generator $\mathcal{L} = -H_0$ where H_0 is the Hamiltonian of the harmonic oscillator in R^d and the invariant measure is the ground state of the harmonic oscillator.

Consider now the hamiltonian

$$H = H_0 + V(x) \tag{18}$$

where the potential V is such that H is self-adjoint, positive and has 0 as isolated eigenvalue. Without loss of generality we the corresponding eigenfunction $\Omega(x)$ to be positive.

Therefore $\int \Omega(x)\Omega_0(x)dx$ is positive and

$$(\Omega, \Omega_0)\Omega = lim_{t\to\infty}e^{-tH}\Omega_0 \tag{19}$$

Define

$$d\mu_T = Z_T^{-1} e^{\int_{-T}^{T} V(\xi(s))ds} d\mu_0 \tag{20}$$

where Z_t is a numerical constant chosen so that $d\mu$ is a probability measure. By construction

$$|e^{2tH}\Omega_0|^{-1}\ (e^{-(t_1-t)H}\Omega_0, f_1 e^{-(t_2-t_1)H} f_2 \dots e^{-(t-t_n)H}\Omega_0) = \int \Pi_k f_k(\xi_k(t_k)) d\mu_t \tag{21}$$

It follows from our assumptions that the left hand side converges when $t \to \infty$. Therefore also the right hand side converges and

$$(\Omega, f_1 e^{-(t_2-t_1)H} f_2 \dots e^{-(t_{n-1}-t_n)H}\Omega) == lim_{t\to\infty} \int \Pi_k f_k(\xi_k) d\mu_t \tag{22}$$

Recall that, before taking the limit, the measure μ_T is defined on continuous functions supported in $(-T, T)$. With a procedure similar to that we used in the case of the Ornstein–Uhlenbeck process one can now define a measure on $\mathcal{D}'$ (or even in a smaller space of distributions). But now this measure is no longer gaussian and it is more difficult to compute the momenta and correlations of the random variables $\phi(f)$.

This limit measure is defined by

$$\mu(\Pi_k f_k(\xi_k)) = lim_{t\to\infty} \int \Pi_k f_k(\xi_k) d\mu_t \tag{23}$$

3 Markov Processes on Function Spaces

We must prove that there is a Markov process over some function space associated to the semigroup with generator $\Delta - V$ and that this process has an invariant measure (given by (23)). As remarked above, this abstract procedure does not guarantee a-priori that there is realization of this process in a space of continuous paths, since the proof relies on convergence *in distribution*: To find the support we must return to the meaning of convergence for the measures μ_T. Define for $f \in D$, $supp(f) \subset [-T, T]$ the *characteristic function*

$$\Phi_T(f) \equiv \int e^{i\phi(f)} d\mu_T \tag{24}$$

It is easy to verify:

(1)

$$f_n \to f \;\; in \;\; \mathcal{D} \leftrightarrow \Phi_T(f_n) \to \Phi_T(f) \tag{25}$$

(2)

$$\sum_{i,j=1}^{N} \bar{c}_j c_i \Phi_T((f_i - \bar{f}_j) \geq 0 \tag{26}$$

for every choice of functions f_i and complex numbers c_i.

(3)

$$\Phi_T(0) = 1 \tag{27}$$

Notice that $\Phi_T(f) = \Phi_S(f)$ if the support of f is contained in $[-T, T] \cap [-S, S]$. Setting $\Phi(f) = lim_{T\to\infty}\Phi_T(f)$ one obtain in this way a functional on $\mathcal{D}$ with the properties (1)–(3) above.

This properties are shared by the limit. We must take $f \in \mathcal{D}$ since we are dealing with functions which have arbitrary but finite support. We are now in condition to apply the following theorem [1–3].

Theorem 1 (Minlos) *Let $\Phi(f)$ be a functional on $\mathcal{D}$ with the properties (1)–(3) described above.*

Then there exists a unique probability measure μ on $\mathcal{D}'$ such that $\Phi_\mu(f) = \int e^{i\xi(f)} d\mu(\xi)$. We shall call Φ_μ characteristic functional *of the measure μ.* ◇

Minlos' theorem is a generalization of a theorem of Bochner that we will now state and prove. Recall that a function f on R^n is Function of positive type *of positive type* if it is bounded, continuous and for any choice of $\lambda_1, \ldots, \lambda_n \in R^n$ the matrix

$$F_{i,j} \equiv f(\lambda_i - \lambda_j) \tag{28}$$

is positive.

Theorem 2 (Bochner) *The cone of functions of positive type coincides with the Fourier transforms of the finite positive measures on R^n.* ◇

Proof (1) If μ is a finite positive measure one has

$$\sum_{i,j} \hat{\mu}(\lambda_i - \lambda_j)\bar{f}_i f_i = \int d\mu(x) |\sum_1^N e^{i\lambda_k x} f_k|^2 \geq 0 \tag{29}$$

(2) Let f be a continuous function of positive type. Let $\mathcal{H}_0$ be the pre-Hilbert space of complex-valued functions that are different from zero only in finite number of points (this is a vector space for point-wise addition) endowed with the scalar product

$$(\phi, \psi)_f = \sum_{x,y \in R^N} \bar{\phi}(x) f(x-y) \psi(y) = (U_t\phi, U_t\psi)_f \qquad \forall t \in R \tag{30}$$

where $U_t\phi(x) = \phi(x-t)$. Let $\mathcal{H}$ be the closure of $\mathcal{H}_0$ under the topology of the scalar product (30). Let Ξ be the ideal of functions for which $(\phi, \phi)_f = 0$. Then U_t is well defined on the quotient $\mathcal{H}/\Xi$.

Since U_t is strongly continuous (since f is continuous) one can use the spectral theorem and Stone's theorem to prove that there exists a family of projection operators P_λ on R^n such that

$$(\phi, U_t\phi)_f = \int e^{it\lambda} d(\phi, P_\lambda \phi)_f \tag{31}$$

Let now $\tilde{\phi}_0$ be the equivalence class in $\mathcal{H}/\Xi$ of the function ϕ_0 defined by:

$$x = 0 \Rightarrow \phi_0 \equiv 1 \qquad x \neq 0 \Rightarrow \phi_0 \equiv 0 \tag{32}$$

so that

$$x = t \Rightarrow U_t\tilde{\phi}_0 = 1 \qquad x \neq t \Rightarrow U_t\tilde{\phi}_0 = 0 \tag{33}$$

Then

$$f(t) = (U_t\tilde{\phi}_0, \tilde{\phi}_0) = \int e^{-it\lambda} d(\tilde{\phi}_0, P_\lambda \tilde{\phi}_0) \tag{34}$$

Therefore $f(t)$ is the Fourier transform of a positive measure. ♡

The process can be realized with paths in a space smaller than $\mathcal{D}'(R)$ (recall that Brownian motion in $[0, T]$ can realized on continuous paths).

As in the case of Brownian motion there is no optimal regularity, but only an upper bound. This bound depends on the regularity properties of the measure $\mu(\Pi_k f_k(\xi_k))$ as a function of the f_k for which the integral is defined and this in turn depends on the regularity of the potential V. In general one can find a Sobolev space on which the process is realized.

4 Processes with (Continuous) Paths on Space of Distributions. The Free-Field Process

The construction we have given of a Markov process using the harmonic oscillator semigroup can be repeated for any self-adjoint operator on $L^2(R)$ provided it has a positivity preserving kernel. If the kernel is positivity improving there is a unique ground state.

When one seeks a generalization to paths in function spaces (e.g. the space of distributions) one should require that the ground state measure be meaningful on the space one considers.

By the theorem of Kolmorogov this is certainly the case if the ground state can be represented as a product measure i.e. the state must be a (infinite) product state in suitable coordinates.

This is the case for gaussian measures which are ground states for positive Hamiltonians which are quadratic in the position-momentum variables. Since we want to have a measure on continuous paths in some distribution space we must require that in the dual space (the space of linear function on the paths) be present also the functions $f(\xi)\delta(t-T)$, where ξ belong to some function space, e.g. some Sobolev space on R^d.

The space X is conventionally called *test function space*; it is in duality with *path space*.

If both the path space and the test function space are spaces of functions that admit Fourier transform one can describe both spaces using Fourier transform.

The Fourier transform in R^{d+1} (with coordinates $x_0 = t, x_1, \ldots, x_d$) of the function $f(x)\delta(t-\tau)$ is $\hat{f}(p)e^{i\tau p_0}$ in "generalized" momentum space with coordinates $k_0, k_1, \ldots, k_d$.

Therefore we can take the covariance for the process (a positive bilinear form over functions of space and time) can be taken to be

$$(F, G)_1 \equiv \int (\hat{F})^*(k_0, k)\hat{G}(k_0, k)\frac{1}{k_0^2 + k^2 + m}dk_0 dk \tag{35}$$

where the parameter m represents the "mass" of the particle associated to the field. We have chosen capital letters to denote function *on space–time*.

Recalling that $\delta(t-\tau) \in H_{-1}$ is continuous in τ in this topology, we see that the definition is consistent with the previous setting if we have $\xi(\delta(t)) = \xi_t$. This infinite-dimensional Ornstein–Uhlenbeck process is gaussian and has expectation and covariance given as follows

$$E(\phi(F) = 0 \qquad E(\phi(F)\phi(G)) = (F, G)_{-1} \tag{36}$$

The Ornstein–Uhlenbeck on distributions can be constructed as an infinite product of one-dimensional Ornstein–Uhlenbeck processes.

For N finite the generator is

$$H = -\frac{1}{2m}\Delta + \frac{1}{2}(x, Ax) - I \qquad A > 0 \tag{37}$$

Since A can be diagonalized, with eigenvalues a_i the resulting process is made of N independent copies of the Ornstein–Uhlenbeck process. Take now N functions $f_1, \ldots, f_N$ in $\mathcal{S}$ and a measure μ on R^d under which they are orthonormal. Call K_N this Hilbert space.

Every element in $f \in K_N$ can be written as $f(x) = \sum c_k f_k(x)$ where c_k are random variables which define a Markov process with continuous trajectories in time (in R^1). We have therefore constructed a gaussian random process on the space of function K_N.

By Kolmogorov's theorem we have constructed a process on the infinite tensor product of copies of $L^2(R^1)$. But the covariances of the component processes decrease to zero (the eigenvalues of the harmonic oscillator increase as N^2) and therefore the densities tend to the function identically equal to zero and the measure of any finite interval tends to zero.

The convergence is in measure and the limit measure of the process may have support "at infinity" [2, 4].

We shall start with the constriction of a gaussian measure on the nuclear space $\mathcal{H}^{-\infty}(R^d)$ (recall that $\mathcal{H}^{-n}$ is defined like the Sobolev space H^n but using the hamiltonian of the harmonic oscillator instead of the Laplacian).

Each $f \in \mathcal{S}$ defines a linear functional (a coordinate) on $\mathcal{H}^{-\infty}(R^d)$

$$\theta \to f(\theta) \equiv \theta(f) = < f, \theta >_0 \tag{38}$$

where the suffix 0 indicates that the duality is made with respect to $L^2(R^d)$. If T is measurable and B is a Borel in R^d the subset of $\mathcal{H}_{-\infty}(R^d)$ defined by

$$\{\theta : (f_1(\theta), \ldots f_n(\theta) \in B\} \tag{39}$$

where B is a *(Borel) cylinder set* and

$$\theta \to F(f_1(\theta) \ldots f_n(\theta)) \tag{40}$$

is a *Borel cylinder function.* In the usual way cylinder sets and functions define the class of measurable sets and functions. For any measure $d\mu$ in $\mathcal{H}_{-\infty}(R^d)$ the bilinear form (f, Cg) on $\mathcal{S}(R^d)$ defined by

$$\mathcal{H}_\infty \ni f, g \to (f, C, g) \int \bar{f}(\theta) g(\theta) d\nu(\theta) \tag{41}$$

is called *covariance* of μ.

We suppose that C is not degenerate (i.e. $(f, Cf)) = 0$ implies $f = 0$ a.e. We will say that the measure ν on $\mathcal{H}_\infty(R^d)$ is gaussian if the restriction to every finite-dimensional space is gaussian.

Let C be a covariance defined on the nuclear space $\mathcal{H}_\infty(R^d)$. Kolmogorov theorem implies that there is a *unique* gaussian measure having Cas covariance.

In the case $d = 0$ (processes over a finite dimensional space) we have seen that one can chose the representation in such a way that the measure be carried by continuous functions.

If $C = (-\Delta + I)^{-1}$ acting on $L^2(R^d)$ then the Ornstein–Uhlenbeck process can be defined on $\mathcal{D}'(R^d)$.

We seek conditions under which the process can be defined on a smaller space. A first step in this direction is a generalization of Bochner theorem. We shall state it without proof.

A distribution $T \in \mathcal{D}'(R^n)$ is Distribution of positive type *of positive type* if for each $\psi \in \mathcal{D}(R^n)$ such that $\bar{\psi}(-x) = \psi(x)$ one has

$$T(\bar{\psi}\,\psi) \geq 0 \tag{42}$$

Theorem 3 (Bochner–Schwartz) *A distribution $T \in \mathcal{D}'(R^n)$ is of positive type if and only if $T \in \mathcal{S}'(R^n)$ and moreover is the Fourier transform of a positive measure* of at most exponential growth. ◇

Notice that the nontrivial part of the theorem is the statement that if the measure is of positive type then there is an equivalent measure which is supported by the *smaller set $\mathcal{S}'(R^n)$*.

Theorem 4 (Minlos) *Let $d > 1$ and let ϕ be a function on $\mathcal{S}(R^d)$. Necessary and sufficient condition for the existence of a measure $d\mu$ on $\mathcal{S}'(R^d)$ that satisfies $\phi(f) = e^{iF(f)}d\mu(F)$ is that*

(1) $\phi(0) = 1$

(2) $F \to \phi(F)$ is continuous in the strong topology

(3) For any $\{f_1, \ldots, f_n \in \mathcal{S}\}$ and $\{z_1, \ldots, z_n \in C\}$ one has

$$\sum_{i,j=1}^{n} z_j z_i \phi(f_1 - f_j) \geq 0 \tag{43}$$

◇

There is in general no canonical measure space. A refinement [3, 5] of the theorem of Minlos proves that if the covariance C can be extended continuously to a Sobolev space $\mathcal{H}^d$ the gaussian measure with covariance C can be realized the space of continuous functions of t with values in a Sobolev space $\mathcal{H}^{n(m,d-1)}$ for a suitable (negative) function $n(m, d)$.

5 Osterwalder Path Spaces

We introduce now, in the infinite dimensional setting, a Markov process which has a relevant role in the *Weyl quantization* of classical fields. Recall that a Markov process is defined by a family of random variables that depend on a parameter $t \in R$ and take value in a (possibly infinite-dimensional) space X.

We start with a somewhat more abstract presentation [4].

Definition 1 A (generalized) path space consists of
(1) a probability space $\{Q, \Sigma, \mu\}$,
(2) a distinguished sub-σ-algebra Σ_0
(3) a one-parameter group $U(t)$ of measure-preserving automorphisms of $L_\infty(Q, \Sigma, \mu)$ which are strongly continuous in measure.
(4) a measure-preserving automorphism R (time-reflection) of $L_\infty(Q, \Sigma, \mu)$ such that

$$R^2 = I \quad RU(t) = U(-t)R \quad RE_0 = E_0R \tag{44}$$

where E_0 is the conditional expectation with respect to Σ_0
(5) Σ is generated by $\cup_{t\in R}\Sigma_t$, where $\Sigma_t = U(t)\Sigma_0$. ◇

We will denote with E_+ the conditional expectation with respect to $\cup_{t\geq 0}\Sigma_t$.

We have used the notation *generalized* because we will be interested in the case in which the space in which the path occurs is a space of distributions.

We will later introduce the dual space, *the space of fields at a fixed time* (linear functionals on the generalized space) and see under which conditions the semigroup structure of the Markov process is reflected in automorphisms of the algebra generated by the fields. In order to prove this connection, we restrict the class of Markov processes and the class of path spaces.

Definition 2 A *Markov path space* is a space of paths which satisfies the further property
(1) $RE_0 = E_0$ (called *reflection invariance* or also *reflection positivity*
(2)

$$E_+E_- = E_+E_0E_- \tag{45}$$

◇

We will be interested in path space that satisfy further conditions. We call them *Osterwalder–Schrader* (O.S) path spaces. O.S. are the initials of K. Osterwlder and R. Schrader that have established [6] the correspondence between a class of Markov Fields (that were analyzed by Symanzik [7] and Nelson [3]) and the relativistic local free field (Wightman) [8].

Definition 3 An O.S. path space in a path space satisfying the following (positivity) condition: $E_0RE_0 \geq 0$ i.e. $(Rf, f) \geq 0 \ \forall f \in L_2(Q, \Sigma, \mu)$. ◇

The O.S. positivity condition plays a major role One proves [9] (Nelson).

Proposition 1 *Every Markov path space is a O.S. path space.* ◇

The proof is as follows:

$$E_+RE_+ = E_+RE_+E_+ = E_+E_-RE_+ = E_+E_0E_-RE_+ = E_+E_0E_- = E_0 \geq 0 \quad (46)$$

♡

Notice that the converse *is not true*.

6 Strong Markov Property

We shall see that O.S. path spaces have the *strong Markov property* i.e. the Markov property *with respect to all time* (and not only with respect to time zero). We begin by proving

Proposition 2 *An O.S. path satisfies* reflection positivity *i.e.* $RE_0 \geq 0$ ◇

Proof One has indeed

$$RE_0 = E_0RE_0 = E_0RE_0 = E_0E_+RE_+E_0 \geq 0 \quad (47)$$

♡

We now prove that to every O.S. path space is associated a semigroup structure.

Theorem 5 ([3, 10, 11]) *Let* $(\{Q, \Sigma, \mu\}, \Sigma_0, U(t), R)$ *be an O.S. path space.*
There is a Hilbert space $\mathcal{H}$ *and a contraction* $K : L^2(Q, \sigma, \mu) \to \mathcal{H}$ *such that*
(1) the range of K *is dense in* $\mathcal{H}$
(2) $S(t)K(F) = K((Ut)F), \ F \in L^2(Q, \Sigma, \mu)$ *defines a strongly continuous self-adjoint contraction semigroup on* $\mathcal{H}$
(3) If we define $\Omega = K(I)$ *then* $\|\Omega\| = 1, P(t)\Omega = \Omega \forall t \geq 0$. ◇

Proof The proof is essentially a G.N.S. construction.

Define the scalar product $< f, g >= (Rf, g) \ f, g \in L^2(Q, \Sigma, \mu)$. By O.S. positivity this defined a positive semi-definite inner product. Let $\mathcal{N}$ be the ideal defined by $\mathcal{N} = \{f \in L^2(Q, \Sigma^*, \mu) \ <f,f>= 0\}$.

The ideal $\mathcal{N}$ is invariant under $U(t)$. This is seen as follows:

$$< U(t)f, U(t)f >) = (RU(t)f, U(t)f) = (U(-t)Rf, U(t)f) = (Rf, U(2t)f)$$
$$=< f, U(2t)f > \leq < f,f >^{\frac{1}{2}} < U(2t)f, U(2t)f >^{\frac{1}{2}} \quad (48)$$

Define $\mathcal{H}$ to be the Hilbert space completion of $L^2(Q, \Sigma^*, \mu)\backslash\mathcal{N}$ in $<, >$ and let $K(f)$ be the canonical projection of f in $\mathcal{H}$. The range of K is dense in $\mathcal{H}$ and $\|K(f)\|_{\mathcal{H}} \leq \|f\|_{L^2}$.

Define $S_t K(f) = K(U(t)f)$. Since $U(t)\mathcal{N} \in \mathcal{N}$ and the product $< . >$ is invariant under $U(t)$ one has

$$S_t S_s K(f) = S_t K(U_s f) = K(U_t U_s f) = K(U_{t+s} f) = S_{t+s} K(f) \tag{49}$$

Therefore the S_t^s form a semigroup which clearly strongly continuous contraction and therefore has a generator. ♡

7 Positive Semigroup Structure

Definition 4 A *positive semigroup structure* $\{\mathcal{H}, T(t), \mathcal{A}, \Omega\}$ consist of
(1) A Hilbert space $\mathcal{H}$
(2) A strongly continuous self-adjoint contraction semigroup $T(t)$ on $\mathcal{H}$ with generator H.
(3) A commutative von Neumann algebra $\mathcal{A} \in \mathcal{B}(\mathcal{H})$ (the algebra generated by the fields at time 0)
(4) a privileged vector $\Omega \in \mathcal{H}$ such that $T(t)\Omega = \Omega, \ \forall t \geq 0$ such that
(5) The vector Ω is cyclic for the algebra generated by $\mathcal{A} \cup T(t), t \geq 0$
(6) for all $f_1, \ldots f_n \in \mathcal{A}^+$ and $t_1, \ldots t_n \geq 0$ one has

$$(\Omega, T(t_1)f_1 T(t_2) \ldots T(t_n)f_n \Omega) \geq 0 \tag{50}$$

We have denoted by $\mathcal{A}^+$ the set of positive elements in $\mathcal{A}$ ◇

Condition (2) means that the union of the subsets

$$T(t_i)f_1 T(t_2)f_2 \ldots T(t_n)f_n \Omega, \qquad t_1 \ldots t_n \geq 0, \quad f_i \in \mathcal{A} \tag{51}$$

is dense in $\mathcal{H}$. One can think of $\mathcal{A}$ as the algebra generated by $e^{i\alpha x}$ and $T_t = e^{iH_0 t}$ where H_0 is the Hamiltonian.

Condition (5) and (6) are certainly satisfied in a theory in which $[H_0, x_k] = C_{k,h} p_h$ (on a suitable domain) and the representation of the Weyl algebra is irreducible.

We begin with a lemma.

Lemma 1 *Let $\{\Omega, \Sigma, \mu), \Sigma_0, U(t), R\}$ be a R.P. path space and let $\mathcal{H}, K, T_t \Omega$ be as in Theorem 5. Then*
(1) if $f \in L^\infty(Q, \Sigma_0, \mu)$ define $\hat{f} K(F) = K(fF)$. Then $\hat{f}$ is a bounded operator on $\mathcal{H}$ and $\|\hat{f}\| = \|f\|_\infty$
(2) $\{\hat{f}\}$ defines a commutative von Neumann algebra of operators on $\mathcal{H}$ with Ω cyclic and separating vector.

(3) For any $0 \le t_1 \le t_2 \cdots \le t_n$ *and for* $f_t = U(t)f_i$ *one has*

$$\int f_{t_1} f_{t_2} \ldots d\mu = (\Omega, \hat{f}_{t_1} P(t_1 - t_2) \ldots P((t_n) - t_{n-1}, \hat{f}_{t_n} \Omega) \tag{52}$$

◇

Proof Notice first that $\forall n, \ \|\hat{f}^n\| \le \|f^n\|_\infty$. Point (1) follows because R is an automorphism of $L_\infty(\mathcal{Q}, \Sigma, \mu)$ and E is the conditional expectation with respect to Σ_0.

Point (2) follows because the restriction K_0 of K to $L^2(\mathcal{Q}, \Sigma_0, \mu)$ is unitary onto its range since $< f, g >= (f, g)$ for $f, g \in L^2(\mathcal{Q}), \Sigma_0, \mu)$ and $\hat{f} = K_0^* f K_0$ so that $\|\hat{f}\|_1 = \|f\|_\infty$.

The restriction of $\mathcal{A}$ to the range of K_0 is therefore a von Neumann algebra with Ω as separating vector, and then $\mathcal{A}$ is a commutative von Neumann algebra of operators on $\mathcal{H}$.

To prove point (3) let for $i = 1, \ldots, n$

$$t_1 \le t_2 \le \ldots \le t_n \qquad f_{t_i} = U(t_i) f_i \quad f_i \in L^2(\mathcal{Q}, \Sigma_0, \mu) \tag{53}$$

It follows that

$$f_{t_1} f_{t_2} \cdots f_{t_n} = U(t_1) f_1 U(t_2 - t_1) f_2 \ldots U(t_n - t_{n-1}) f_{t_n} \tag{54}$$

Statement (3) follows then from the fact that

$$K(U(s_1) g_1 U(t_2 - t_1) g_2 \ldots g_n 1 = T_{s_1} g_1 T_{s_2} \ldots g_n \Omega \tag{55}$$

for $g_1, g_2 \ldots g_n \in L^2(\mathcal{Q}, \Sigma_0, \mu)$ and $s_1, s_2 \ldots s_n \ge 0$. ♡

In what follows the von Neumann algebra $\mathcal{A}$ is taken to be an algebra of functions on path space and therefore we will use the symbol f instead of the symbol a to indicate a generic element.

We prove now that a R.P. path corresponds to a positive semigroup structures We use the same strategy that we used in Book I to show that in Quantum Mechanics conditioning of a one parameter group of unitary operators to a sub-σ algebra leads to a positivity preserving semigroup.

One can recover the group by the Stinespring construction (reconstruction formula). Let $\mathcal{H}$, T_t, $\mathcal{A}$, Ω be the semigroup structure defined in Lemma 1.

Theorem 6 *Let* $\{\Omega, \Sigma, \mu, \Sigma_, U(t), R\}$ *be an O.S. path space and let* $\{\mathcal{H}, T(t), \mathcal{A}, \Omega\}$ *be the associated semigroup structure.*

Then $\{\mathcal{H}, T_t, \mathcal{A}, \Omega\}$ *is a* positive *semi-group structure.* ◇

Proof Let $\{\mathcal{Q}, \Sigma, \mu\Sigma_0, U(t), R\}$ be an O.S. path space and $\mathcal{H}, T(t), \mathcal{A}\Omega$ the associated semigroup structure. Conditions (1)–(5) of Definition 4 are clearly satisfied.

Condition (6) follows because 1 is a cyclic vector in $L_2(\mathcal{Q}, \Sigma_+, \mu)$ for $L_\infty(\mathcal{Q}, \Sigma_0, \mu) \cup \{U(t), t \ge 0\}$ since σ_+ is generated by $\cup_{t \ge 0} U(t)\Sigma_0$. ♡

Notice the role that the O.S. positivity condition has in the proof.

It can be proven that the converse of the statement of Theorem 6 also holds. For this one remarks that $\mathcal{A}$ is isomorphic to $C(Q_0)$ where Q_0 is the spectrum of $\mathcal{A}$. Therefore the proof is similar to the proof of the same statement for the Ornstein–Uhlenbeck process. Define $Q = \otimes_{t\in R} U(t)Q_0$ and the action of $U(t)$ and R on Q by

$$U(t)q(s) = q(t-s) \qquad Rq(s) = q(-s) \tag{56}$$

Define $F(q) = f(q(0)$ for $f \in C(Q_0)$. The difficult point, which we don't discuss here, is to construct a measure on the σ under which $\{\Omega, \Sigma, \mu, \Sigma, U(t), R\}$ is a path space. We do not give here the details.

Proposition 3 *Let $T(t)$ be a positivity preserving semigroup on $L^2(M, \mu)$.*

Then $\{L^2(M, \mu), T(t), L_\infty(M), I\}$ form a positive semigroup structure with I as cyclic vector. Conversely let $\{\mathcal{H}, T(t)\mathcal{A}, \Omega\}$ be a semigroup structure, with Ω cyclic for $\mathcal{A}$.

There exists a probability space $\{M, \mu\}$ and a positivity preserving semigroup $\hat{T}(t)$ on $L^2(M, \mu)$ such that

$$\{\mathcal{H}, T(t), \mathcal{A}\Omega\} \simeq \{L^2(M, \mu), \hat{T}(t), L_\infty M, I\} \tag{57}$$

◇

Proof The first part follows form the definitions. We prove the converse. Let Ω be a cyclic vector for $\mathcal{A}$. It follows that $\mathcal{A}$ is maximally abelian and therefore there is a Baire measure ν on the spectrum Q_0 of $\mathcal{A}$ such that

$$\mathcal{H} \simeq L^2(Q_0, \nu), \qquad \mathcal{A} \simeq L_\infty(Q_0, \nu) \qquad \Omega \simeq 1 \tag{58}$$

Let $\hat{T}(t)$ correspond to $T(t)$ under this isomorphism. Then $\hat{T}(t)$ is a positivity preserving semigroup on $L^2(Q, \nu)$ The proof is the same as in the case of the O.U. process. ♡

Theorem 7 *Let $\{\Omega, \Sigma, \mu, U(t), R\}$ a O.S. path space, and let $\{\mathcal{H}, T_t, \mathcal{A}, \Omega\}$ be the associated semigroup structure. Then $\{\Omega, \Sigma, \mu, U(t), R\}$ is Markov if and only of Ω is cyclic for $\mathcal{A}$.* ◇

Proof One has $E_+RE_+ = E_0$ since the path space is Markov. Therefore forall $f \in L^2(Q, \Sigma_+, \mu)$ one has $(Rf, f) = (E_0 f, E_0 f)$ and therefore

$$\mathcal{H} \simeq L^2(Q)\sigma_0), \mu \qquad T_t \simeq E_0 U(t) E_0, \qquad \Omega \simeq 1 \tag{59}$$

Conversely if Ω is cyclic for $\mathcal{A}$ and T_t is a positivity preserving semigroup, it follows $T_t\mathcal{A}^+\Omega \subset \mathcal{A}^+$ for all $t \geq 0$ and then by polarization $T_t\mathcal{A}\Omega \subset \mathcal{A}$. The Markov property follows then considering

$$F(q) = f_n(q(t_n)) \ldots f_1(q(t_1)),\ f_k \in L_\infty(Q.\Sigma : 0, \mu), \quad t_1 \leq t_2 \cdots \leq t_n \tag{60}$$

then $E_+E_-F = E_+F$ and therefore it is measurable with respect to Σ_0. It follows $E_+E_- = E_+E_0E_-$. ♡

Remark that Theorems 6 and 7 imply that Markov path spaces correspond to semigroup structures in which Ω is cyclic for $\mathcal{A}$. As a consequence for all $f, g \in \mathcal{A}^+$ and $T \geq 0$ one has $(f\Omega, T(t)g\Omega) \geq 0$.

Definition 5 Let M be a probability space. A strongly continuous self-adjoint contraction semigroup $T(t)$ on $L^2(M, d\mu)$ is *positivity preserving* if
(1) $T(t)I = I \forall t \geq 0$
(2) $f \geq 0 \to T(t)f \geq 0$ ◇

Theorem 8 *Let $\{\mathcal{H}, T(t), \mathcal{A}, \}$ be a semigroup structure, and Ω be a cyclic vector. There exist a probability space $\mathcal{M}$, μ and a positivity preserving semigroup $\hat{T}(t)$ on $L^2(\mathcal{M}, \mu)$ such that*

$$\{\mathcal{H}, T(t)\mathcal{A}, \Omega\} \equiv \{L_2(\mathcal{M}, \mu), \hat{T}(t), L^\infty(\mathcal{M}, \iota)\} \tag{61}$$

as positive semigroup structure. ◇

8 Markov Fields. Euclidian Invariance. Local Markov Property

We have seen the construction of O.S. and Markov processes with trajectories continuous in time with values in function spaces over R^d. One may try to have a structure that is more symmetric in space and time and construct *Markov fields* over R^{d+1}.

One may ask that for these fields the Markov property be valid in a generalized sense: given any domain $\Omega \in R^{d+1}$ with interior Ω^0 and smooth boundary $\partial\Omega$ the random field in the complement Ω^c of Ω is completely determined (in law) by a field on the boundary.

Recall that a stochastic process indexed by a set X is a function from X to a probability space $\{\Omega, \mathcal{S}, \mu\}$ where $\mathcal{S}$ are the measurable sets and μ is the measure. If X is a topological space, a *linear process* $f \to \phi(f)$ over X is a stochastic process indexed by X and such that $f_n \to f$ in X implies $\phi(f_n) \to \phi(f)$ in measure where defines the process.

Denote by S the σ-algebra generated by $\{\phi(f),\ f \in \mathcal{S}\}$ ($\mathcal{S}$ is Schwarz space) If Λ is open in E^{d+1} define $\Theta(\Lambda)$ the sigma-algebra generated by functions with support in Λ.

$\mathcal{H}^d$ is the harmonic space of order -1 constructed similarly to the Sobolev space but with the operator $-\Delta + x^2 + 1$ in place of Δ If Λ is a subset of E^d define

$$\Theta(\Lambda) = \cap_{\Lambda \subset \Lambda',\ \Lambda' open} |\Theta \Lambda' \tag{62}$$

Denote by $E\{.;\ \Theta(\Lambda)$ the conditional expectation. Then a Markov field over E^{d+1} is a linear process such that for every measurable f and regular Λ one has

$$E(f|\Theta(\Lambda^c) = E(f|\Theta(\partial\Lambda) \tag{63}$$

where $\partial\Lambda$ is the border of Λ. In particular denote by $\mathcal{D}(R^d)$ the space of C^∞ function on R^n with compact support. A linear process over $\mathcal{D}(R^d)$ is called *random field.*

In the case of the Ornstein–Uhlenbeck process, we have a random process over $\mathcal{H}^{-1}(R^d)$.

Since the injection of $\mathcal{D}$ in $\mathcal{H}^{-1}$ is continuous, a random process over $\mathcal{H}^{-1}(R^d)$ defines a *random field* over $\mathcal{H}^{-1}(R^{d+1})$.

By construction functions (distributions) of the type $f(x)\delta(t - t_0$ with $f \in S$ are defined on the process (are *test functions* i.e *random variables*).

We now require that our random field *euclidian invariant* i.e. invariant under the natural action of the inhomogeneous Euclidian group in R^{d+1}. This implies that it has the Markov property with respect with respect to any choice of a d-dimensional hyperplane.

Recall that the Euclidian group E^{d+1} of R^{d+1} is the inhomogeneous orthogonal group (including reflections) i.e. the group of linear transformations which preserve $|x - y|$. A *representation of the Euclidian group* on a probability space is a homeomorphism $\eta \to T_\eta$ of E^d into the group of measure preserving transformations T_η which continuous in the sense that if $\eta_n \to \eta \in E$ implies $T_{\eta_n} \to T_\eta$ in measure.

On R^{d+1}, $d \geq 1$ introduce coordinates (t, x), $x \in R^d$, $t \in R$ and let Y_s the hyperplane $t = t_0$. Let Y_s be the half-space $t \leq s$ and let $\rho(s)$ the reflection with respect to Y_s. We call $\eta(s)$ be the translation $(x, t) \to (x, t + s)$.

An *Euclidian field* over $\mathcal{H}^{-1}(R^{d+1})$ is by definition a Markov field over $\mathcal{H}^{-1}(R^{d+1})$ and a *representation* T of the Euclidian group on the underlying probability space of ϕ such that, for $f \in \mathcal{H}^{-1}$ and $\eta \in E^d$ the following holds

$$T_\eta\phi(f) = \phi(f \cdot \eta^{-1}) \tag{64}$$

This property is called *Euclidian covariance.*

Any convex bounded domain in R^{d+1} with regular boundary can be seen as the envelope of hyper-planes and therefore we require that the Markov field has the *local Markov property* i.e. for any convex bounded domain $\Omega \subset R^{d+1}$ with regular boundary $\partial\Omega$ the field in the interior $\dot{\Omega}$ and the field in the exterior $\bar{\Omega}^c$ is determined (as a probability space) by the restriction of the field to $\partial\Omega$.

We define therefore a *local Markov field* [3, 7, 9] as follows. If E is a set in R^{d+1} let $O(E)$ be the sigma algebra generated by the $\phi(f)$ with $f \in \mathcal{H}^{-1}(R^d)$ and support (f) $\subset E$.

Let $U \subset R^{d+1}$ with smooth boundary ∂U and denote by U' the complement of U. A *Markov field over* $\mathcal{H}^{-1}(R^{d+1})$ is a random field over $\mathcal{H}^{-1}(R^{d+1})$ with the property that for all open sets $U \subset R^{d+1}$ if u is a positive random variable in $O(U)$ then the following *Markov property* holds

$$E[u|O(U')] = E[u|O(\partial U)] \tag{65}$$

A Markov field is *real* if $\phi(f) = \phi^*(f)$. Notice that if $O \equiv \{x|x_1 = t\}$ (the first coordinate is time) this property corresponds to the Markov property of diffusions.

This condition may be too restrictive; a weaker condition is

$$\forall \epsilon > 0 \quad E[u|O(U')] = E[u|O(\partial U_\epsilon)] \tag{66}$$

where ∂U_ϵ is an ϵ-neighborhood of ∂U.

This allows to consider also derivatives of the random field; it allows to describe the observable *momentum* and it implies that the random fields are in the domain of the Hamiltionian.

9 Quantum Field

We associate a *Euclidian Quantum Field* θ to the Euclidian field ϕ. The quantum field leaves in a Hilbert space $\mathcal{H} = O(R^d) \cap L^2(R^d)$ which is *much smaller* than the Hilbert space $\mathcal{K} \equiv O(R^{d+1}) \cap L^2(\Omega, \mathcal{S}, \mu)$ in which the Euclidian field is defined.

Let E_0 be the orthogonal projection of $\mathcal{K}$ onto $\mathcal{H}$. Define $T(t)u,\ 0 \leq t < \infty$ for $u \in \mathcal{H}$ as

$$T(t)u = E_0 T(\eta(t))u \tag{67}$$

Then

Theorem 9 (Nelson [3, 11, 12]) *Let ϕ be the Euclidian field over $\mathcal{H}^{-1}$ and let $T(t)$ and $\mathcal{H}$ defined as above. There is a* unique *self-adjoint* positive *operator H on $\mathcal{H}$ such that*

$$P_t u = e^{-tH} \qquad 0 \leq t < \infty \tag{68}$$

◇

Proof One has $Y_0 = R^{d-1}$

$$\begin{aligned} P_tP_s &= E\{T_\eta(s)P_tu|O(Y_0)\} \\ &= E\{T_{\eta(s)}[E\{T_{\eta(t)}u|O(Y_0)]|O(Y_0)\} = E\{[E\{T_{s+t}u|Y_s)]|O((Y_s)\}]|O(Y_0)\} \\ &= E\{[E\{T_{s+t}u|Y_s)]|O((Y_s)\}]|O(Z_0)\} = E\{[E\{T_{s+t}u|Y_s)]|O((Z_s)\}]|O(Z_0)\} \\ &= E\{T_\eta(s+t)u\}|O(Z_0) = E\{T_\eta(s+t)u\}|O(Y_0) = P_{s+t}u \end{aligned} \quad (69)$$

The first and second identities are the definition of the operators P_s and P_t, the third is Euclidian covariance. The fourth is Markov property, the fifth is inclusion, the sixth is again Markov property and the last is the definition of the operator P_{t+s}. Let $u \in \mathcal{H}$. Then as $t \to 0$ we have $T(\eta(t))u \to u$ in measure and since $T(\eta(t))$ are unitary, $|T(\eta(t))u| = |u|$. Therefore $P_tu \to u$ as $t \to 0$ and since $\|P_t\| \leq 1$ the family $P(t)$ forms a continuous contraction semigroup on $\mathcal{H}$.

To conclude the proof we show that each P_t is a self-adjoint operator on $\mathcal{H}$. Let ρ the reflection in the hyperplane R^{d-1}. We will prove that T_ρ is the identity on $\mathcal{H}$.

We call this property *reflection property*. Assuming the reflection property we conclude the proof that P_t is self-adjoint. For $u, v \in \mathcal{H}$ one has

$$(v, P_tu) = (v, E_0T(\eta)(t))u) = (v, T(\eta(t))u) = E(\bar{v}, T(\eta t)u) = (P_t, v, u) \quad (70)$$

Notice that $T_{\frac{t}{2}}RT_{\frac{t}{2}} = id$. Since $\eta(t)^{-1}\rho(\frac{t}{2}) = \rho$ (the refection in R^d) the reflection property implies

$$T_{\rho(\frac{t}{2})}\bar{v} = \bar{v} \qquad T_{\rho(\frac{t}{2})}T_{\eta(t)}u = u \quad (71)$$

Using euclidian covariance it follows

$$(v, P_tu) = ET_{\rho(\frac{t}{2})}v, T_{\eta(t)}u) = (T_{\eta(t)}v, u) = (E_0T_{\eta(t)}v, u) = (P_tv, u) \quad (72)$$

We prove now the reflection property i.e. that T_ρ is the identity in $\mathcal{H}$ and f have support in R^d. Since the kernel of the operator $(-\Delta+1)^{-1}$ is positive the potential

$$(-\Delta+1)^{-1}\delta(x) \qquad x \in R^d \quad (73)$$

can be approximated arbitrary well by a positive element in $\mathcal{H}^{-1}(R^d)$. But this positive element is a measure and therefore is invariant under T_ρ. It follows that $T(\rho)$ leaves $\Phi(f)$ fixed and consequently $T(\rho)$ is the identity on $\mathcal{H}$.

We remark that this procedure allows the construction of fields at any time t and the Hamiltonian, generator of the semigroup.

Notice that *the hamiltonian is not a function of the fields at fixed time*. Moreover $e^{-tH} \cup \mathcal{A}_0$ generate the algebra of fields at all times.

It follows also that in a field theory in R^{d+1} in which field at a fixed time cannot be defined (i.e. the distribution $\delta(x_0 - a)f(x)$ $x \in R^d, f \in S$ is not a test function) cannot be obtained from an Euclidian Markov field. This is the case for the models of relativistic field theory that have so far been constructed in space dimension $d \geq 2$. ♡

Let E^{d+1} be the inhomogeneous euclidian group. A representation $T(\xi)$, $\xi \in E^{d+1}$ is a homeomorphism of E^d on the group of measure-preserving of the measure algebra associated to $\{\Omega, S, \mu\}$. An Euclidian (random) field is a Markov field together with a representation of E^d such that for every $f \in S(E^d)$ and ξ one has (covariance)

$$T(\xi)\phi(f) = \phi(f \circ \xi^{-1}) \tag{74}$$

and moreover (reflection positivity)

$$T(\theta)\alpha = \alpha, \qquad \alpha \in \Theta(E^{d-1}) \tag{75}$$

where $\Theta(E^{d-1})$ is reflection with respect to a co-dimension one hyperplane.

We assume moreover

(1) $\forall f \in \mathcal{S}(E^{d+1})\ \phi(f) \in L^p(\Omega,\ \mu)$

(2) the map $S_n(f_1, \dots f_n) \to E(\pi(f_1) \dots \phi(f_n))$ is continuous.

Then one has

Theorem 10 *Let $\phi(f)$ be an euclidian field. The the distributions S_n*

(a) are tempered distributions i.e. $S_n \in \mathcal{S}'(R^d)$ with $S_0 = 1$

(b) are covariant under the Euclidian group $S_n(f) = S_n(f \circ \xi^{-1})$

(c)

$$\sum_{n,m} S_{n+m}(\Theta f_m^* \cdot f_n) \geq 0 \qquad \forall f_n \in S^{0,\infty}(R^{dn}) \tag{76}$$

where

$$S^{0,\infty}(R^{dn}) \equiv f \in \mathcal{S}(R^{dn} \mid f((x_1, \dots x_n) = 0 \ \ unless\ 0 < x_{1,d} < \cdots x_{n.d} < \infty$$

(temporally well ordered)

(d) $S_n(f) = S_n(\Pi(f))$ where Π is a permutation. ◇

We shall not give here the proof of this theorem [5].

10 Euclidian Free Field

We give now an example of euclidian field, the *Euclidian free field* in R^d. Let $m > 0$ and let $\mathcal{H}$ the Hilbert space completion of $\mathcal{S}_R(E^d)$ in the scalar product $(g, (-\Delta + m^2)f)$. Denote by ϕ the real gaussian process on $\mathcal{H}$. When restricted to S this is a random process on $\mathcal{S}(E^d)$.

Theorem 11 *ϕ is an euclidian field that satisfies assumptions 1 and 2 above.* ◇

Proof Let Λ be open in E^d with regular boundary. Define
$\mathcal{U} \equiv \{f \in \mathcal{H},\ suppf \subset \Lambda\}$
$\mathcal{M} \equiv \{f \in \mathcal{H}\ suppf \subset \Lambda^c\}$
$\mathcal{N} \equiv \{f \in \mathcal{H}\ suppf \in \partial\Lambda\}$
$\mathcal{L} \equiv \mathcal{M} \cap \mathcal{M}'$

Let $f \in \mathcal{U}$ and let h the orthogonal projection of f on $\mathcal{M}$ We prove that $h \in \mathcal{N}$. Since Δ is a local operator we have

$$(g, (-\Delta + m^2)h) = (g, (-\Delta + m^2)f) \tag{77}$$

and this implies $(-\Delta + m^2)h = (\Delta + m^2)f$ as distributions on Λ_0^c. It follows that $h = f$ as distributions on Λ_0^c. But $f = 0$ on Λ_0^c. Therefore $supph \subset \Lambda^c - \Lambda_0^c = \partial\Lambda$.

Let $\mathcal{K}$ be a closed subspace of $\mathcal{H}$ and let $\tilde{\mathcal{K}}$ be the sigma-algebra generated by $\phi(f)$, $f \in \mathcal{K}$. If $\mathcal{K}_n \downarrow \mathcal{K}$ it follows that $\tilde{\mathcal{K}}_n \downarrow \tilde{\mathcal{K}}$.

Considering a sequence of open sets Λ_n such that $\Lambda_n \downarrow \Lambda^c$ one derives that $\Theta(\Lambda^c) = \tilde{\mathcal{M}}$. In the same way one establishes

$$\Theta(\partial\Lambda) = \tilde{\mathcal{N}} \qquad \Theta(\Lambda) = \tilde{U} \tag{78}$$

Since $\mathcal{U} \perp \mathcal{L}$ the sigma-algebras $\tilde{\mathcal{U}}$ and $\tilde{\mathcal{L}}$ are independent (as random variables) and $\tilde{\mathcal{M}}$ is the sigma algebra generated by $\tilde{\mathcal{N}}$ and $\tilde{\mathcal{L}}$. It follows that for any function α positive, measurable and integrable in $\tilde{\mathcal{U}}$ one has

$$E(\alpha\ |\tilde{\mathcal{M}}) = E(\alpha\ |\tilde{\mathcal{N}}) \tag{79}$$

Therefore ϕ is a Markov field. To verify Euclidian covariance, notice that defining an action of the euclidian group on $\mathcal{H}$ as $U(\xi)f = f\xi^{-1}$ and then an action on $(\Omega, \mathcal{S}, \mu$ as $\xi \to \Gamma(U(\xi))$ (here Γ is the functor of second quantization) one has

$$T(\xi)\phi(f) = \phi(f \circ \xi^{-1}) \tag{80}$$

The integrability conditions are satisfied because $(\phi(f))^n$ is integrable for every $n \in \mathcal{N}$. It remains to prove that reflection with respect to the hyper-plane E^{d-1} is implemented by unitary operators that leave $\Theta(E^{d-1})$ invariant.

Let as before

$$\mathcal{H}_0 \equiv \{f \in \mathcal{H},\ \ suppf \in E^{d-1}\} \tag{81}$$

If $f \in \mathcal{H}_0$ its Fourier transform $\hat{f}$ is in $L^2(R^{d-1}, \frac{dk}{k^2+m^2})$ and has the form $\hat{f}(k) = f_0(\mathbf{k})$ $\mathbf{k} = k_1, \ldots k_{d-1}$. It follows $T(\Theta)\alpha = \alpha$ for $\alpha \in \Theta(E^{d-1})$. ♡

The Markov field we have described corresponds to the solutions of the Klein–Gordon equation for a scalar particle of mass m. Indeed the evolution of this field is characterized in Fourier transform by

$$\phi(t,p) = \frac{1}{\sqrt{2}}[e^{i(p^2+m^2)t}a(p) + e^{-i(p^2+m^2)t}a^*(p)] \tag{82}$$

One has therefore

$$\int \phi(t,p)\phi(0,p)dp = \int |\phi(0,p)|^2 e^{i(p^2+m^2)t}dp \tag{83}$$

Following the unitary evolution up to time T and conditioning to time zero one obtains a semigroup which corresponds to a gaussian stochastic process with covariance $\frac{1}{k^2+m^2}$.

11 Connection with a Local Field in Minkowski Space

There is a connection [5, 7, 12] between a Markov field in E^d which satisfies euclidian covariance and reflection positivity and a local field in Minkowski space–time $R^{d-1} \times R$ with positive energy. The connection is by analytic continuation of the correlation functions (marginals) through a wedge in the product of the complexified euclidian space and the complex euclidian group.

This wedge has as edges on one side the product of E^d and the euclidian group and on the other side the product of Minkowski space and the Lorenz group. The euclidian correlation function of the Markov field are defined on the *Euclidian edge*.

Using covariance, reflection positivity and regularity they can be continued through the wedge and their image on the *Minkowski edge* are the *Wightman functions* Wightman functions of a relativistic field with energy-momentum spectrum contained in the forward light cone.

Positivity of the energy, space-like commutativity and analyticity of the representation of the Poincaré group are sufficient to prove that this continuation is reversible (edge-of-the-wedge Theorem) [5, 7, 8]. Notice that the wedge has not in general a simple structure [5] and therefore the continuation *is not simply a rotation in the complex plane*.

Things simplify in the case of the Euclidian free field. Since the underlying process is gaussian the covariance (two-point function) determines completely the process.

It is therefore sufficient to prove the strong Markov property for the covariance. The claim now follows form the explicit form $E(\phi, C\phi) = (\phi, \frac{1}{-\Delta+m}\phi)$.

Notice that in this case the connection with the Minkowski free field (of mass m) is particularly simple since the function has an analytic extension to the full Minkowsky plane and therefore the continuation is made simply a linear transformation ($t \to it$).

We stress that this is not the case when an euclidian invariant interaction is introduced. In fact to the present time in a relativistic theory only two types of interactions have been described that have a Markov counterpart, the positive polynomial one and the quadratic negative exponential one [5, 11–13].

In both case the Markov process has the strong local Markov property (in particular the fields at fixed time exists). Here we do not discuss this point.

12 Modifications of the O.U. Process. Modification of Euclidian Fields

For the O.U. measure one has

$$\int e^{iq(f)} d\phi_0 = e^{-\frac{1}{2}(f,f)_1} \tag{84}$$

(the full Gaussian property) where

$$(f,f)_1 = \int |\hat{f}|^2(p) \frac{1}{|p|^2 + m} dp \tag{85}$$

m is a positive parameter.

Also in this infinite dimensional setting one may try to modify the O.U. process adding to the Hamiltonian a "potential" (a function of the fields) and to obtain a corresponding "Feynman–Kac formula". This can be done as in the finite-dimensional case, by adding to the measure a *multiplicative functional* that plays the role of e^{-tV}.

This procedure present difficulties if one insists that the functional be a *local* function of the Markov field since the *points* of the measure space are distributions and it is in general not possible to take their point-wise product (the singular sets may overlap). Remark that the terminology: $\Phi(x)$ *is the field at the point* x is descriptive but incorrect.

Only in two space–time dimensions the product of fields at the same point can be reasonably defined an even then after an accurate procedure (renormalization).

So far this attempt has had success for polynomial and exponential interactions and only in the case of space dimension one. In the lectures we shall not describe this theory.

Definition 6 A real random variable is *additive* with respect to the euclidian field Θ if for any open covering Λ_i of E^d there exist random variables $\alpha_i \in \Theta(\Lambda_i)$ such that $\alpha = \sum_i \alpha_i$ A random variable is *multiplicative* with respect to Θ if for every open covering there exist random variables β_i such that $\beta = \Pi_i \beta_i$. The random variable α is additive if and only if $\beta \equiv e^{\alpha}$.

Theorem 12 *Let ϕ be a Markov field on $S(E^d)$ with probability space $\{\Omega, S, \mu\}$. Let β a multiplicative random variable with expectation one. Then ϕ is also a Markov field on $S(E^d)$ with probability space $\{\omega, S, \beta d\mu\}$.* ◇

Proof We prove the Markov property with respect to the new probability space. Remark that if if $\mathcal{A}$ and $\mathcal{B}_n$ are complete measure sigma algebras on a probability space, with $\mathcal{B}_n$ monotonically decreasing, the following relation holds

$$\cap_n (\mathcal{A} \cup \mathcal{B}_n = \mathcal{A} \cup (\cap_n \mathcal{B}_n) \tag{86}$$

We notice also that if ϕ is a Markov field on $S(E^d)$, Λ is an open subset of E^d ad Λ' is a closed subset of Λ^c which contains $\partial\Lambda$ then

$$E(\Theta(\Lambda \cup \Lambda') \mid \Lambda') = \Theta(\Lambda') \tag{87}$$

Define now a new measure β defining for every μ-measurable function α

$$E_\beta(\alpha) = E_\mu(\beta\alpha) \tag{88}$$

Let Λ be open in E^d and let $\alpha \in \Theta(\Lambda)$ be positive and μ-measurable. We must prove

$$E_\beta\{\alpha \mid \Theta(\Lambda^c\} = E_\alpha\{\alpha \mid \Theta(\partial\Lambda)\} \tag{89}$$

Notice that by the Radon–Nikodym theorem there is a unique random variable $\tilde{\alpha}$ in $\Theta(\Lambda^c)$ such that

$$E_\beta(\gamma\alpha) = E_\beta(\gamma\tilde{\alpha}) \tag{90}$$

There exist random variables $\beta_1 \in O(\Lambda)$ $\beta_2 \in O(\Lambda_0)$, $\beta_3 \in O(\Lambda_0^c)$ such that

$$E(\alpha\gamma\beta_1\beta_2\beta_3) = E(\tilde{\alpha}\gamma\beta_1\beta_2\beta_3) \tag{91}$$

of $\gamma \in O(\Lambda_0^c)$. Since γ is arbitrary

$$E(\alpha\beta_1\beta_2 \mid \Theta(\Lambda^c)) = \tilde{\alpha}E(\beta_1\beta_2 \mid \Theta(\lambda^c)) \tag{92}$$

One has $\Lambda' = \Lambda_0 \cap \Lambda^c$ and therefore $\tilde{\alpha} \in \Theta(\bar{\Lambda}_0)$. Since Λ_0 is an arbitrary open set which contains $\partial\Lambda$ we conclude $\tilde{\alpha} \in O(\partial\Lambda)$.

References

1. B. Gnedenko, *The Theory of Probability* (Mir publishers, Moscow, 1976)
2. Y. Berezansky, Y. Kondriatev, *Spectral Methods in Infinite-Dimensional Analysis* (Kluwer, Dordrecht, 1995)
3. E. Nelson, J. Funct. Anal. **11**, 211–219 (1972)
4. H.H. Kuo, *White Noise Distribtion Theory* (CRC press, New York, 1996)
5. J. Glimm, A. Jaffe, J. Math. Phys. **13**, 1568–1584 (1972)
6. K. Osterwlader, R. Schrader, Comm Mat. Phys. **31**, 83–112; Commun. Math. Phys. **42**(1075), 281–305 (1973)
7. K. Symanzik, Euclidean quantum field theory, in *Local Quantum Theory*, ed. by R. Jost (Academic Press, New York, 1969)
8. A. Wighman, L. Garding, Ark. Physik **28**, 129–194 (1965)
9. E. Nelson, J. Funct. Anal. **12**, 211–227 (1973)
10. A. Klein, J. Funct. Anal. **27**, 277–291 (1978)
11. J. Froelich, Ann. Inst. H. Poincaré **21**, 271–317 (1974)
12. B. Simon, Helv. Phys. Acta **46**, 686–696 (1973)
13. S. Albeverio, R. Höegh-Krohn, Commun. Math. Phys. **68**, 95–128 (1979)
14. J. Doob, *Probability Theory* (Wiley, New York, 1953)

Lecture 9: Modular Operator. Tomita–Takesaki Theory Non-commutative Integration

We review in this Lecture some basic elements of the *modular theory* and its connections with the theory of Tomita–Takesaki which we treated very briefly in Volume I of these Lecture Notes.

There the theory was discussed in the context of the theory of C^* algebras and their one-parameter groups of automorphisms.

In this Lecture we take a slightly different approach, which has some connection with Friedrichs extension of a symmetric positive form on a separable Hilbert space and in general with transforming a Hermitian matrix to diagonal form.

Recall that a complex Hilbert space $\mathcal{H}$ with an involution J is said to be in *standard form* if

$$\mathcal{H}_{real} \cap \mathcal{H}_{Im} = 0 \qquad \mathcal{H}_{real} \cup \mathcal{H}_{Im} = \mathcal{H}$$

An example of a Hilbert space that is not in standard form is the domain (with the graph norm) of a symmetric positive operator A which is not essentially self-adjoint. In this case the missing space is the deficiency space of A. The domain of the Friedrichs extension is in standard form . In this context, the Tomita–Takesaki theorem says that the Hilbert space generated by a von Neumann factor with a cyclic and separating vector Ω is in standard form.

We shall give a presentation of this theory [4, 5] which takes advantage from this point of view. Notice that proving that a closed positive quadratic form is associated to a self-adjoint operator (and therefore to its spectral decomposition) is the infinite dimensional analogue of finding a base in which the matrix which represents the quadratic form is diagonal.

We recall that modular theory, and the corresponding theory of the modular operator, has deep connections with the K.M.S. condition (at finite temperature). It plays a major role in Quantum Statistical Mechanics and in relativistic (algebraic) Quantum Field Theory. It has also relevance in the theory of non-cummutative integration.

We will recall later the basic facts about the K.M.S. condition. We shall also give some elements of an extension to the non-commutative setting of the classic Radon–Nikodym theorem about equivalence of measures.

G. Dell'Antonio, *Lectures on the Mathematics of Quantum Mechanics II: Selected Topics*, Atlantis Studies in Mathematical Physics: Theory and Applications 2, DOI 10.2991/978-94-6239-115-4_9

In Quantum Mechanics and in Algebraic Field Theory the role of positive (normalized) measures is taken by the states of a von Neumann algebra (or of a C^* algebra) and the problem will be the equivalence of the representations associated to the states via the G.N.S. construction.

We have seen in Volume I that in a representation of on a von Neumann algebra which satisfies the K.M.S condition with respect to a one-parameter group of automorphisms σ_t there exists t_0 for which $\phi(a\alpha_{t_0}(b)) = \phi(ba)$ for a dense set $\mathcal{M}$. If this relation holds with $t = 0$, one has a tracial state invariant for the dual action of the automorphism group. In this case the cone of positive states corresponds to the cone of positive measures in the commutative case.

In this lecture we shall also mention briefly the theory of *dual cones* which is strictly connected to the Tomita–Takesaki theory but has an independent interest since it is an extension to the non-commutative setting of the classic Radon–Nikodym theorem about equivalence of measures.

If for a von Neumann algebra $\mathcal{M}$ which is a factor ($\mathcal{M} \cap \mathcal{M}' \equiv \{cI\}$) admits a tracial state (a normal state σ such that $\sigma(ab) = \sigma\phi(ba)$) then there exists a *natural* isomorphism between $\mathcal{M}$ and $\mathcal{M}'$ that can be used to set up a non-commutative integration theory.

For a von Neumann factor that admits a trace one can construct a non-commutative version of the classical integration theory for spaces of finite measure. The foundation of this theory was given by I. Segal [1] and E. Nelson [3] with relevant contribution by D. Gross [2].

1 The Trace. Regular Measure (Gage) Spaces

In this non-commutative case we define *non-commutative space with finite regular gage* a triple $\{\mathcal{H}, \mathcal{A}, \mu\}$ where $\mathcal{H}$ is a complex Hilbert space, $\mathcal{A}$ is a von Neumann algebra and μ a non-negative function defined on the projections of $\mathcal{A}$ and such that
(i) μ is completely additive: if $\mathcal{S}$ is a collection of mutually orthogonal projection in $\mathcal{A}$ with upper bound P, then $\mu(P) = \sum_{Q\in\mathcal{S}} \mu(Q)$.
(ii) μ is invariant under unitary transformations
(iii) μ is finite ($\mu(I) < \infty$)
(iv) μ is regular (if P is not zero $\mu(P)$ is strictly positive).

Under these assumption one can extend linearly μ to the entire $\mathcal{M}$ as a norm-continuous function. The function so extended is called *trace*; we shall use the symbol $Tr(A),\ A \in \mathcal{A}$.

If $A \in \mathcal{A}$ as operator on $\mathcal{H}$ has spectral decomposition $A = \int \lambda dE(\lambda)$ then $Tr(A) = \int \lambda d\mu(\lambda)$. If $A \geq 0$ then $Tr(A) \geq 0$. The trace is *central* if ($Tr(AB) = Tr(BA)$).

If $A \in \mathcal{A}$ is a closed operator we define $|A| = (A^*A)^{\frac{1}{2}}$. For $1 \leq p < \infty$ define $\|A\|_p = (Tr(|A|^p)^{\frac{1}{p}}$ and $\|A\|_\infty = |A|$. With this definition $\|A\|_p$ is a norm for each p in $[1, \infty]$.

We denote by $L^p(\mathcal{A})$ the completion of $\mathcal{A}$ in $\mathcal{B}(\mathcal{H})$ in the $\|.\|_p$ norm. It is easy to see that $L^\infty(\mathcal{A}) \equiv \mathcal{A}$ as normed spaces. In [1] (see also [2, 3]) one proves that for $1 \leq p \leq \infty$ one can identify $L^p(\mathcal{A})$ with a suitable space of bounded operators on $\mathcal{H}$. In particular one can identify a positive element in $L^p(\mathcal{A})$ with a self-adjoint operator.

Remark that if $\mathcal{A}$ is a type I factor, in particular if $\mathcal{A} = \mathcal{B}(\mathcal{H})$, the space $L^1(\mathcal{A})$ is the space of trace-class operators and $L^p(\mathcal{A})$ is the space of operators of Schatten class p. With this notation the *L^p non-commutative theory* is developed in complete parallelism with that of Lebesgue integration spaces.

If $A \in L^1(\mathcal{A})$ the function $Tr(A)$ defines a linear continuous functional on $L^1(\mathcal{A})$ and Hölder's inequalities hold as well as interpolation formulas. In particular if $\{\mathcal{H}, \mathcal{A}, m\}$ is a non-commutative measure space and $a,\ b \in \mathcal{A}$ one has

$$\|ab\|_p \leq \|a\|_\infty \|b\|_p \qquad \|ba\|_p \leq \|a\|_\infty \|b\|_p \tag{1}$$

It follows that right and left multiplication by $a \in \mathcal{A}$ extends to a bounded operator on $L^p(\mathcal{A})$. We shall denote R_a and L_a these operators. By construction R_b and L_a commute for any choice of $a,\ b$.

The relevance for Physics of the regular measure spaces is due to the fact that for these space one has theorems similar to the theorems of Frobenius for matrices which give existence and uniqueness of the lowest eigenvalue of a positive matrix. They are also relevant to establish a theory of non-commutative Markov processes.

A bounded operator A on $L^2(\mathcal{A})$ is said to *preserve positivity* if AB is a non-negative element of $L^2(\mathcal{A})$ when B is non-negative. Let $\{\mathcal{H}, \mathcal{A}, \mu\}$ be a non commutative space with regular gage and let π be a projection in $\mathcal{A}$. We shall call *Pierce subspace* associated to π the range of $P_\pi \equiv L_\pi R_\pi$ as operator on $L^2(\mathcal{A})$.

The role of the support of a function is now taken by the Pierce subpaces.

We shall not give here a treatment of the general aspects of this non commutative integration theory. We only notice that it has an important role in the theory of fields of spins and of fermions on a lattice and elements of this application are given in Lecture 16.

We quote an important theorem [1–3].

Theorem 1 *Let $\mathcal{H}, \mathcal{A}, \mu$ be a space with a finite regular gage and let A be an Hermitian bounded operator on $L^2(\mathcal{A})$ which is positivity preserving. If $\|A\|$ is an eigenvalue of A and if A does not leave invariant any proper Pierce subspace, then the eigenvalue $\|A\|$ has multiplicity one and one can choose the associated eigenvalue to be non-negative and cyclic for $\mathcal{A}$.* ◇

The theory of Takesaki–Tomita extends this non-commutative integration theory to normal states which do not define a trace but satisfy, for some value $t_0 \neq 0$ of the parameter, the K.M.S. condition relative to a modular group of automorphisms associated to the state. In a sense this represents the non-commutative version of the integration theory in a compact $\Omega \subset R^d$ with respect to a finite measure which is absolutely continuous with respect to Lebesgue measure.

An important feature of the Tomita–Takesaki theory is that it connects an analytic property (to be analytic in a strip with a suitable relation of the values at the boundary) with a one-parameter group of automorphisms that leave invariant the algebra of observables. The group of automorphisms may be the group of time-translations, the sub-group of *boosts* in the Lorenz group, ...

We now recall some basic elements about the K.M.S. condition.

2 Brief Review of the K-M-S. Condition

As we saw in Volume one of these Lecture Notes, the K.M.S. condition is a generalization of the Gibbs condition for the equilibrium of a system in Classical Statistical Mechanics. In this theory a state of a classical Hamiltonian system with Hamiltonian $H \geq 0$ at temperature T is represented by a Liouville distribution in phase space that can be written modulo normalization as $e^{-\frac{H}{T}}$ so that at temperature 0 the system in phase space is localized on the minima of H.

The same is assumed to be true in Quantum Statistical Mechanics but the observables are operators and the integration over phase space is substituted with a non-commutative integration given by taking the trace. The expectation value of the observable A at equilibrium at temperature T is now

$$Tr\left(Ae^{-\frac{H}{T}}\right) \tag{2}$$

If one considers the evolution of the correlations under the Hamiltonian H one must study the function

$$\Phi_{A,B}(t) \equiv Tr\left(AB(t)e^{-\frac{H}{T}}\right) = Tr\left(Ae^{itH}Be^{-itH}e^{-\frac{H}{T}}\right)$$

Since the operators A and B in general do not commute with H this expression is not invariant under interchange of A with B. But the right-hand side can be written as

$$Tr\left(Ae^{itH}Be^{-i[t+i\frac{1}{T}]H}\right) \tag{3}$$

Since H is positive, the function $\Phi_{A,B}(t)$ can be continued for $T > 0$ as analytic function in the strip $0 < Imt < \frac{1}{T}$ continuous at the boundary. The same is true for the function $\Phi_{B,A}(t)$.

One verifies easily, by the cyclic property of the trace, that the analytic function $\Phi_{A,B}(z)$, $0 < Imz < \frac{1}{T}$ can be continued as a continuous function to the boundary of the strip and satisfies at the boundary

$$\Phi_{A.B}(x + i0) = \Phi_{B,A}\left(x + i\frac{1}{T}\right)$$

In the case of an infinite-dimensional thermodynamic system the operation *trace* is not defined in phase space, but in an algebraic formulation one may have a function with the property of the trace. Therefore it is natural to state the condition of being in equilibrium at temperature T as the condition that for all element $A,\ B$, the function $Tr_T(AB(t))$ has the same property as in the finite-dimensional case.

This was the proposal of the physicists Kubo, Martin and Schwinger and since than this condition is known under the acronym K-M-S.

Given a dynamical system $\{\mathcal{A}, \alpha_t\}$ one says that the state ρ_β *satisfies the K.M.S. condition* for the group α_t at the value β of the parameter $(0 < \beta < \infty)$ (in short, ϕ *is a* β*-K.M.S. state)* if for every $x \in \mathcal{A}$ and every $y \in \mathcal{A}$ the following holds

$$\rho_\beta(y\ \alpha_{\xi+i\beta}(x) = \rho_\beta(\alpha_\xi(x)\ y), \qquad \xi \in R \tag{4}$$

We extend this definition to cover also the cases $\beta = 0$ and $\beta = \infty$. We will say that ρ_0 satisfies the K.M.S. condition for the group α_t at $\beta = 0$ if

$$\rho_0(y\ \alpha_\zeta(x)) = \rho_0(\alpha_\zeta(x)\ y) \quad \forall x \in \mathcal{A}\ \ y \in \mathcal{A} \tag{5}$$

We will say that ρ_∞ satisfies the K.M.S. condition for the group α_t at infinity if for any $x \in \mathcal{A}^a$ and every $y \in \mathcal{A}$ the analytic function $f(\zeta) \equiv \rho_\infty(y\ \alpha_\zeta(x))$ satisfies

$$|f(\zeta)| \leq \|x\|\ \|y\| \quad if \quad Im\zeta \geq 0. \tag{6}$$

In this case the state ρ_∞ is said to be *ground state* relative to the automorphisms group α_t. The origin of this name is clear from the finite -dimensional case (in that case it is the state with minimum energy) and for the general case it will be clearer later. An important result is the following, that we have described in Volume I of these Lecture Notes.

Let $\mathcal{A}$, α_t be a C^* dynamical system and let $0 \leq \beta \leq \infty$. The following conditions on a state ρ are equivalent

(1) ρ is β-K.M.S. state

(2) ρ satisfies the α_t K.M.S. condition for a dense set of elements $x \in \mathcal{A}^a$.

(3) For any pair $x,\ y\ \in \mathcal{A}$ there exists a function $f_\rho(\zeta)$ bounded continuous in the strip

$$\Omega_\beta \equiv \{\zeta \in C,\ 0 \leq Im\zeta \leq \beta\} \tag{7}$$

holomorphic in the interior of Ω and satisfying the boundary conditions

$$f_\rho(t) = \rho(y\ \alpha_t(x)), \quad f_\rho(t + i\beta) = \rho(\alpha_t(x)\ y) \tag{8}$$

If $\beta = \infty$ the last condition takes the form

$$f_\rho(t) = \rho(y\ \alpha_t\ (x)) \quad t \in R \quad \|f_\rho\| \leq \|x\|\ \|y\| \tag{9}$$

For the proof we refer to Vol. I Lecture 9.

3 The Tomita–Takesaki Theory

We recall now briefly the main points of this theory. If the von Neumann algebra $\mathcal{M}$ on a Hilbert $\mathcal{H}$ has a cyclic and separating vector Ω, one can associate to this vector a positive operator Δ (called *modular operator*) and an anti-linear isometry j such that

$$j\Omega = \Omega, \quad j\Delta^{\frac{1}{2}} a\ \Omega = a^*\ \Omega \quad \mathcal{M}\Omega \subset D\left(\Delta^{\frac{1}{2}}\right) \tag{10}$$

$$j\mathcal{M}j = \mathcal{M}' \quad \Delta^{it}\mathcal{M}\Delta^{-it} = \mathcal{M} \qquad \forall t \tag{11}$$

The *modular group* associated to the state Ω is the group of inner automorphisms with generator $log\Delta$. The state satisfies the K.M.S. condition with respect to this group. The case $\Delta = I$ corresponds to a tracial state and in this case the existence of the anti-linear isometry j follows from $(\Omega, a^*b\Omega) = (\Omega, ba^*\Omega))$.

We begin with some preliminary result and the connection to the Friedrichs extension of symmetric strictly positive operators.

Remark that the Friedrichs extension can be interpreted in the following way: given a closed strictly positive quadratic form q in a complex Hilbert space $\mathcal{H}$ consider the subspace X for which $q(\phi, \psi)$ takes real values for every $\phi, \psi \in X$. It is a *real* vector space closed in the topology induced by the quadratic form.

On the other hand, every positive self-adjoint operator A determines a real subspace Y, closed in the graph topology of A, which has the property that for any pair of vectors $\phi, \psi \in D(A)$ the number $(\phi, A\psi)$ is real. This defines Y as a *real subspace*. The construction of the Friedrichs extension (an operator) can be interpreted as construction of Y starting with the subspace X i.e. as a natural closed map $X \to Y$.

If $\mathcal{H}$ is finite dimensional, therefore isomorphic to $C^n \equiv R^n \oplus R^n$ the operator A is represented by a strictly positive matrix with eigenvalues $\lambda_1, \ldots \lambda_n$. In this case it is possible to transform A into diagonal form (in fact into the identity) in the real Hilbert space $R^n \oplus R^n$ by means of *complex linear transformation* in C^n consisting in a rotation followed by a dilation by a factor $\sqrt{\lambda_k}$ in the direction of the eigenvectors. For compact operators there is a similar procedure but the proof of closure is more demanding.

The construction of the Friedrichs extension can be seen as an extension of this construction to the case of quadratic forms which correspond to operators with (partly) continuous spectrum. This clarifies the important role of the following structure.

Let $\mathcal{H}$ be a complex Hilbert space, and define

$$< \psi, \phi >= Re(\psi, \phi) \quad \phi, \ \psi \in \mathcal{H} \tag{12}$$

With this definition $\mathcal{H}$ acquires the structure of a real Hilbert space, that we denote by $\mathcal{H}_r$, with scalar product

$$(\psi, \phi) =< \psi, \phi > +i < i\psi, \phi > \tag{13}$$

(in our notations, (ψ, ϕ) is *linear* in ϕ and *anti-linear* in ψ).

Let us assume that there exists a closed *real* subspace $\mathcal{K} \in \mathcal{H}$ with the following properties
(a)

$$\mathcal{K} \cap i\mathcal{K}^{\perp} = \emptyset, \quad (\mathcal{K} + i\mathcal{K})^{\perp} = \emptyset \tag{14}$$

If this is the case,we say that the space $\mathcal{K}$ is *in standard form*. A large part of the Tomita–Takesaki theory is related to the fact that if a representation of a von Neumann algebra $\mathcal{A}$ has a cyclic and separating vector Ω, then $\mathcal{A}_r\Omega$ and $\mathcal{A}'\Omega$ compose a standard form [4, 5].

The following construction defines uniquely a self-adjoint operator Δ (called *modular operator* associated to the subspace $\mathcal{K}$) and an anti-linear isometry j. If $\mathcal{K}$ is the real subspace X associated to a strictly positive quadratic form q the operator we obtain is the Friedrichs extension of q.

The case of interest for us is that in which $\mathcal{K}$ is generated by the self-adjoint elements of a von Neumann $\mathcal{M}$ acting on a cyclic and separating vector Ω. In this case we will prove that the modular operator has the properties indicated above. Notice that in this case $\mathcal{K}$ is generated by the convex cone which is obtained by applying to Ω the *positive* elements of $\mathcal{M}$. This will lead to the Tomita–Takesaki duality theory and to the equivalence theory for representations of C^* algebras.

We shall use later the following result

Lemma 1 *Let ρ be a state of a von Neumann algebra $\mathcal{M}$ and let τ be a linear positive functional on $\mathcal{M}$ satisfying $\tau \leq \rho$. There exist $h \in \mathcal{M}_1^+$ and λ, $Re\lambda \geq \frac{1}{2}$such that for any $a \in \mathcal{M}$*

$$\tau(a) = \lambda \ \rho(h \ a) + \bar{\lambda} \ \rho(a \ h) \tag{15}$$

If the representation induced by ρ is irreducible, there is a unique operator h with this property. ◇

Proof We could reduce ourselves to the case $\mathcal{M} \subset \mathcal{B}(\mathcal{H})$, and ρ defined by a projection operator π_ϕ, $|\phi| = 1$, and $a = b^*b$. In this case $\rho(a) = Tr(\pi_\phi \ a) = (b\phi, b\phi)$, and $\tau(b^*b) = Tr(\sigma b^*b)$ for a suitable density matrix σ. Lemma 1 follows then the from elementary inequalities.

A more algebraic proof is as follows. Let Ξ the convex compact set in $(\mathcal{M}^*)_{s.a}$ defined by

$$\tau \in \Xi_\rho \Leftrightarrow \exists \lambda, \ \ Re\lambda \geq \frac{1}{2}, \ \ \exists h \in \mathcal{M}_+^1 \ : \ \tau(a) = \lambda\rho(ha) + \bar{\lambda}\rho(ah) \ \ \forall a \in \mathcal{M}. \tag{16}$$

We must prove that if $0 \leq \tau \leq \rho$ then $\tau \in \Xi_\rho$. Suppose this is false. By the Hahn-Banach separation theorem there exist $a \in \mathcal{M}_{s.a.}$ and $t \in R^+$ such that $\tau(a) > t$, $\rho(a) \leq t$. Set $a = a_+ - a_-$, $h = [a_+]$ (the projection on the support of a_+). Then

$$\tau(a_+) \geq [\tau(a_+) - \tau(a_-)] > t \geq 2Re\lambda\tau(a_+) \geq \tau(a_+) \tag{17}$$

a contradiction. ♡

Corollary *If τ is faithful and*

$$\tau(a) = \lambda\rho(k\ a) + \bar{\lambda}\rho(a\ k) \quad k \in \mathcal{M}_{s.a.} \tag{18}$$

then $k = [a_+]$. ◇

Proof It is easy to verify that (18) holds for $[a_+]$. Suppose it be true for h. One has

$$(\lambda+\bar{\lambda})((h-[a_+])^2) = \lambda h(h-[a_+])+\bar{\lambda}(h-[a_+])h-\lambda[a_+]((h-[a_+])-\bar{\lambda}(h-[a_+])[a_+] \tag{19}$$

If (19) holds for h, then

$$2Re\lambda\rho((h-[a_+])^2) = \tau(h-[a_+]) - \tau(h-[a_+]) = 0 \tag{20}$$

Therefore $h = [a_+]$. ♡

Let now $\mathcal{M}$ be a von Neumann algebra on $\mathcal{H}$ with a cyclic and separating vector Ω. It is easy to verify that Ω is cyclic and separating also for $\mathcal{M}'$. Remark that if ρ is a normal faithful state the representation associated to ρ through the G.N.S. construction provides an isomorphism and therefore we can identify $\mathcal{M}$ with $\Pi_\rho(\mathcal{M})$. Let $\mathcal{K}$ be the closure of $\mathcal{M}_{s.a.}\ \Omega$. Define

$$< \phi, \psi >= Re(\phi, \psi) \quad \phi, \ \psi \in \mathcal{H} \quad (\phi, \psi) =< \phi, \psi > +i < i\phi, \psi > \tag{21}$$

With the scalar product $< .,. >$ the space $\mathcal{H}$ becomes a real Hilbert space $\mathcal{H}_r$ and $\mathcal{K}$ can be regarded a subspace of $\mathcal{H}_r$.

Proposition 1 *$\mathcal{K}$ is in standard form (i.e. has the properties in (14)).* ◇

Proof Property (a) follows from the fact that Ω is cyclic. To prove (b) notice that $\mathcal{M}'_{s.a.}\Omega$ is orthogonal in $\mathcal{H}_r$ to $\mathcal{M}_{s.a.}\Omega$. Indeed if $a' \in \mathcal{M}'_{s.a.}\Omega$ and $a \in \mathcal{M}_{s.a.}$ one has

$$(a'\Omega, ia\Omega) = -(ia\Omega, a'\Omega) \quad \Rightarrow < a'\Omega, ia\Omega >= 0 \tag{22}$$

It follows $\mathcal{M}'_{s.a.}\Omega \subset (i\mathcal{K})^{\perp}$; similarly $i\mathcal{M}'_{s.a.}\Omega \subset (\mathcal{K})^{\perp}$. Therefore

$$\mathcal{M}'\Omega \subset (\mathcal{K} \cap i\mathcal{K})^{\perp} = \mathcal{K}^{\perp} + (i\mathcal{K}))^{\perp} \tag{23}$$

and from the density of $\mathcal{M}'\Omega$ it follows $\mathcal{K} \cap i\mathcal{K} = \emptyset$. ♡

Before giving the general construction of the modular operator associated to a subspace $\mathcal{K}$ in standard form, we give the proof [4] of a property that will be useful in what follows.

Proposition 2 *Let A_0 be a closed symmetric operator with domain dense in $\mathcal{H}$ and suppose $(x, A_0x) \geq 0$ for every $x \in D(A_0)$. Let A_0 be affiliated to a von Neumann algebra $\mathcal{M} \in B(\mathcal{H})$ in the sense that for every $b \in \mathcal{M}'$ and for every $x \in D(A_0)$*

$$(bx, A_0x) = (A_0x, bx) \tag{24}$$

holds (notice that A_0 is only symmetric and we have not a spectral representation).

Let A the Friedrichs extension of A_0. Then A is affiliated to $\mathcal{M}$. ◇

Proof The statement is trivial if $\mathcal{M}'$ consists only of multiples of the identity. Let V be unitary in $\mathcal{M}'$. Then VAV^* is a positive extension of VA_0V^*. Denoting by D' the closure of $D(A_0)$ with respect to the scalar product defined by

$$< u, v >\equiv ((A_0 + I)u, v)) \tag{25}$$

From the construction of the Friedrichs extension the identity map on $D(A_0)$ has a unique self-adjoint extension ι which satisfies $D(VAV^*) \subset V.\iota(D')$. Since A_0 is affiliated to $\mathcal{M}$ one has $VA_0V^* = A_0$. It remains to prove that $V(\iota(D') \subset \iota(D')$.

Let $z \in \iota(D')$ such that $\iota(z') = z$ with $z' \in D'$. There exists a sequence $\{x_n\} \in D(A_0)$ which converges to z'. Since A_0 is affiliated to $\mathcal{M}$ one has $Vx_n \in D(A_0)$. The sequence $\{x_n\}$ converges in D' and therefore

$$\begin{aligned} lim_{n,m\to\infty} \| V\, x_n - V\, x_m \|^2 &= lim_{n,m\to\infty}((A_0 + I)V(x_n - x_m), V(x_n - x_m)) \\ &= lim_{n,m\to\infty} < x_n - x_m, x_n - x_m >= 0 \end{aligned} \tag{26}$$

It follows that $\{Vx_n\}$ converges in D' to an element u' and $\{Vx_n\}$ converges in $\mathcal{H}$ to $\iota(u')$. Since $\{x_n\}$ converges to z in $\mathcal{H}$ one has that $\{Vx_n\}$ converges to Vz. Therefore $Vz = \iota(u') \in \iota(D')$ and $V\iota(D') \subset \iota(D')$. ♡

4 Modular Structure, Modular Operator, Modular Group

Given $\mathcal{K}$ in standard form we construct now an invertible anti-isometry j and a self-adjoint operator Δ (*modular operator*). In the case in which $\mathcal{K}$ is constructed from a von Neumann algebra that has a cyclic and separating vector we shall see that the isometry intertwines the algebra and its commutant (which are therefore equivalent) and the modular operator is the generator of a group of inner automorphisms which satisfies the K.M.S. condition for $\beta = 1$.

The construction we give shows that the modular operator is defined, *independently from the theory of von Neumann algebras,* starting from a real subspace of a complex Hilbert space with a procedure which is similar to the one followed in the construction of the Friedrichs extension of a closed positive quadratic form. Assume that the subspace $\mathcal{K}$ of the real Hilbert space $\mathcal{H}_r$ satisfies condition (14). Let P and Q the orthogonal projectors of $\mathcal{H}_r$ on $\mathcal{K}$ and $i\mathcal{K}$. Define

$$A = P + Q, \quad jB = P - Q \tag{27}$$

where jB is the polar decomposition of $P - Q$ in $\mathcal{H}_r$.

Proposition 3 ([4]) *The operators A, B, P, Q, j satisfy*
(i) A and B are linear complex and $0 \leq A \leq 2I \quad 0 \leq B \leq 2I$
(ii) A , (2I - A) e B are injective and $B = \sqrt{A(2I - A)}$
(iii) j is an anti-linear isometry, $j^2 = I$. *If* $\phi, \psi \in \mathcal{H}$ *then* $(j\phi, \psi) = (\phi, \bar{j}\psi)$
(iv) B commutes with A,P,Q , j
(v) $jP = (I - Q)j, \quad jQ = (I - P)j \quad jA = (2I - A)j$ ◇

Proof (i) It is easy to show that $iP = Qi$. It follows that $a \equiv P + Q$ is linear over the complex field and $B \equiv P - Q$ is anti-linear. From $B^2 = (P - Q)^2$ one derives that B^2 and therefore also B is linear. Therefore j is anti-linear. The operators A and B are positive in $\mathcal{H}_r$ and from (27) it follows that they are self-adjoint and positive also in $\mathcal{H}$. The bounds $\|A\| \leq 2$, $\|B\| \leq 2$ are obvious from the definition.
(ii) If $A\phi = 0$ one has

$$\|P\phi\|^2 + \|Q\phi\|^2 =< P\phi, \phi > + < Q\phi, \phi >=< A\phi, \phi >= 0 \tag{28}$$

It follows $\phi \in K^{\perp} \cap (iK)^{\perp}$ and therefore $\phi = 0$. This proves that A is injective. Similarly one shows, analyzing $I - P$ e $I - Q$, that $2I - A$ is injective since P e Q are idempotents and $B^2 = A(A - 2I)$.
(iii) j is self-adjoint in $\mathcal{H}_r$ and it is an injective isometry since B is injective. Therefore $j^2 = I$. One has

$$(j\psi, \phi) =< j\psi, \psi > +i < ij\psi, \phi >=< \psi, j\phi > -i < i\psi, j\phi >= (\psi, \bar{j}\phi) \tag{29}$$

(iv) B commutes with A, P, Q. Since $P - Q$ is self-adjoint in $\mathcal{H}_r$ it follows that it commutes with j.

(v) We have $BjP = (P - Q)P = (I - Q)(P - Q) = (I - Q)Bj = B(I - Q)j$ Since j is injective, $jP = (I - Q)j$. Taking adjoints and summing one obtains $jA = (2I - A)j$. ♡

We can now introduce the modular operator.

Definition 1 (*modular operator*) We call the operator $\Delta \equiv \frac{2I-A}{A}$ *modular operator* associated to the subspace $\mathcal{K}$ in standard form ◇

Proposition 4 ([5]) *The operator Δ is self-adjoint, positive, and $\Delta^{-1} = j\Delta j$. Moreover $\mathcal{K} + i\mathcal{K} \subset D(\sqrt{\Delta})$ and for any pair $\phi, \quad \psi \in \mathcal{K}$ one has*

$$j\sqrt{\Delta}(\phi + i\psi) = \phi - i\psi \tag{30}$$

◇

Proof Since $0 < A < 2I$ both A and $2I - A$ are injective and therefore Δ is positive and injective. The equality $\Delta^{-1} = j\Delta j$ follows from point (v) of Proposition 3.

If ϕ and ψ are in $\mathcal{K}$ one has

$$(2I - P - Q)\phi = (P - Q)\phi, \quad (2I - P - Q)(i\psi) = -(P - Q)(i\psi) \tag{31}$$

and therefore $(2I - A)(\phi + i\psi) = jB(\phi - i\psi)$ and for every $\xi \in D(A^{-1})$

$$(\phi + i\psi, \Delta\xi) = ((2I - A)(\phi + i\psi), A^{-1}\xi) \quad = (jB(\phi - i\psi), A^{-1}\xi) = (j(\phi - i\psi), \sqrt{\Delta}\xi) \tag{32}$$

In the last equality we have used point (ii) of Proposition 3 and $D(\Delta) \subset D(\sqrt{\Delta})$. In particular one has

$$|(\phi + i\psi, \sqrt{\Delta}\,(\sqrt{\Delta}\xi)) \leq \|\phi - i\psi\| \|\sqrt{\Delta}\xi\| \tag{33}$$

A density argument shows that $\phi + i\psi \in D(\sqrt{\Delta})$ and $\sqrt{\Delta}(\phi + i\psi) = j(\phi - i\psi)$. ♡

We have seen that if a von Neumann algebra $\mathcal{M}$ has a cyclic and separating Ω, then the subspace generated by the action on Ω of the self-adjoint elements of $\mathcal{M}$ satisfies the condition for the existence of a modular operator Δ, which in general depends on the subspace and therefore on Ω.

Definition 2 (*modular group*) The unitary group generated by Δ is called *modular group*

Proposition 5 ([4, 5]) *The unitary group $t \to \Delta^{it}$* (modular group *associated to the subspace $\mathcal{K}$) commutes with j and leaves $\mathcal{K}$ invariant.* ◇

Proof Proposition 5 follows from Proposition 2, but can also be seen as follows. From the definition of Δ one has

$$\Delta^{it} = (2I - A)^{it} A^{-it} \tag{34}$$

It follows from Proposition 3 that $j\, A^{it} = (2I - A)^{-it} j$ (keeping into account that j is anti-linear). From this one concludes $j\Delta^{it} = \Delta^{it} j$. Therefore Δ^{it} commutes with A, B, j and in particular

$$\Delta^{it}\mathcal{K} = \Delta^{it} P\mathcal{H}_r = P\Delta^{it}\mathcal{H}_r = P\mathcal{H}_r = \mathcal{K} \tag{35}$$

♡

It is easy to see that the analytic vectors for the group of automorphisms generated by Δ^{it} are dense in $\mathcal{K}$.

We return now to the case of a modular group associated to a cyclic and separating vector state of a von Neumann algebra $\mathcal{M}$.

Proposition 6 *If the modular group is associated to a cyclic and separating vector Ω of a von Neumann algebra $\mathcal{M}$ over a Hilbert space $\mathcal{H}$ (and therefore $\mathcal{K}$ is the closure of $\mathcal{M}_{s.a.}\Omega$), then the* closed operator $j\sqrt{\Delta}$ *extends the map*

$$a\Omega \to a^*\Omega, \quad a \in \mathcal{M} \tag{36}$$

which is densely defined in $\mathcal{H}$. ◇

Proof We have seen that the closure of $(a + a^*)\Omega; \quad a \in \mathcal{M}$ has the properties required for the space $\mathcal{K}$. From Proposition 4 applied to $a\Omega$ one sees that for all $a \in \mathcal{A}$ one has $j\sqrt{\Delta}\, a\Omega = a^*\, \Omega$. ♡

We remark that one can prove in a simpler way the existence of the modular operator, but not the property to generate a one-parameter group that intertwines $\mathcal{M}$ with $\mathcal{M}'$. Indeed the anti-linear operator $S_0 : a\Omega \to a^*\Omega, \quad a \in \mathcal{M}$ is densely defined (since Ω is cyclic for $\mathcal{M}$) and closable since $S_0 \subset F_0$ where F_0 is defined by

$$F_0 \,:\, b\Omega \to b^*\Omega, \quad b \in \mathcal{M}' \tag{37}$$

It is easy to verify that $S_0 \subset F_0^*$ and since F_0 is densely defined S_0 is closable. Denote by S the closure of S_0; the polar decomposition gives $S = J\, \Delta^{\frac{1}{2}}$ where $\Delta = S^*S$ is self-adjoint and J is anti-unitary. From $J^2\Delta^{\frac{1}{2}} = J\Delta^{-\frac{1}{2}}J$ one derives $J^2 = I \quad \Delta^{\frac{1}{2}} = J\Delta^{-\frac{1}{2}}J$.

As a further remark notice that in general if a, b are self-adjoint in $\mathcal{M}$ the operator ba is not self-adjoint. Therefore in general a self-adjoint element of $\mathcal{M}$ leaves K invariant but its action *does not commute in general with the conjugation*. The role of the modular operator is to quantify this non-commutativity. If the state Ω is tracial, the modular operator is the identity. ♣

5 Intertwining Properties

We now prove that the isometry j intertwines $\mathcal{A}$ and $\mathcal{A}'$ ($j\mathcal{A}j = \mathcal{A}'$). This relation will also be at the basis of the duality theory for positive cones.

In view of its independent interest, we give first the proof when there is a faithful tracial state ($\tau(ab) = \tau(ba)$)$\forall a, b \in \mathcal{M}$).

Denote by $\Pi_\tau \mathcal{M}$ the $G.N.S.$ representation associated to τ; we shall identify it with $\mathcal{M}$. Any normal state ω of $\mathcal{M}$ can be written as $\omega(a) = \tau(\rho\, a) = \tau(\sqrt{\rho} a \sqrt{\rho})$ where $\rho \in \mathcal{M}$ is a positive operator.

Suppose ρ invertible. Then $\sqrt{\rho}$ regarded as element of $\mathcal{H}$ is cyclic for the algebra of left multiplication $\mathcal{N}$

$$\mathcal{N} = \{L_a,\ a \in \mathcal{M}\} \tag{38}$$

It is easy to see that $\mathcal{N}'$ is the algebra of right multiplication R_a. One has

$$S : a\sqrt{\rho} \to a^* \sqrt{\rho}, \quad Ja = a^* \quad \Delta = L_\rho\, R_{\rho^{-1}} \tag{39}$$

An easy calculation leads to

$$\Delta^{it} L_a \Delta^{-it} b = L_{\rho^{it} a \rho^{-it}} b \qquad a, b \in \mathcal{N} \qquad \Delta^{it} \mathcal{N} \delta^{-it} \subset \mathcal{N} \tag{40}$$

On the other hand $JL_a Jb = ba^+ = R_{a^*} b$ and therefore $j\mathcal{M}j \in \mathcal{M}'$. We shall now treat the general case.

Proposition 7 ([4]) *One has*

$$Q\Omega = P\Omega = A\Omega = B\Omega = j\Omega = \Delta\Omega = j\Omega = \Omega \tag{41}$$

Moreover for every $a' \in \mathcal{M}'_{s.a.}$ there exists $a \in \mathcal{M}_{s.a.}$ such that

$$jba'\Omega = a\Omega \tag{42}$$

◇

Proof By definition $\Omega \in K$ and since $\mathcal{M}'\Omega \in (iK)^\perp$ one has also $\Omega \in K^\perp$. Therefore $P\Omega = Q\Omega = \Omega$ and $j\Omega = \Delta\Omega = \Omega$.

To prove (42) assume first that b is a positive element of $\mathcal{M}'$ which satisfies $0 \leq b \leq I$. Then the functional $\psi \in \mathcal{M}_*$ defined by

$$\psi(a) = (b\Omega, a\Omega) \tag{43}$$

is positive and dominated by ϕ_Ω (notice that $b^* a = (b^*)^{\frac{1}{2}}\ a\ b^*)^{\frac{1}{2}}$). Using this property and restricting ψ to the self-adjoint elements of $\mathcal{M}$ one can show that there exists a positive $c \in \mathcal{M}$ such that $\psi(a) = (a\Omega, c\Omega)$. Therefore $a\Omega = P(b\Omega)$. The identity (42) follows then from $Q\Omega = 0$. ♡

We shall extend now Proposition 8 to obtain a relation between elements $\mathcal{M}$ and those of $\mathcal{M}'$. We shall do this viewing (42) as a relation between a and a' that contains the modular group Δ^{it} and later use the invariance of $\mathcal{M}$ under the modular group.

Proposition 8 *For each $a' \in \mathcal{M}'$ and complex number λ, $Re\ \lambda > 0$ there exists $a \in \mathcal{M}$ such that*

$$bja'jb = \lambda(2I - A)aA + \bar{\lambda}Aa(2I - A) \tag{44}$$

◇

Proof By linearity we can assume a' positive and $a < I$ The functional $b \to (b\Omega, x'\omega)$, $b \in \mathcal{M}$ is positive and dominated by ϕ_Ω; there exists therefore $a \in e\mathcal{M}_+$ such that

$$(b\Omega, a'\Omega) = ((\lambda ab + \bar{\lambda}ba\Omega, \Omega), \quad \forall b \in \mathcal{M} \tag{45}$$

Substituting c^*b for b , $c \in \mathcal{M}$ one obtains

$$(b\Omega, a'\Omega) = \lambda(b\Omega, ca\Omega) + \bar{\lambda}(b\ a\Omega, c\Omega) \tag{46}$$

Given $b',\ c' \in \mathcal{M}'$ choose $b,\ c \in \mathcal{M}$ satisfying Proposition 5. Substituting $b\Omega$ with $jBb'\Omega$ and $c\Omega$ with $jBc'\Omega$ one has

$$(Bja'jBc'\Omega, b'\Omega) = \lambda(jBb'\Omega, ca\Omega) + \bar{\lambda}(ba\Omega, jBc'\Omega) \tag{47}$$

Using $a\Omega = j\Delta^{\frac{1}{2}}a^*\Omega$ which holds for every $a \in \mathcal{M}$ (48) can be rewritten as

$$\begin{aligned}(Bja'jBc'\Omega, b'\Omega) &= \lambda(jBb'\Omega, j\Delta^{\frac{1}{2}}ac\Omega) + \bar{\lambda}(j\Delta^{\frac{1}{2}}bc'\Omega, Bc'\Omega)\\ &= \lambda(ajBc'\Omega, (2I - a)b'\Omega) + \bar{\lambda}((2I - A)c'\Omega, a\, j\, Bb'\Omega)\end{aligned} \tag{48}$$

We now recall that $A - jB = 2Q$ and $Q\mathcal{M}'\Omega = 0$; it follows

$$(Bja'jBc'\Omega, b'\Omega) = [(\lambda(2I - A)aA + \bar{\lambda}Aa(2I - A)]c'\Omega, b'\Omega) \tag{49}$$

The elements b' and c' are generic elements in $\mathcal{M}'$ and Ω is cyclic for $\mathcal{M}'$. We have therefore obtained the identity

$$bja'jb = \lambda(2I - A)aA + \bar{\lambda}Aa(2I - A) \tag{50}$$

♡

We will now transform (51) in a relation that contains a, a', j and the modular group. We do so using the following lemma; the proof is obtained [4] e.g. considering the function $g(z) = \pi\frac{e^{i\theta z}}{sen(\pi z)}f(z)$ and applying the formula that gives the residue at $z = 0$ as an integral along a suitable boundary.

Lemma 2 ([4]) *If $Re\lambda > 0$ and $f(z)$ is bounded and analytic in the strip $\{z \in C, |Rez| \leq \frac{1}{2}\}$ then setting $\lambda = e^{i\frac{\theta}{2}}, |\theta| < \pi$ one has*

$$f(0) = \frac{1}{2} \int e^{-\theta t} \frac{1}{\cosh(\pi t)} \left[\lambda f\left(it + \frac{1}{2}\right) + \bar{\lambda} f\left(it - \frac{1}{2}\right) \right] dt \tag{51}$$

◇

Using the previous Lemma we can prove

Proposition 9 ([4]) *If a, a' λ satisfy Proposition 6 and $\lambda = e^{i\frac{\theta}{2}}, \quad |\theta| < \pi$ one has*

$$a = \frac{1}{2} \int \Delta^{it} j a' j \Delta^{-it} \frac{e^{-\theta\, t}}{cosh\ (\pi t)} dt \tag{52}$$

◇

Proof Let $\phi, \psi \in K$ be analytic vectors of δ^{it}. Consider the analytic function

$$f(z) = (BaB\ \Delta^{-\bar{z}}\phi, \Delta^{z}\psi) \tag{53}$$

It is bounded in every strip and therefore we can use Lemma 9.8. From (51)

$$f\left(it + \frac{1}{2}\right) = (\Delta^{-it}(2I - A)aA\Delta^{it}\phi, \psi); \quad f\left(it - \frac{1}{2}\right) = (\Delta^{-it}Aa(2I - A)\Delta^{it}\phi, \psi) \tag{54}$$

Making use of Proposition 6

$$\lambda f\left(it + \frac{1}{2}\right) + \bar{\lambda} f\left(it - \frac{1}{2}\right) = (\Delta^{-it}Bja'jB\Delta^{it}\phi, \psi) \tag{55}$$

An application of Lemma 1 provides

$$\begin{aligned}(BxB\phi, \psi) &= \frac{1}{2} \int \frac{e^{-\theta\, t}}{cosh(\pi\ t)} (\Delta^{-it}Bja'jB\Delta^{it}\phi, \psi)dt \\ &= \frac{1}{2} \int \frac{e^{-\theta t}}{cosh(\pi t)} (\Delta^{-it}ja'j\Delta^{it}B\phi, b\phi)dt\end{aligned} \tag{56}$$

Proposition 9 follows because $\mathcal{K}$ generates $\mathcal{H}$ and the range of B is dense. ♡

We prove now [5]

Proposition 10 *For every $t \in R$ and $a' \in \mathcal{M}'$ one has $\Delta^{it}ja'j\Delta^{-it} \in \mathcal{M}$.* ◇

Proof Let $b' \in \mathcal{M}'$ and $\phi, \quad \psi \in \mathcal{H}$. Define

$$g(t) = ([\Delta^{-it}ja'j\Delta^{it}\ b' - b'\ \Delta^{-it}ja'j\Delta^{it}]\phi, \psi) \tag{57}$$

From Proposition 9 for $|\theta| < \pi$

$$\int g(t)\frac{e^{-\theta t}}{cosh(\pi t)}dt = 0 \tag{58}$$

The function $h(z) = \int g(t)\frac{e^{-zt}}{cosh(\pi t)}dt$ is holomorphic in the upper half plane and is zero for z real. Therefore $\int g(t)e^{-ist}\frac{1}{cosh(\pi t)}dt = 0$. Uniqueness of Fourier transform implies $g \equiv 0$. Hence $\Delta^{-it}ja'j\Delta^{it} \in \mathcal{M}'' = \mathcal{M}$. ♡

Theorem 2 *Let $\mathcal{M}$ be a von Neumann algebra on a Hilbert space $\mathcal{H}$ and let the vector $\Omega \in \mathcal{H}$ be cyclic and separating. There exists a positive self-adjoint operator Δ (called* modular operator *with respect to Ω) and an anti-linear isometry j such that $j\mathcal{M}j = \mathcal{M}'$ and $\Delta^{it}\mathcal{M}\Delta^{-it} = \mathcal{M}$ for every real t. One has $j\,\Omega = \Omega$, $\mathcal{M}\Omega \in D(\sqrt{\Delta})$ and*

$$j\,\sqrt{\Delta}\,a\,\Omega = a^*\Omega \quad \forall a \in \mathcal{M} \tag{59}$$

◇

Proof Equation (59) follows from the definition of j. To prove the remaining part of the theorem, let $\mathcal{K}$ be the closure of $\mathcal{M}_{s.a.}\Omega$.

We have seen that this linear space satisfies the conditions which allow the construction of the modular operator. From Proposition 10 (for $t = 0$) we know that $j\mathcal{M}'j \in \mathcal{M}$. The thesis of Theorem 2 could then be obtained by proving that the modular operator Δ' associated to the real subspace $\mathcal{K}'$ which is the closure of $\mathcal{M}'_{s.a.}\Omega$ satisfies $\Delta'\,\Delta = I$ (the conjugations satisfy $j' = j$).

This provides the inclusion $j\mathcal{M}j \in \mathcal{M}'$. A direct proof is as follows. Let $a,\ b$ self-adjoint in $\mathcal{M}$. Since $j\Omega = \Omega$ one has

$$(bjaj\Omega, \Omega) = (\Omega, ajbj\Omega) \tag{60}$$

This linear relation extends to all elements of $\mathcal{M}$. Choose $b' \in \mathcal{M}'$ and remark that $bjb'j \in \mathcal{M}$. Substituting $jb'j$ in place of b one obtains

$$(b(jb'j(jaj)\Omega, \Omega) = (\Omega, aj(bjb'j)j\Omega) \tag{61}$$

From this one derives $(ajbj\Omega, b'\Omega) = (jbja\Omega, b'\Omega)$. Since $\mathcal{M}'\Omega$ is dense in $\mathcal{H}$ it follows $ajbj\Omega = jbja\Omega$. This is a linear equation valid for every $a \in \mathcal{M}$. Substituting a with $ac\quad a,\ c \in \mathcal{M}$ one obtains

$$jbjac\Omega = acjbj\Omega = ajbjc\Omega \tag{62}$$

and therefore $jbja = ajbj$ because of the density of $\mathcal{M}\Omega$ in $\mathcal{H}$. Hence $jbj \in \mathcal{M}'$. ♡

6 Modular Condition. Non-commutative Radon–Nikodym Derivative

Given a Hilbert space $\mathcal{H}$ and a closed real subspace $\mathcal{K}$ of $\mathcal{H}_r$ we shall say that the unitary group $\{U_t\}$, $t \in R$ *satisfies the modular condition with respect to* $\mathcal{K}$ if for any pair of vectors $\phi, \psi \in \mathcal{K}$ there exists a bounded continuous function $f_{\phi,\psi}$ defined on the strip

$$\mathcal{S}_{-1} = \{z \in C : \ -1 \leq Imz \leq 0\} \tag{63}$$

holomorphic in the interior and satisfying the boundary conditions

$$f(t) = (U_t\phi, \psi) \qquad f(t-i) = (\psi, U_t\phi) \quad t \in R \tag{64}$$

◇

Proposition 11 ([4, 5]) *Let $\mathcal{M}$ be a von Neumann algebra with cyclic and separating vector Ω.*

The unitary group $t \to \Delta^{it}$ satisfies the modular condition with respect to the closure of $\mathcal{M}\Omega$ and is the unique unitary representation with these properties. ◇

We shall now give the relation between the modular group and the K.M.S. condition. Consider a C^* dynamical system which we will denote by $\{\mathcal{A}, \ \alpha_t\}$. An element $x \in \mathcal{A}$ is *analytic* for α_t if the map $t \to \alpha_t(x)$ has an extension to an entire analytic function $\zeta \to \alpha_\zeta(x)$ $\zeta \in \mathcal{C}$.

If $x \in \mathcal{A}$ define

$$x_n \equiv \frac{\sqrt{n}}{\sqrt{\pi}} \int \alpha_t(x) e^{-n^2 t} dt \tag{65}$$

For any integer n the element x_n is analytic for α_t and that $lim_{n\to\infty}|x_n - x| = 0$. Therefore the set $\mathcal{A}^a$ of analytic vectors in norm-dense in $\mathcal{A}$ and in fact it is a *-subalgebra of $\mathcal{A}$.

The same conclusions are reached if one considers a W^*-dynamical system or a dynamical system with values in a von Neumann algebra. An important property of the K.M.S. condition the following that we have already noted in Volume 1 of this Lecture notes.

Let $\{\mathcal{A}, \alpha_t\}$ be a C^* dynamical system and let ρ_β be a state which satisfies the α_t-K.M.S. condition for a value β of the parameter $(0 \leq \beta \leq \infty)$. Then ρ_β is invariant for the automorphisms group α_t.

Proposition 12 *Let $\mathcal{A}, R, \{\alpha_t\}$ be a C^* dynamical system. Suppose that a state ρ satisfy the K.M.S. condition at $\beta = 1$. Let $(\Pi_\rho, U_t^\rho, \mathcal{H}_\rho \Omega_\rho)$ the cyclic covariant representation associated to ρ by the G.N.S. construction and let $\mathcal{K}$ the closure of $\Pi_\rho(\mathcal{M}_{s.a}\ \Omega_\rho)$. Then U_t^ρ satisfies the modular condition with respect to $\mathcal{K}$.* ◇

Proof Since the state is $\alpha-$ invariant, the representation is covariant. It is also easy to see that U_t^ρ leaves for every t invariant the subspace $\mathcal{K}$. For every $\psi \in \mathcal{K}$ we can choose a sequence a_n b_n $\in \mathcal{M}_{s.a.}$ such that $\Pi_\rho(a_n)\Omega_\rho$ converges to ϕ and $\Pi_\rho(b_n)\Omega_\rho$ converges to ψ.

By assumption, there exists functions f_n bounded and continuous in the strip

$$\mathcal{S}_1 = \{z \ : \ 0 \leq Im\, z \leq 1\} \tag{66}$$

holomorphic *in the interior* and which satisfy the boundary conditions

$$f_n(t) = (U_t^\rho \ \Pi_\rho(a_n)\Omega_\rho, \Pi_\rho \ (b_n)\Omega_\rho), \quad f_n(t+i) = (\ \Pi_\rho(b_n)\Omega_\rho, U^\rho \Pi_\rho \ (a_n)\Omega_\rho) \tag{67}$$

Since the f_n are uniformly bounded and uniformly convergent by the Phragmen–Lindelhof theorem the functions f_n converge to a function f which is holomorphic in the interior of the strip and satisfies the boundary conditions

$$f(t) = (U_t^\rho \phi, \psi) \qquad f(t+i) = (\psi, U_t^\rho \phi) \tag{68}$$

Setting $g(z) = \bar{f}(\bar{z})$ one sees that g satisfies the modular condition with respect to $\mathcal{K}$.

In order to show that to $\mathcal{K}$ corresponds a modular structure we must show that the conditions in (67) are satisfied. The second condition is trivially satisfied since Ω_ρ is cyclic. Let us prove that $\mathcal{K} \cap i\mathcal{K} = \emptyset$.

Let $\phi \in \mathcal{K} \cap i\mathcal{K}$ and $\psi \in \mathcal{K}$. Since U^ρ satisfies the modular condition there exist functions f_1 and f_2 holomorphic in the strip $\{z : \ -1 \leq Im\, z \leq 0\}$ which satisfy the boundary conditions

$$f_1(t) = (U_t^\rho \phi, \psi), \quad f_1(t-i) = (\psi, U_t^\rho \phi) \quad f_2(t) = (U_t^\rho i\phi, \psi) \quad f_1(t-i) = (\psi, iU_t^\rho \phi) \tag{69}$$

One has $if_1(t) = f_2(t)$, $\quad -if_1(t-i) = f_2(t-i)$; this implies $if_1(z) = f_2(z)$, $-if_1(z) = f_2(z)$ in the interior of the strip, and therefore $f_1 = f_2 \equiv 0$. This holds for every $\psi \in \mathcal{K}$; since $\mathcal{K}$ generates $\mathcal{H}$ over the complex field, it follows $\phi = 0$. ♡

As a consequence of Proposition 12 we can prove Theorem 3

Theorem 3 *For every normal faithful state ρ of a von Neumann algebra $\mathcal{M}$ there exists a unique W^* dynamical system (which will be denoted by $(\mathcal{M}, \alpha_t, \rho)$) such that ρ satisfies the K.M.S. condition with respect to α_t. We shall call* modular group associated to *ρ (denoted by σ_t^ρ) the group of automorphisms of this dynamical system.* ◇

Proof Consider the cyclic representation associated to ρ by the G.N.S. construction. Since ρ is normal and cyclic, we can identify $\mathcal{M}$ with its image in Π_ρ. Since Ω_ρ is separating we can construct the modular operator and define

$$\sigma_t(a) = \Delta^{it} a \Delta{-it} \quad a \in \mathcal{M} \quad t \in R \tag{70}$$

By construction the map $t \to \Delta^{it}$ satisfies the modular condition with respect to the closure of $\mathcal{M}_{s.a.}\Omega_\rho$. It is easy to see that the modularity condition implies the K.M.S. condition with respect to $\{\sigma_t\}$ at the value 1.

To prove the converse, let $(\mathcal{M}, R, \alpha)$ satisfy the K.M.S. at the value one of the parameter and let $U(t)$ be the family of unitary operators that implements α_t in the Hilbert space $\mathcal{H}_\rho$. Using Proposition 12 for the dynamical system $(\mathcal{M}, R, \alpha)$ it is easy to see that for every $t \in nR$ and $a \in \mathcal{M}$ one has

$$\alpha_t(a) = U_t^\rho a U_{-t}^\rho = \Delta^{it} a \Delta^{-it} = \sigma_t(a) \tag{71}$$

♡

As immediate consequence of Theorem 3 one has

Lemma 3 *If σ_t^ρ is the modular group associated to the normal faithful state ρ of a von Neumann algebra $\mathcal{M}$ and α is an automorphism of $\mathcal{M}$ then $\{\alpha^{-1} \cdot \sigma_t^\rho \cdot \alpha\}$ is the modular group associated to the state $\rho \cdot \alpha$.* ♢

Proof Choose $a, b \in \mathcal{M}$. Using condition K.M.S. for $\alpha(a), \alpha(b)$ one can construct two function holomorphic in the interior of the strip $\mathcal{S}_1$ which at the boundary coincide with

$$\rho(\alpha(b)\ \sigma_t^\rho(\alpha(x))) \equiv (\rho \cdot \alpha)(b(\alpha^{-1} \cdot \sigma_t^\rho \cdot \alpha(a))) \tag{72}$$

and with

$$\rho(\sigma_t^\rho(\alpha(a)\ \alpha(b)) = (\rho \cdot \alpha)(\alpha^{-1} \cdot \sigma_t^\rho \cdot \alpha(a))b) \tag{73}$$

Lemma 3 follows then from Proposition 12. ♡

We shall now briefly study the relation among faithful normal states in term of their modular operators. We begin by constructing the analog of a Radon–Nikodym derivative in the commutative case.

Proposition 13 *Let ρ be a normal faithful state of a von Neumann algebra $\mathcal{M}$ and let σ_t^ρ be the corresponding modular group. If $\rho' \in \mathcal{M}_*$ satisfies $0 \le \rho' \le \rho$ and ρ' is invariant under the dual action of $\{\sigma_t^\rho\}$ then there exists unique an element $h \in \mathcal{M}_{s.a.}$ such that $\rho'(a) = \rho(ha) = \rho(ah)$. Moreover h is invariant under σ^ρ.* ♢

Proof Lemma 1 guarantees the existence of a unique $h \in \mathcal{M}$ for which

$$\rho' = \frac{1}{2}[\rho(h\ .) + \rho(.\ h)] \tag{74}$$

The element h is invariant because both ρ and ρ' are invariant and h is unique. We show that this implies $\rho(a\ h) = \rho(h\ a)$ for all $a \in \mathcal{M}$ (in fact one can show that the two statement are equivalent). For each $a\ \ h \in \mathcal{M}$ there exists a function f holomorphic in Ω_1 and such that

$$f(t) = \rho(a\ h), \qquad f(t+i) = \rho(h\ b) \tag{75}$$

If h is invariant, f is a constant. Therefore $\sigma(a\ h) = f(0) = f(i) = \sigma(h\ a)$. From (74) one derives $\rho(.) = \rho'(h\ .) = \rho(.\ h)$. ♡

We study now the properties of states *that have the same modular operator.*

Proposition 14 (non-commutative Radon–Nikodym derivative) *Let ρ and ρ' be faithful normal states of the von Neumann algebra $\mathcal{M}$. If they have the same modular operator there exists a unique positive injective operator h affiliated to $\mathcal{M} \cap \mathcal{M}'$ such that $\rho'(a) = \rho(ha)$ for every $a \in \mathcal{M}$. The element h plays therefore the role of* non-commutative Radon–Nikodym derivative *of ρ' with respect ρ.* ◇

Proof Consider first the case $\rho' \leq \rho$. From the previous lemma $\rho'(.) = \rho(h\ .)$ where h is invariant under σ_t^ρ.

Let u be unitary and a arbitrary in $\mathcal{M}$. Using the K.M.S. condition for ρ we obtain two functions f, g continuous in Ω_1 and holomorphic in the interior which satisfy

$$f(t) = \rho'(u^* \sigma_t^\rho(h\ u\ a)), \qquad f(t+i) = \rho'(\sigma_t^\rho(h\ u\ a)\ u^*)$$

$$g(t) = \rho'(u^*\ \sigma_t^\rho(u\ a)) \qquad g(t+i) = \rho'(\sigma_t^\rho(u\ a)u^*) \tag{76}$$

From $h \in \mathcal{M}_{s.a.}$ it follows $f(t+i) = f(t+i)$ and therefore $f = g$. Evaluating this function at zero

$$\rho(u^*\ h\ u\ a) = \rho(h\ a) \tag{77}$$

From the uniqueness of h follows $u^*hu = h$. This must be true for every unitary in $\mathcal{M}$ and therefore $h \in \mathcal{M} \cap \mathcal{M}'$. In the general case, we remark that σ is also the modular group for $\rho + \rho'$; therefore there exist operators $h, h' \in \mathcal{M} \cap \mathcal{M}'$ such that

$$\rho(a) = (\rho + \rho')(ha), \qquad \rho'(a) = (\rho + \rho')(h'a) \tag{78}$$

Since ρ and ρ' are faithful both h and h' are injective. Hence $k = h\ (h')^{-1}$ is affiliated to $\mathcal{M}$ and satisfies $\rho'(a) = \rho(ka)$. ♡

We turn now to the case of two states ρ and ρ' whose modular groups *commute.*

Proposition 15 ([5]) *Let ρ and τ be two normal faithful states of $\mathcal{M}$ and let σ_t^ρ and σ_t^τ be their modular groups. The following conditions are equivalent*
(1) ρ is invariant under the action of σ^τ
(2) τ is invariant under the action of σ^ρ
(3) σ^τ and σ^ρ commute
(4) there exists a unique positive injective operator h affiliated to $\mathcal{M} \cap \mathcal{M}'$ such that $\tau(a) = \sigma(h\ a)\quad \forall a \in \mathcal{M}$.
(5) there exists a unique positive injective *operator k affiliated to $\mathcal{M} \cap \mathcal{M}'$ such that $\sigma(a) = \tau(k\ a)\quad \forall a \in \mathcal{M}$.* ◇

Proof (1) $\leftrightarrow$ (3) and (2) $\leftrightarrow$ (3)

According to Proposition 13 the modular group for $\rho \cdot \sigma_t^\tau$ is $\rho[\sigma_{-s}^\tau \cdot \sigma_t^\rho \cdot \sigma_s^\tau]$. If ρ is invariant under σ^τ one has

$$\rho \cdot \sigma_s^\tau = \{\sigma_{-s}^\tau \cdot \sigma_t^\rho \cdot \sigma_s^\tau = \sigma_t^\rho \tag{79}$$

and therefore σ^τ and σ^ρ commute.

Conversely if the modular groups commute one derives $\rho \cdot \sigma^t au(a) = \rho(ha)$ where h is a positive operator affiliated to a $\mathcal{M} \cap \mathcal{M}'$. Uniqueness of h implies $h_{ns} = h_s^n$ for every integer n and thus $h_s = I$ for every s and ρ is σ_t^τ invariant.

(2) $\leftrightarrow$ (4) and (1) $\leftrightarrow$ (5)

Straightforward

(1)$\leftrightarrow$ (4)

Consider the state $\xi = \frac{1}{2}(\rho + \tau)$ and denote σ^ξ its modular group. Since ξ is σ^τ invariant, from (1) $\leftrightarrow$ (2) follows that τ is σ^ξ invariant.

From $\tau \leq 2\xi$ there exists $0 \leq k \leq 2I$, σ^ξ invariant, such that $\tau(a) = \xi(ka)$.

Uniqueness and invariance of ρ and τ imply that also k is invariant.

Since $\rho(a) = \xi((2I - k)a)$ and both k and $2I - k$ are injective (both ρ and τ are faithful) one concludes that $h \equiv \frac{k}{2I-k}$ is a positive injective operator affiliated to $\mathcal{M} \cap \mathcal{M}'$. And $\rho(a) = \tau(ha)$. ♡

Proposition 15 is a non-commutative Radon–Nikodym theorem and the operator h plays the role of Radon–Nikodym derivative. To see this analogy notice that, by a theorem of Gelfand and Neumark, every abelian von Neumann algebra can be faithfully represented by the algebra $\mathcal{A} \equiv L^\infty(X)$ of multiplication by complex valued essentially bounded functions on a locally compact space X. In this case $\mathcal{A}' = \mathcal{A}$.

If $X = L^\infty(T^d)$ the normal states are represented by positive measurable functions $f(x)$ on T^d with integral one (more precisely by the measures $f(x)dx$). The cyclic and separating states are represented by strictly positive functions. The state ϕ_f on $L^\infty(T^d)$ is defined by

$$\phi_f(a) = \int a(x)f(x)dx, \quad a \in L^\infty \tag{80}$$

In this case the operator j is complex conjugation and Δ is the identity.

7 Positive Cones

The positive cone $\mathcal{C}_f$ defined by f coincides the positive cone $\mathcal{C}_f'$ and is represented by the positive integrable functions. Given an element $g \in \mathcal{A}$ the functional ϕ_g is positive iff g is positive and is such that $\phi_g(a) = \int a(x)\frac{g(x)}{f(x)}f(x)dx$. Therefore $\frac{g(x)}{f(x)}$

is the Radon–Nikodym derivative of the state ϕ_g (i.e. of the measure $g(x)dx$) with respect to the state ϕ_f (i.e. of the measure $f(x)dx$. ♣

Remark that in the commutative case, if the total measure is one, the function one is a cyclic and separating vector, and Eq. (80) can be interpreted as follows: given an element of $L^\infty \equiv \mathcal{A}'$ the linear functional $a \to \phi_b(a)$ is positive iff b belongs to the positive cone of $\mathcal{A}$.

There exists therefore a duality, originated by the state Ω, between the positive cone in $\mathcal{A}'$ and the positive cone in $\mathcal{A}$. This duality is elementary in the commutative case and holds for any cyclic and separating state. The formalism described here allows for an extension of this duality to the non-commutative case (*Tomita duality*)

Let be a von Neumann algebra $\mathcal{M}$ (on a Hilbert space $\mathcal{H}$) with a cyclic and separating vector Ω., with corresponding modular operator Δ and invertible anti-linear isometry j. Denote by $S_0 \in \mathcal{M}\Omega$ the densely defined operator $S_0\ a\Omega = a^*\Omega \quad a \in \mathcal{M}$ and by $F_0 \in \mathcal{M}'\Omega$ the densely defined operator $F_0\ a\Omega = a^*\Omega, a \in \mathcal{M}'$.

Denote by S and F their closures. One has the polar decomposition $S = j\sqrt{\Delta}$ con $\Delta = S^*S$. If $x \in \mathcal{H}$ denote by ϕ_x on $\mathcal{M}$ the linear functional defined by

$$\phi_x(a) = (a\Omega, x) \tag{81}$$

Similarly denote by ϕ'_x on $\mathcal{M}'$ the linear functional

$$\phi_x(a') = (a'\Omega, x) \tag{82}$$

Definition 3 We will say that x is $\mathcal{M}_\Omega$-positive if the functional ϕ_x is positive. ◇

Denote by $\mathcal{C}_\Omega$ the cone of $\mathcal{M}_\Omega$-positive vectors. Similarly denote by $\mathcal{C}'_\Omega$ the cone of $\mathcal{M}'_\Omega$-positive vectors.

Theorem 4 ([4]) *The functional $\phi_{x'}$ on $\mathcal{M}'$ is positive iff there exists a self-adjoint operator h affiliated to $\mathcal{M}$ such that $x' = h\Omega$. This is also the condition under which the G.N.S. representation $\Pi_x\mathcal{M}$ of $\mathcal{M}$ generated by the state ϕ_x is equivalent to to the representation $\Pi_\Omega(\mathcal{M}) \equiv \mathcal{M}$.*

Conversely the functional ϕ_x on $\mathcal{M}$ is positive iff there exists a positive self-adjoint operator h affiliated to $\mathcal{M}'$ such that $x = h\Omega$. ◇

We remark that this theorem poses a duality between the cone $\mathcal{C}_\Omega$ and the cone of positive elements in $\mathcal{M}'$ and also between the cone $\mathcal{C}'_\Omega$ and the cone of positive elements in $\mathcal{M}$. We do not give the proof of Theorem 4.

The results we have described must be placed in the context of the theory of positive dual cones by Tomita and Takesaki. A rather detailed analysis can be found e.g. in [5].

References

1. I. Segal, Ann. Math. **57**, 401–457, **58**, 595–596 (1953)
2. E. Nelson, J. Funct. Anal. **15**, 103–116 (1974)
3. L. Gross, J. Funct. Anal. **10**, 52–109 (1972)
4. K. Pedersen, *Lecture Notes in Mathematics*, vol. 731 (Springer, Berlin, 1979)
5. R. Kadison, R. Ringrose, *Operator Algebras*, vol. III, IV Graduate Studies in Mathematics vol 15, 16 American Math. Soc. (1997)

Lecture 10: Scattering Theory. Time-Dependent Formalism. Wave Operators

Scattering theory, in Quantum as in Classical Mechanics, describes those effects of the interaction of a system of N particles which can be measured when the components of the system have become spatially separated so that they can be uniquely identified and the mutual interactions have become negligible.

In this lecture we shall limit ourselves to a system of two quantum particles which interact through a potential force that is invariant under translation. In this case the problem can be reduced to that of one particle in interaction with a potential force.

This problem is by far simpler than the corresponding N-body problem in which several *channels* may be present and the final state may contain bound states of some of the particles. In this lecture we shall analyze the *time-dependent formalism* in which the motion in time is explicitly considered.

In the next Lecture we shall study the same problem through a study of the relation between the eigenfunction of the interacting system and of a reference system, which we take to be free. The latter procedure is called *time independent scattering theory* to stress that only the relation between eigenfunction is considered.

In the time-dependent formulation scattering theory in the one-body problem with forces due a potential V is essentially the comparison of the asymptotic behavior in time of the system under two dynamics given by two self-adjoint operators H_1 and H_2.

We shall treat in some detail the case in which the ambient space is R^3, both systems are described in cartesian coordinates, and the reference hamiltonian is the free hamiltonian; in this way the reference motion has a simple description. We have

$$H_1 = -\frac{\hbar^2}{2m}\Delta \qquad H_2 = -\frac{\hbar^2}{2m}\Delta + V(x) \tag{1}$$

where m is the mass of the particle and $V(x)$ is the interaction potential. In the formulation of the general results we leave open the choice of the reference

G. Dell'Antonio, *Lectures on the Mathematics of Quantum Mechanics II: Selected Topics*, Atlantis Studies in Mathematical Physics: Theory and Applications 2, DOI 10.2991/978-94-6239-115-4_10

hamiltonian so that the formalism can be applied more generally e.g. in presence of an electromagnetic field we can use as reference the hamiltonian

$$H_1 = -\frac{1}{2m}(\nabla - A(x))^2 \tag{2}$$

In general we shall choose units in which $2m = \hbar = 1$. We shall make stringent assumptions on $V(x)$, and in particular that the potential $V(x)$ is time-independent and Kato small so that $-\frac{\hbar^2}{2m}\Delta + V$ is (essentially) self-adjoint. We will assume also that $V(x)$ vanishes sufficiently fast at infinity, e.g. $lim_{|x|\to\infty}|x|^p\ V(x) = 0$ for a suitable value of $p > 1$.

The theory can also be applied when H_1 is periodic in space; this is the case if one describes scattering of a particle by a crystal.

Notice that the same comparison problem can be posed when the potential depends on time and in particular if it is *periodic in time* with period T (and sufficiently regular as a function of the space variables).

We shall not treat this case.

1 Scattering Theory

We shall formulate scattering theory as comparison between the asymptotic behavior for $t \to \pm\infty$ of a generic element $\phi \in \mathcal{H}$ that evolves according the dynamics given H_2, i.e. $\phi(t) = e^{-itH_2}\phi$, and the behavior of two elements $\phi_\pm(t)$ which evolve according to H_1 and differ very little from $\phi(t)$ when $t \to \pm\infty$. In general we will consider only the case $\mathcal{H} \equiv L^2(R^d)$, $d = 3$.

The cases $d = 2$ can be treated along the same lines with an extra care due to the weaker decay in space-time of the solution of the free Schrödinger equation.

We assume

$$lim_{t\to\pm\infty}|\phi(t) - \phi_\pm(t)| = lim_{t\to\pm\infty}|e^{-iH_2t}\phi - e^{-iH_1t}\phi_\pm|_2 = 0 \tag{3}$$

Remark that in this equation it is required only that the limit of the difference exists, while in general the limit of each term does not exist in the topology of $\mathcal{H}$. For example if

$$H_1 = \Delta, \quad H_2 = \Delta + V(x), \quad V \in C_0^\infty \quad V(x) > 0 \tag{4}$$

each of the two dynamics has a *dispersive property* in the following sense: for $t \to \pm\infty$ one has, for ϕ in the orthogonal complement of the discrete spectrum of H_k, $k = 1, 2$

$$lim_{t\to\pm\infty}\ sup_x|\phi(x,t)| \equiv lim_{t\to\pm\infty}\ sup_x|(e^{itH_k}\phi_\pm)(x)| = 0 \tag{5}$$

and therefore we would compare functions which for $t \to \pm\infty$ *tend to be infinitesimal everywhere*. Of course the rate of vanishing will be in general different in different directions, but the comparison would become difficult.

One way to overcome this problem could be (and this is the approach of Enss, that we shall discuss later) to use *time-dependent scales in space* which increase suitably with time so that on the new scales the functions have little dispersion. In this way one can compare the asymptotic effects of the interaction *in different directions*.

Another method consists in noticing that both dynamics are unitary, and therefore equation (4) is equivalent to

$$lim_{t\to\pm\infty} e^{-itH_1} \, e^{itH_2}\phi = \phi_\pm \qquad lim_{t\to\pm\infty} e^{itH_2} \, e^{-itH_1}\phi_\pm = \phi \tag{6}$$

In the domain of existence we will define the *wave operators*

$$W_\pm(H_2, H_1) = lim_{t\to\pm\infty} e^{itH_2} \, e^- itH_1 \tag{7}$$

Let us remark that, whenever defined, the wave operator satisfies

$$W_\pm(H_2, H_1) e^{itH_1} = e^{itH_2} W_\pm(H_2, H_1) \tag{8}$$

The wave operators on their domain of definition intertwine the two dynamics. In particular the domain of $W_\pm(H_2, H_1)$ is invariant under the flow of H_1.

Let us exemplify (3) and (4) in the case of main interest for us, namely $H_1 = -\Delta$ e $H_2 = -\Delta + V$ where V has suitable regularity and decay properties.

The existence of $W_\pm(-\Delta + V, -\Delta)$ answers the question whether a state which evolves almost freely at $t \simeq -\infty$ after the interaction with the potential V(x) will have an almost free evolution at $t \to +\infty$.

The existence of $W_\pm(-\Delta, -\Delta + V)$ answers the question whether for a given initial datum the evolution $-\Delta + V$ is asymptotic for $t \to +\infty$ or $t \to -\infty$ to free evolution.

It is evident that if the initial datum corresponds to a bound state the answer to this second question will be negative. Therefore the domain of the operator $W_\pm(-\Delta, -\Delta + V)$ is contained in the orthogonal complement of the bound states of the hamiltonian $-\Delta + V$. The purpose of the analysis in this lecture is find conditions under which this is the only subspace excluded, and any free asymptotic behavior can be approximated by choosing the initial datum in the complementary subspace.

This implies that the range of $W_\pm(-\Delta + V, -\Delta)$ is the entire Hilbert space. [1, 2] Notice, for comparison, that in Classical Dynamics there are bounded regular potentials which have no bound states and for which the limit does not exists for some initial datum, or it exists only in one direction of time.

2 Wave Operator, Scattering Operator

Definition 1 (*Wave Operator*) If the spectrum of the operator H_1 is absolutely continuous (as is for $H_1 = -\Delta$) we shall define *Wave Operator* relative to the pair $H_2,\ H_1$ the operator

$$W_\pm(H_2, H_1) = s - lim_{t\to\pm\infty} e^{itH_2}\ e^{-itH_1} \tag{9}$$

If the spectrum of H_1 is not absolutely continuous the definition of wave operator is suitably generalized. Denote by $\mathcal{H}_{1,ac} \subset \mathcal{H}$ the (closed) subspace of absolute continuity for H_1 defined by

$$\mathcal{H}_{1,ac} \equiv \{\phi \in \mathcal{H}\ :\ (E_1(\lambda)\phi, \phi) \in C_{a.c.}\} \tag{10}$$

where $E_1(\lambda)$ is the spectral family of H_1 and $C_{a.c.}$ is the space of absolutely continuous functions. We define *Generalized Wave Operators* the limit (if it exists)

$$W_\pm(H_2, H_1) \equiv s - lim_{t\to\pm\infty} e^{itH_2}\ e^{-itH_1}\ \Pi_1 \tag{11}$$

where Π_1 is the orthogonal projection on $\mathcal{H}_{1,ac}$. ◇

Remark that if $H_1 = \Delta$ the spectrum is absolutely continuous; in this case $\Pi_1 = I$ and definition coincides with that in (11).

If $H_1 = -\Delta + V$ the spectrum of the operator H_1 can have a singular continuous part as well as a discrete one; in this case we must refer to (13) for the definition of wave operator. It follows from the definition that

$$W^*_\pm(H_2, H_1)W_\pm(H_2, H_1) = \Pi_1 \tag{12}$$

where Π_1 is the orthogonal projection on the absolutely continuous part of the spectrum of H_1.

Definition 2 (*Scattering Operator*) On the elements in $\phi_- \in D(W_-(H_1, H_2))$ such that $W_-(H_1, H_2)\phi_- \in D(W_+(H_1, H_2))$ we define the *Scattering Operator* the map (H_2, H_1) defined by

$$\phi_- \to \phi_+ \equiv S\ \phi_- \tag{13}$$

In the case $H_1 = -\frac{\hbar^2}{2m},\quad H_2 = -\frac{\hbar^2}{2m} + V$ the operator $S(H_2, H_1)$ is usually called *Scattering Matrix.*

From the definition one has $e^{itH_2}S = Se^{itH_2}$ Notice that the operator S is the map $\phi_- \to \phi_+$ and represents the probability amplitude that a given free motion at $t = -\infty$ gives to a definite free motion at $+\infty$. The adjoint S^* is defined on the domain of $W_+(H_1, H_2)$ and on suitable domains the following identities hold

$$S = W_+^* \, W_- \qquad S^* = W_-^* \, W_+ \tag{14}$$

For the physical interpretation (which we will give later by the introduction of a flux across surfaces) the operators S must be unitary. This implies that it must have as domain and range the entire Hilbert space

$$Range W_+(H_1, H_2) = \mathcal{H} = Range W_-(H_1, H_2) \tag{15}$$

In the case of scattering by a potential the assumptions we shall make on $V(x)$ have the purpose to guarantee the existence of the Wave Operators and the validity of (17). Notice that it is convenient to formulate the scattering problem with reference to two Hamiltonians H_1 and H_2 rather than to a free and interacting ones.

This underlines the symmetric role of the two Hamiltonians and allows the formulation of the *chain rule* which permits to deduce the existence of the wave operator $W_\pm(H_3, H_1)$ from the existence of $W_\pm(H_3, H_2)$ and $W_\pm(H_2, H_1)$.

We have now formulated the two fundamental problem of scattering theory in Quantum Mechanics:

(i) *Existence of the Wave Operator*
(ii) *Asymptotic completeness*: $Range W_- = Range W_+$

Another interesting question refers to the inverse scattering problem. Given the unitary operator S and the operator H_1 prove existence and uniqueness of an operator H_2 which satisfies (8). For a general introduction to this class of problems one can consult [1–3].

A simple example, due to G. Schmidt, show that the dispersive properties of the dynamics are important for uniqueness. Let

$$H_1 = i\frac{d}{dx}, \qquad H_2 = i\frac{d}{dx} + V(x), \qquad \mathcal{H} = L^2(-\infty, +\infty) \tag{16}$$

Then

$$(e^{-itH_1}\phi)(x) = \phi(x-t) \quad H_2 = U^{-1} \, H_1 \, U, \quad U = e^{i\int_0^x V(y)dy} \tag{17}$$

It follows

$$(e^{itH_2} \, e^{-itH_1})\phi(x) = e^{i[V(x+t)-V(x)]}\phi(x) = e^{i\int_x^{x+t} V(y)dy}\phi(x) \tag{18}$$

In this example $W_\pm$ are multiplication operators

$$W_\pm(H_2, H_1) = e^{i\int_x^{\pm\infty} V(y)dy} \tag{19}$$

(they exist if $V \in L^1$) and S is the operator of multiplication by the phase factor $S = e^{-i\int_{-\infty}^{\infty} V(x)dx}$ In this case the inverse scattering problem does not have a unique solution.

On the contrary for the Schrödinger equation (a dispersive one) one can prove that for short range potentials the potential is uniquely determined by the S matrix. We shall give an outline of the proof of this statement in the next chapter.

3 Cook–Kuroda Theorem

A first result in scattering theory is the following theorem, first proved by Cook and then improved by Kuroda [4]

Theorem 1 (Cook, Kuroda) *Suppose that there exists a dense set $D \in \mathcal{H}_{1,ac}$ on which the following properties are satisfied*

(a) for $\phi \in \mathcal{D}$ there exists t_0 (which may depend on ϕ) such that

$$e^{-itH_1}\phi \in D(H_1) \cap D(H_2), \qquad t_0 \leq t < +\infty \tag{20}$$

(b) $(H_2 - H_1)e^{-itH_1}\phi$ is continuous in as a function of t for $t \in (t_0, \infty)$
(c)

$$\int_{t_0}^{\infty} |(H_2 - H_1)e^{-itH_1}\phi|_2 dt < \infty \tag{21}$$

Under these assumptions $W_+(H_2, H_1)$ exists. The same is true for $W_-(H_2, H_1)$. ◇

Proof For $\phi \in \mathcal{D}$ $t, s \geq t_0$

$$\frac{d}{dt}(e^{itH_2}\, e^{-itH_1}\phi) = ie^{itH_2}\, (H_2 - H_1)\, e^{itH_2} \tag{22}$$

Therefore for $t > t_0$

$$e^{itH_2} e^{-itH_1}\phi = e^{it_0H_2} e^{-it_0H_1}\phi + i\int_{t_0}^{t} e^{i\tau H_2}\, (H_2 - H_1)\, e^{-i\tau H_1}\phi d\tau \tag{23}$$

If $\phi \in \mathcal{D}$ under assumptions (b), (c) the integral on the right hand side converges when $t \to \infty$. Therefore the limit $lim_{t\to+\infty} e^{itH_2}\, e^{-itH_1}\phi$ exists for $\phi \in \mathcal{H}_{1,ac}$. ♡

If $\mathcal{H} \equiv L^2(R^3)$ $H_1 \equiv -\Delta$ $H_2 \equiv -\Delta + V$ (17.8) reads

$$\int_{t_0}^{\infty} |V(x)e^{it\Delta}\phi|_2 dt < \infty \tag{24}$$

In this case $\mathcal{D}$ can be chosen to be the collection of functions with Fourier transform in C_0^{∞}. For sufficiently regular potentials Eq. (26) follows from *dispersive estimates* for the functions $e^{ik^2t}\hat{\phi}(k)$. For $t \neq 0$ the integral kernel of e^{-itH_1} is

$$(e^{-itH_1}\phi)(x) = \left(\frac{1}{4\pi it}\right)^{\frac{3}{2}} \int_{R^3} e^{\frac{|x-y|^2}{4it}} \phi((y)\, dy \tag{25}$$

Therefore

$$|(e^{-itH_1}\ \phi)(x)| \leq \left(\frac{1}{4\pi t}\right)^{\frac{3}{2}} \int |\phi(y)|dy \tag{26}$$

and then

$$\int_1^\infty |(H_2 - H_1)e^{-itH_1}\phi|_2 dt \leq \int_1^\infty \frac{\|\phi\|_1 \|V\|_2}{(4\pi t)^{\frac{3}{2}}} dt \ = \ C \int_1^\infty \frac{dt}{t^{\frac{3}{2}}} < \infty \tag{27}$$

Therefore condition (c) is satisfied if $V \in L^2(R^3)$ by taking as dense domain $L^2(R^3) \cap L^1(R^3)$. It is easy to show that also conditions (i) and (ii) are satisfied if $V \in L^2(R^3)$ and therefore in this case the wave operators exist. Making use of Hölder inequality in (29) instead of Schwartz inequality and because $t^{-\alpha} \in L^1(1, \infty)$ one verifies that if $\alpha > 1$ the condition on V can be weakened to

$$\int_{R^3} \frac{|V(x)|}{(1 + |x|)^{1-\epsilon}} dx \ < \infty \qquad \epsilon > 0 \tag{28}$$

Remark that from (27) one derives that, as a function of x, $(e^{it\Delta}\phi)(x)$ goes to zero when $t \to \infty$. One refers to this fact by saying that the Schroedinger equation with hamiltonian H_0 has a *dispersive property* (contrary e.g. to the wave equation). Under very mild assumption on $V(x)$ one can prove that also the solutions of the Schroedinger equation with potential $V(x)$ have a dispersive property.

From the proof of the Cook–Kuroda theorem one sees that *the dispersive property* plays an important role in the proof of the existence of the scattering operator. For scattering theory in dimension 3 it is also important to prove that, a part from the common factor $t^{-\frac{3}{2}}$, the rate of decay to zero is not uniform in different spacial directions so that a trace remains of the initial datum.

In particular one can show that if $\phi \in L^2(R^3)$ is sufficiently regular one has

$$lim_{t\to\infty} t^{\frac{3}{2}} |e^{it\Delta}\phi - \phi^{asint}(t)|_2 = 0 \tag{29}$$

where

$$\phi^{asint}(t) \equiv \frac{m}{(it)^{\frac{3}{2}}} e^{i\frac{mx^2}{2t}} \hat{\phi}(\frac{mx}{t}) \tag{30}$$

($\hat{\phi}$ is the Fourier transform of ϕ).

If $\hat{\phi}$ has support in a very small neighborhood of k_0 and one multiplies $e^{-itH_0}\phi$ by a factor $t^{\frac{3}{2}}$ one obtains a function which has essential support in a very narrow cone with vertex in the origin and axis $\hat{k}_0 \equiv \frac{k_0}{|k_0|}$. Therefore at this scale the asymptotic state describes a particle which moves freely in the direction k_0.

We shall come back to this point when we shall discuss the method of V. Enss [5].

4 Existence of the Wave Operators. Chain Rule

In what follows we shall use the symbol $W_\pm$ for the operator $W_\pm(H_2, H_1)$.

Lemma 1 *Set* $H_2 - H_1 \equiv A \in B(\mathcal{H})$ *and* $W(t) \equiv e^{itH_2}e^{-itH_1}$.

If W_+ *exists one has, for every* $\phi \in \mathcal{H}_{1,ac}$

$$|W_+\phi - W(t)\phi|^2 = -2\, Im \int_t^\infty (e^{isH_1}\; W_+^*\; Ae^{-isH_1}\phi, \phi)ds \tag{31}$$

◇

Proof By definition

$$(W_+ - W(t))\phi = i \int_t^\infty e^{isH_1}\; A\; e^{-isH_1}\phi\; ds \tag{32}$$

By unitarity $|(W_+ - W(t))\phi|^2 = 2Re\;((W_+ - W(t))\;\phi, W_+\phi)$. Eq. (31) follows from this together with (32). ♡

From the existence of the wave operators one derives some unitary equivalences. In particular

Theorem 2 (Dollard, Kato [6]) *If the operator* $W_+(H_2, H_1)$ *exists, it is a partial isometry with domain* $\mathcal{H}_{1,ac}$ *and range* $\mathcal{M}_+ \equiv W_+\;\mathcal{H} \subset \mathcal{H}_{2,ac}$.

The orthogonal projection $\mathcal{E}_+$ *on* $W_+\;\mathcal{H}$ *commutes with* H_2. *The restriction of* H_1 *to* $\mathcal{H}_{1,ac}$ *is unitary equivalent to the restriction of* H_2 *to* $W_+\mathcal{H}$.

In particular the absolutely continuous spectrum of H_1 *is contained in the absolutely continuous spectrum* H_2. *Analogous results hold for* W_-.

If both W_+ *and* W_- *exist, then* $S \equiv W_+^*\; W_-$ *commutes with* H_1.

◇

Proof From the definition one has $W_+^*\; W_+ = \Pi_1$ and $W_+\; W_+^* = \mathcal{E}_+$.

On the other hand

$$e^{isH_2}\; W_+ = s - lim_{t\to\infty} W(t+s)e^{isH_1} = W_+e^{isH_1} \tag{33}$$

Multiplying both terms by e^{-izs}, $Imz < 0$ and integrating over s from 0 to $+\infty$ (i.e. taking Laplace transform) one obtains

$$(H_2 - z)^{-1} \, W_+ = W_+ \, (H_1 - z)^{-1} \tag{34}$$

From this follows $W_+ \, H_1 \subset H_2 \, W_+$ and by duality $W_+^* \, H_2 \subset H_1 \, W_+^*$.

$$\mathcal{E}_+ \, H_2 = W_+ \, W_+^* \, H_2 \subset W_+ \, H_1 \, W_+^* = H_2 \, \mathcal{E}_+ \tag{35}$$

This proves that $\mathcal{E}_+$ commutes with H_2 and therefore $\mathcal{M}_+$ reduces H_2.And also that $\mathcal{E}_+ \, H_2 \, \mathcal{E}_+ = H_2 \, \mathcal{E}_+$. Therefore equality holds in (35).

Multiplying with $\mathcal{E}_+$ on the right

$$H_2 \, W_+ = \mathcal{E}_+ \, H_2 \, \mathcal{E}_+ W_+ = W_+ \, H_1 \, \Pi_1 15 \tag{36}$$

From (36) one sees that $H_{1,a.c.}$ is unitarily equivalent to the part $H_{2,a.c;}$ that acts on M_+ and in particular that $\sigma_{a.c.}(H_1) \subset \sigma_{a.c.}(H_2)$.

Analogous results hold for W_- when this operator exists. ♡

Corollary *If $W_+(H_2, H_1)$ exists, one has the following strong convergence properties when $t \pm \infty$*

$$e^{itH_2} \, e^{-itH_1} \Pi_1 \to_s W_+, \quad e^{itH_1} \, e^{-itH_2} \mathcal{E}_+ \to_s W_+^* \tag{37}$$

$$e^{-itH_2} \, W_+ - e^{-itH_1} \, \Pi_1 \to_s 0, \quad e^{itH_1} \, e^{-itH_2} \, W_+ \to_s \Pi_1 \tag{38}$$

$$(W_+ - 1)e^{-itH_1} \, \Pi_1 \to_s 0, \qquad (W_+^* - 1)e^{-itH_1} \, \Pi_1 \to_s 0 \tag{39}$$

$$e^{itH_1} \, W_+ \, e^{-itH_1} \to_s \Pi_1, \quad e^{itH_1} \, W_+^* \, e^{-itH_1} \to_s \Pi_1 \tag{40}$$

$$(1 - \mathcal{E}_+) \, e^{-itH_1} \, \Pi_1 \to_s 0, \quad (1 - \Pi_2) \, e^{-itH_1} \, \Pi_1 \to_s 0 \tag{41}$$

◇

We now prove the *chain rule*.

Theorem 3 (Chain rule) *If both $W_+(H_2, H_1)$ and $W_+(H_3, H_2)$ exist then the operator $W(H_3, H_1)$ exists and one has the chain rule*

$$W_+(H_3, H_1) = W_+(H_3, H_2).W_+(H_2, H_1) \tag{42}$$

◇

Proof The strong limit of a sequence of products of bounded closed operators coincides with the strong limit of the sequence of the factors. Therefore

$$W_+(H_3, H_2).W_+(H_2, H_1) = s - lim_{t\to\infty} e^{itH_3} e^{-itH_2} \ \Pi_2 e^{itH_2} e^{-itH_1} \Pi_1 \tag{43}$$

Since Π_2 commutes with H_2 it follows from (45) that

$$W_+(H_3, H_2).W_+(H_2, H_1) = s - lim_{t\to\infty} e^{itH_3} \ \Pi_2 e^{-itH_1} \Pi_1 \tag{44}$$

On the other hand

$$W_+(H_2, H_1) = s - lim_{t\to+\infty} e^{-itH_3} e^{itH_1} \Pi_1. \tag{45}$$

It is therefore sufficient to prove

$$s - lim_{t\to\infty} e^{itH_3} \ (I - \Pi_2) \ e^{-itH_1} \ \Pi_1 = 0 \tag{46}$$

Due to unitarity of $e^{itH_3} e^{-itH_2}$ it is equivalent to prove (H_2 and Π_2 commute)

$$s - lim_{t\to\infty} (I - \Pi_2) \ e^{itH_2} \ e^{-itH_1} \ \Pi_1 = 0 \tag{47}$$

But $Range W_+(H_2, H_1) \subset \mathcal{H}_{2,ac}$. Therefore $(I - \Pi_2) W_+(H_2, H_1) = 0$

♡

5 Completeness

Definition 3 The wave operator $W_+(H_2, H_1)$ is *complete* if

$$range \ W_+(H_2, H_1) = \mathcal{H}_{2,ac} \tag{48}$$

◇

If both W_+ and W^- exist and are complete, then

$$Range \ W_+(H_2, H_1) = Range \ W_-(H_2, H_1) = \mathcal{H}_{2,ac} \tag{49}$$

Therefore

$$S(H_2, H_1) \ \equiv W_+^*(H_2, H_1) \ W_-(H_2, H_1) \tag{50}$$

is a unitary operator from $\mathcal{H}_{1,ac}$ to $\mathcal{H}_{1,ac}$ A simple corollary of the chain rule is the following

Proposition 1 *If both $W_+(H_2, H_1)$ and $W_+(H_1, H_2)$ exist, then they are complete. The same is true for W_-.*

◇

Notice that in the analysis of the example we have given we have used the explicit form of the integral kernel of e^{itH_1}, or equivalently of the *generalized eigenfunctions* of H_1 i.e. the solution of $H\psi_E = E\psi_E$ which do not belong the the Hilbert space $\mathcal{H}$. To prove the existence of $W_\pm(H_1, H_2)$ it is therefore convenient to have a good control of the generalized eigenfunctions of $H = -\Delta + V$.

A result which can be proved without a detailed analysis of the generalized functions, and at the same time is general enough to cover many physically interesting cases is the following

Theorem 4 (Birman, de Branges, Kato [1, 7]) *The generalized wave operators $W_\pm(H_2, H_1)$ exists and are complete if $(H_2 - z)^{-1} - (H_1 - z)^{-1} \equiv A$ is of trace class for $Imz \neq 0$.*

◇

Proof Let us recall that a trace class operator A con be written as

$$A\phi = \sum_{n=1}^{\infty} c_n (f_n, \phi) f_n, \qquad \sum_n |c_n| < \infty \tag{51}$$

where $\{f_n\}$ is an orthonormal complete basis. Denote by A_N the sum of the first N terms and let $H^N \equiv H_0 + A_N$ so that $H^N - H^{N-1}$ is a rank one projection. The *chain rule* suggests to give first the proof when A is a rank N operator and then consider the limit $N \to \infty$.

In Proposition 2 we shall give the proof for the case of rank one. The chain rule shows then that operator $W_{N,\pm} = W_\pm(H^N, H_0)$ exists for every N.

Recall that by Weyl theorem the absolutely continuous spectrum of H^N does not depend on N. From (47) one derives

$$(W_{n,\pm} - e^{itH^n} e^{-itH^{n-1}})\phi = i \int_t^\infty e^{isH^{n-1}} (H^n - H^{n-1}) e^{-isH^{n-1}} \phi \, ds \tag{52}$$

where $H^n - H^{n-1} = |f_n><f_n|, f_n \in \mathcal{H}$.

Therefore, with $g_n \equiv (W_n^+)^* f_n$

$$|(W_{n,+} - W_n(t))\phi|^2 \leq \int_t^\infty |(e^{-isH^{n-1}} \phi, f_n)|^2 ds]^{1/2} \int_t^\infty |(e^{-isH_1} \phi, g_n)|^2 ds]^{1/2} \tag{53}$$

By iteration one has

$$|[W_\pm(H_2, H_1) - e^{itH_2} e^{-itH_1}]\phi|_2^2$$

$$\leq 2 \left[\sum_{n=1}^{\infty} |c_n| \int_t^\infty |(e^{-isH^{n-1}} \phi, f_n)|^2 ds \right]^{1/2} \left[\sum_{k=1}^{\infty} |c_k| \int_t^\infty |(e^{-isH^{n-1}} \phi, g_n)|^2 ds \right]^{1/2} \tag{54}$$

For ϕ in a dense subset of $\Pi_1 \mathcal{H}$ the right-hand side of (56) is bounded; indeed for any self-adjoint operator H

$$\int_{-\infty}^{\infty} |(e^{-isH_1}\phi, f)|^2 ds \leq 2\pi \; |\phi|_2^2 \; |f|_2^2 \tag{55}$$

where $||\phi||^2 = ess.sup_\lambda \; \frac{d\,(E(\lambda)\phi,\phi)}{d\lambda}$ This follows Parseval's theorem because $\int_{-\infty}^{\infty} e^{-it\lambda}(d(E(\lambda)\phi, f)dt$ is the Fourier transform of $\frac{d}{d\lambda}(E(\lambda)\phi, f)$. If $|\phi|_2 < \infty$ it follows from (57)

$$|(W_{n,+} - W_n(t))\phi|_2 \leq |\phi|_2^{1/2}(8\pi\|A\|_1)^{1/4} \; \eta(t,\phi)^{1/2} \tag{56}$$

where we have set

$$\eta(t,\phi) \equiv \sum_{k=1}^{\infty} |c_k| \int_t^{\infty} |(e^{-isH_1}\phi, f_k)ds \leq 2\pi||\phi|| \; \|A\|_1 \tag{57}$$

From (56) and the triangular inequality it follows

$$|(W_n(\tau) - W_n(t'))\phi|_2 \leq ||\phi||^{1/2}(2\pi\|A\|)^{1/4}\left[\eta(t',\phi)^{1/2} + \eta(\tau,\phi)^{1/2}\right] \tag{58}$$

We can now to the limit $N \to \infty$. Since $\|A - A_n\| \to 0$ one has norm convergence of e^{itH^n} to e^{itH_2} and from (58) it follows

$$|(W(t) - W(\tau))\phi|_2 \leq |\phi|^{\frac{1}{2}}(8\pi|A|_1)^{1/4} \; \left[\eta(t,\phi)^{1/2} + \eta(\tau,\phi)^{1/2}\right] \tag{59}$$

This inequality proves that the $lim_{t\to\infty}$ exists if $||\phi|| < \infty$. Notice now that the set of $\phi \in \Pi_1\mathcal{H}_1$ for which $||\phi|| < \infty$ is dense in $\mathcal{H}_{1;a.c.}$.

Since the collection $W(t)$ is uniformly bounded it follows that the limit exists for $\phi \in \mathcal{H}_{1,ac}$. We conclude that the limit $lim_{t\to\infty} W(t) \; \Pi_1 \equiv W_+$ exists. In the same way one proves the existence of $W_-(H_2, H_1)$. Exchanging the role of H_1 e H_2 one proves the existence of $W_\pm(H_1, H_2)$.

This concludes the proof of Theorem 4 *under the condition* that the following Proposition 2 holds. ♡

Proposition 2 *The generalized wave operators $W_\pm(H_2, H_1)$ and $W_\pm(H_1, H_2)$ exists and are complete if $(H_2 - z)^{-1} - (H_1 - z)^{-1} \equiv A$ is a rank-one operator for $Imz \neq 0$.* ◇

Proof We shall give the proof in several steps.

Step (a)

As first step we shall give the proof in the case $\mathcal{H}$ is identified with $L^2(R, dx)$, H_1 is multiplication by x and the operator $A \equiv H_2 - H_1$ is the rank-one operator

$(A\,u)(x) = (u, f)\,f(x)$ where $f(x)$ is regular and fast decreasing at $\pm\infty$. In this case one has

$$|A\,e^{-itH_1}u|_2 = |f|\,|\int_{-\infty}^{\infty} e^{-itx}u(x)\bar{f}(x)dx| \tag{60}$$

If $u(x)$ is regular and decreases fast enough the integral in (60) is finite. Since the functions with the required properties are dense in $L^2(R, dx)$, the sufficient conditions in Theorem 1 are satisfied. This proves existence.

Step (b)

To extend the proof to the case $f \in L^2(R, dx)$ remark that by (62) and Schwartz's inequality one has

$$|W_+\,u - W(t)\,u|^2 \le 2\left[\int_t^{\infty} |(e^{-isH_1}u, f)|^2 ds\right]^{1/2} \left[\int_t^{\infty} |(e^{isH_1}\,W_+^* f,\,u)|^2 ds\right]^{1/2} \tag{61}$$

This integrals are finite, as one sees using Parseval's inequality

$$\int_{-\infty}^{\infty} |(e^{isH_1}\,W_+^*\,f,\,u)|^2 ds \le 2\pi|f|^2|u|_\infty^2 \tag{62}$$

(by assumption $|u|_\infty$ is finite). Since W_+^* is an isometry we can bound by this term by $C\,|u|_\infty$. From (61) we obtain

$$|W_+\,u - W(t)u| \le (8\pi)^{1/4}|u|_\infty^{1/2}\left(\left[\int_t^{\infty} |(e^{-isH_1}u, f)|^2 ds\right]^{1/4} + \left[\int_T^{\infty} |(e^{-isH_1}u, f)|^2 ds\right]^{1/4}\right) \tag{63}$$

Inequality (63) depends only on the L^2 norm of f and therefore extends to all functions in $L^2(R, dx)$.

Step (c)

Proposition 2 holds true if H_1 is a self-adjoint operator in a Hilbert space $\mathcal{H}$ and $H_2 = H_1 + (., f)f$ with $f \in \mathcal{H}$. To see this, Π_1 be the orthogonal projection on the absolutely continuous part of the spectrum of H_1. Set $f = g + h$, $g = \Pi_1 f$. By assumption $g \in \mathcal{H}_{1,a.c.}$ and therefore g can be represented by a function $g(x)$ on the spectrum.

Consider first the case in which $g(x)$ is regular and rapidly decreasing at infinity. In this case, we can proceed as in case (a), substituting g to f. If $g(x)$ is not regular and/or does not have fast decrease one can proceed by approximation, as in case (b) above since the convergence extends by continuity to $H_{1,a.c.}$.

Step (d)

Consider next the case $H_2 = H_1 + A$ with A of rank one. We treat first the case $\mathcal{H} \equiv L^2(S, dx)$ where S is a Borel set in R^1 and H_1 is multiplication by x.

Let H_1' be the maximal extension of the operator defined as multiplication by x. Then the subspace $\mathcal{H}$ reduces H' and this reduction coincides H.

Let $H_2' = H_1' + (.,f)f$. Also H_2' is reduced by $\mathcal{H}$ and the reductions of $e^{-itH_2'}$ and of $e^{-itH_1'}$ coincide respectively e^{-itH_2} and e^{-itH_1}.

From the existence of $W_+' = s - lim_{t\to\infty} e^{-itH_2'} e^{itH_1'}$ the existence of $W_+(H_2, H_1)$ follows by reduction. In the case in which the spectrum of H_1 is not absolutely, consider as before the projection of f on the absolutely continuous part of the spectrum of H_1.

Step (e)

Let us consider the general case, without assumptions on the structure of $\mathcal{H}$. Let H_1 be self-adjoint and let

$$H_2 = H_1 + (.,f)f, \quad f \in \mathcal{H} \tag{64}$$

Denote by $\mathcal{H}_0$ the smallest subspace of $\mathcal{H}$ which contains f and reduces H_1. Let Π_0 be the orthogonal projection on $\mathcal{H}_0$.

The subspace $\mathcal{H}_0$ can be characterized as the closure of the set of elements in $\mathcal{H}$ that have the form $E_1(\lambda)f$ for all real λ ($E_1(\lambda)$ is the spectral set of H_1).

It follows that also H_2 is reduced by $\mathcal{H}_0$ and

$$H_2\ \Pi_0\ u = H_1\ \Pi_0\ u + (\Pi_0 u, f), \quad f \in H_0 \tag{65}$$

Denote by $\mathcal{H}_0^\perp$ the subspace of $\mathcal{H}$ which is orthogonal to $\mathcal{H}_0$. The subspace $\mathcal{H}_0^\perp$ reduces both H_1 and H_2; if $u \in \mathcal{H}_0^\perp$ one has $H_2\ u = H_1\ u$. To prove existence of $W_+(H_2, H_1)$ it is therefore sufficient to consider only vectors in $\mathcal{H}_0$ and therefore to the case in which $\mathcal{H}_0 \equiv \mathcal{H}$.

Let

$$f = g + h \quad g = \Pi_1 f, \quad h = (I - \Pi_1)f \tag{66}$$

where as before Π_1 is the projection on the absolutely continuous part of the spectrumé of H_1. From the construction we have made we deduce that $H_{1,a.c.}$ is spanned by vectors of the form $E(\lambda)\ g$.

Therefore $\mathcal{H}_{1,a.c.}$ is the closure of vectors of the form

$$\phi(H_1)g = \left[\int_{-\infty}^{\infty} \phi(\lambda)\ dE(\lambda)\right]\ g \tag{67}$$

But

$$(\phi_1(H_1), \phi_2(H_1)) = \int_S (\psi_1(\lambda), \psi_2(\lambda))d\lambda \tag{68}$$

where $k = 1,\ 2$

$$\psi_k(\lambda) \equiv \phi_k(\lambda)\ \rho(\lambda)^{1/2}, \qquad \rho(\lambda) = \frac{d((E_1(\lambda)\ g, g)}{d\lambda} \tag{69}$$

and we have denoted with S the Borel set of all λ for which $\frac{d((E_1(\lambda)\,g,g)}{d\lambda}$ exists and is positive (recall that $g \in \mathcal{H}_{1,a.c.}$). If ϕ spans all measurable bounded functions, then $\psi(\lambda)$ spans a dense subset of $L^2(S)$.

Therefore we can identify $\mathcal{H}_{1,a.c.}$ with $L^2(S)$ through the map $\phi(H_1) : \; g \to \psi$. In this representation of $H_{1,a.c.}$ the operator H_1 is multiplication by x.

We have therefore reduced the problem to the particular cases which we have considered before. This concludes the proof of Proposition 2. ♡

Theorem 4 is important because can be used also in the case of potential scattering with localized impurities. It is enough to choose

$$H_1 \equiv -\Delta + V_{per}, \qquad H_2 = H_1 + W(x) \tag{70}$$

with $V_{per} \in L^2_{loc}$ and W such that $|W(x)|^{1/2}(1-\Delta)^{-1}$ is of trace class.

The spectrum of H_1 is absolutely continuous (and composed in general by bands). Therefore the wave operators $W_\pm(H_2, H_1)$ exist and are complete. Their domain is the entire Hilbert space and the range is the subspace of absolute continuity of H_2.

The wave operators are unitary. ♣

6 Generalizations. Invariance Principle

When we will analyze the time-independent scattering theory we shall see that the assumption $H_2 - H_1 \in J_1$ can be replaced by the weaker one $H_2 - H_1 \in J_2$ (Hilbert-Schmidt class). For this it will be enough e.g. $V(x) \in L^1 \cap L^2$.

It is convenient to generalize the previous results and study the existence of the wave operators $W_\pm$ for hamiltonian that are suitable functions of H_1 and H_2. This will lead to weaker conditions for the existence of the wave operators. A class of allowed functions can be obtained by using the following Lemma.

Lemma 2 *Let $\phi(\lambda)$ be a function on R of locally bounded variation with the property that it is possible to subdivide R in a finite number of open sub-intervals I_k (excluding therefore a locally finite number of points) such that in each of these intervals the function ϕ is differentiable with continuous derivative.*

Under this assumption for every $w \in L^2(R, dx)$ one has

$$2\,\pi\,|w|^2 \geq \int_0^\infty |l.i.m. \int_{-\infty}^\infty e^{-it\lambda - is\phi(\lambda)} w(\lambda) d\lambda|^2 dt \tag{71}$$

where $l.i.m.$ denotes limit in the mean. Moreover the right hand side converges to 0 when $s \to \infty$. ◇

Proof Let $H\ u(x) = x\ u(x)$ and let $\mathcal{F}$ denote Fourier transform. The right hand side of (71) is

$$(2\ \pi)^{1/2} |\eta_{t\geq 0}\ \mathcal{F} e^{-is\phi(H)} w|^2 \tag{72}$$

($\eta_{t\geq 0}$ is the indicator function of the negative semi-axis). Inequality in (73) follows immediately and convergence to zero is equivalent to $s\!-\!lim_{t\to\infty} \Theta_{t\geq 0}\ U\ e^{-is\phi(H)} = 0$.

We can limit therefore to prove convergence to zero for functions that belong to a domain on which H is essentially self-adjoint, for example to indicator functions of finite interval. We can moreover restrict attention to intervals $(a,\ b)$ in which the function ϕ is continuously differentiable. One has

$$v(t,s) \equiv \int_a^b e^{-it\lambda - is\phi(\lambda)} d\lambda = i \int_a^b (t + s\ \phi'(\lambda))^{-1} \frac{d}{d\lambda} e^{-it\lambda - is\phi(\lambda)} d\lambda \tag{73}$$

Under the assumption we have made on ϕ if $t,\ s > 0$ the function $\psi(\lambda) \equiv (t + s\ \phi'(\lambda))^{-1}$ is positive and of bounded variation. Its total variation in $[a, b]$ is such that

$$\int_a^b |d\ \psi(\lambda)| \leq M \frac{s}{(t + c\ s)^2} \leq \frac{M}{c((t + c\ s)} \tag{74}$$

where M is the total variation of $\phi'(\lambda)$ in $[a, b]$ and c is the minimum value of $\phi'(\lambda)$ in the same interval. Integrating by parts the right hand side in (74) one obtains

$$|v(t,s)| \leq \psi(a) + \psi(b) + \int |d\psi(\lambda)| \leq \frac{2\ c + M}{c(t + c\ s)} \tag{75}$$

It follows

$$\int_0^\infty |v(t,s)|^2 dt \leq \frac{2(c + M)^2}{c^2\ s} \tag{76}$$

♡

Using Lemma 2 we shall now prove the following *invariance principle*.

Theorem 5 *Let $H_2,\ H_1$ be self-adjoint operators such that $H_2 - H_1 \in J_1$. Let ϕ be a function on R with the properties described in Lemma 2. Then the generalized wave operators $W_\pm(\phi(H_2),\ \phi(H_1))$ exist, are complete and are independent of ϕ.*

In particular they are equal to $W_\pm(H_2, H_1)$ as one sees choosing $\phi(\lambda) = \lambda$.

◇

Proof We have previously shown that

$$|W_+\ u - W(t)\ u| \leq ||u||(8\ \pi\ \|A\|_1) \tag{77}$$

if u is in the subspace of absolute continuity of H_2 ($\|A\|_1$ is the trace norm of A) and

$$||u||^2 = ess.sup._\lambda \frac{d(E(\lambda u, u)}{d\lambda} \tag{78}$$

With $v \equiv e^{-is\phi(H_1)}\ u$ one has $||v|| = ||u||$. Setting $t = 0$ from (79) we obtain

$$|(W_+ - 1)e^{-is\phi(H_1)}\ u| \leq ||u||(8\ \pi\ \|A\|_1)^{1/4}\ \eta(0, e^{-is\phi(H_1)}\ u)^{1/4} \tag{79}$$

with

$$\eta(0, e^{-is\phi(H_1)}\ u) = \sum_k |c_k| \int |(e^{-itH_1 - is\phi(H_1)}\ u, f_k)|^2 dt \tag{80}$$

The integrals in (79) and (80) have the same structure as the integrals (73) of the previous Lemma if we substitute $\frac{d(E_1(\lambda,u,f_k)}{d\lambda}$ with $w(\lambda)$

Remark that this function belongs to L^2 and its L^2 norm is not larger than $||u||$. Due to Lemma 2 each term in the sum (81) converges to zero when $s \to \infty$. Since the series is dominated uniformly in s by the convergent series $\sum_k |c_k|||u||^2 \equiv |A|_1\ ||u||$ it follows that the entire series converges to zero.

The set of u with $||u|| < \infty$ is dense in $\Pi_1 \mathcal{H}$ and therefore

$$s - lim_{s\to\infty}(W_+ - 1)e^{-is\phi(H_1)}\ \Pi_1 = 0 \tag{81}$$

From $e^{is\phi(H_1)} = \int_{-\infty}^{\infty} e^{-is\phi(\lambda)}\ d\lambda$ it follows

$$W_+\ e^{-is\phi(H_1)} = e^{-is\phi(H_2)}W_+ \tag{82}$$

Multiplying to the left (81) by $e^{is\phi(H_2)}$ one obtains

$$s - lim_{s\to\infty}e^{is\phi(H_2)}\ e^{-is\phi(H_1)}\ \Pi_1 = W_+\ \Pi_1 = W_+ \tag{83}$$

Therefore we prove that $W_+(\phi(H_2), \phi(H_1))$ exists coincides with $W_+(H_2, H_1)$ if we prove that the space of absolute continuity of $\phi(H_1)$ and of H_1 coincide. For the proof we make use of the properties of the function $\phi(\lambda)$.

Let $\{F_1(\lambda)\}$ be the spectral family of $\phi(H_1)$. For any Borel set $S \in R$ one has $F_1(S) = E_1(\phi^{-1}(S))$

If $|S| = 0$ the properties of ϕ imply $|\phi^{-1}(S)| = 0$ and therefore $F_1(S)u = 0$ if $u \in H_{1,a.c.}$. On the other hand $F_1(\phi(S)) = E_1(\phi^{-1}\ [\phi(S)]) \geq E_1(S)$.

If $|S| = 0$ then $|\phi(S)| = 0$ and therefore if u is absolutely continuous with respect to $\phi(H_1)$ one has $|E_1(S)\ u| \leq |F_1(\phi(S))\ u| = 0$. This shows that the absolutely continuous spectrums of H_1 and of $\phi(H_1)$ coincide and concludes the proof of Theorem 5. ♡

Specializing the function ϕ one obtains useful criteria for the existence of the wave operators and for asymptotic completeness. In particular

Theorem 6 *Let H_2 and H_1 be strictly positive operators on a Hilbert space H.*

If for some $\alpha > 0$ the difference $H_2^{-\alpha} - H_1^{-\alpha}$ is trace-class, then the wave operators $W_\pm(H_2, H_1)$ exist, are complete and coincide with $W_\mp(H_2^{-\alpha}, H_1^{-\alpha})$. ♢

Proof Let γ be the smallest between the lower bound of the spectra of H_2 and H_1. Consider the function defined as $\phi(\lambda) \equiv -\lambda^{-\frac{1}{\alpha}}$ for $\lambda \geq \gamma$ and by $\phi(\lambda) = \lambda$ for $\lambda < \gamma$. It is easy to verify that this function satisfies the requirements of Lemma 2. ♡

We shall use now Theorem 6 to prove asymptotic completeness of the wave operator for the system

$$\mathcal{H} = L^2(R^3), \qquad H_1 = -\Delta, \qquad H_2 = H_1 + V \tag{84}$$

where V is the operator of multiplication by $V(x) \in L^1 \cap L^2$. We shall use a particular case of the following theorem

Theorem 7 (Kato) *Let H_1 be self-adjoint and bounded below. Let V a symmetric operator relatively bounded wit respect to H_1 with bound less then one. Assume that V can be written as $V = V_1 \ V_2$ with $V_k \ (H_1 + z)^{-1}$, $k = 1, 2$ of Hilbert-Schmidt class when z is negative and smaller than the lower bound of the spectrum of H_1. Then the wave operators $W(H_2, H_1)$ and $W(H_1, H_2)$ exist and are complete.* ♢

Proof There is no loss of generality in assuming that H_1 and H_2 are strictly positive; therefore one can choose $z = 0$. By assumption $V_k \ H_1^{-1} \in J_2, \quad k = 1, 2$. To this class belongs also $V_k \ H_2^{-1}$ since J_2 is a bilateral and $(H_1 + c\ I)\ (H_2 + c\ I)^{-1}$ is bounded. One has

$$\frac{1}{H_2} - \frac{1}{H_1} = \frac{1}{H_2} V \frac{1}{H_1} = \frac{1}{H_2} V_1 \ V_2 \frac{1}{H_1} \in J_1 \tag{85}$$

and the thesis of the theorem follows from Theorem 6. ♡

Theorem 7 can be used to prove asymptotic completeness when $V(x) \in L^1 \cap L^2$. Notice that $V \in L^2(R^3)$ implies that V is infinitesimal relative to $-\Delta$.

Therefore in order to apply Theorem 8 it suffices to prove $V\ (-\Delta + c)^{-1} \in J_2$ per $c > 0$.

The integral kernel of this operator is

$$|V(x)|^{1/2} \frac{e^{-c|y-x|}}{4\pi|x-y|} \tag{86}$$

and this is of Hilbert–Schmidt class because

$$\int\int |V(x)|e^{-2c|x-y|}\ |x-y|^{-2} dx\, dy \leq \int |V(x)| dx \int e^{-2|y|}\ |y|^{-2} dy < \infty \tag{87}$$

Let us consider now the continuity of the dependence of $W_\pm(H_2, H_1)$ on H_2 e H_1. We shall prove continuity at least for perturbations of trace class.

Theorem 8 *Let H_2 and H_1 be self-adjoint and such that $W_\pm(H_2, H_1)$ exist. Then for each $A \in J_1$ the wave operator $W_\pm(H_2 + A, H_1)$ and $W_\pm(H_2, H_1 + A)$ exist, and when A converges to zero in J_1 one has, in the strong operator topology*

$$W_\pm(H_2 + A, H_1) \to W_\pm(H_2, H_1), \qquad W_\pm(H_2, H_1 + A) \to W_\pm(H_2, H_1) \tag{88}$$

◇

Proof Existence follows from Theorem 7. Moreover from the chain rule

$$W_\pm(H_2 + A, H_1) = W_\pm(H_2 + A, H_2)\ W_\pm(H_2, H_1) \tag{89}$$

It is therefore sufficient to consider the case $H_2 = H_1$. From the estimates obtained in the proof of Theorem 7 one has

$$|W_\pm(H_1 + A, H_1)u \ - \ u| \leq ||u||(4\pi||A||_1)^{1/2} \tag{90}$$

The thesis of the theorem follows then from the density of $\{u, \ : \ ||u|| < \infty\}$ in $\Pi_1\mathcal{H}$. ♡

Stronger continuity results can be obtained from Theorem 7. It can be proved e.g. that if A_n is a sequence of operators which converge to zero in strong resolvent sense, i.e. if for any $z_0 \notin R$ one has

$$lim_{n\to\infty}|(H_2 + A_n - z_0)^{-1} - (H_2 - z_0)^{-1}| = 0 \tag{91}$$

then $s - lim W_\pm(H_2 + A_n, H_1) = W_\pm(H_2, H_1)$.

For a detailed analysis of asymptotic completeness in quantum scattering theory one can usefully consult [6].

References

1. V. Amrein, J. Jauch, K. Sinha, *Scattering Theory in Quantum Mechanics* (V.Benjamin, Reading Mass, 1977)
2. D. Yafaev, Trans. Am. Math. Soc. **189**, 255–285
3. I.M. Gel'fand, B.M. Levitan, Trans. Am. Math. Soc. **54**, 253–304 (1955)
4. S. Kuroda, Nuovo Cimento **12**, 431–454 (1959)
5. V. Enss, *Chapter Notes in Mathematics* vol 1159 (1984)
6. S. Agmon, Spectral properties of schroedinger operators and scattering theory. Ann Scuola Normale Pisa **2**, 151–218 (1975)
7. M. Reed, B. Simon, *Methods of Modern Mathematical Physics*, vol II (Academic Press, 1975)
8. J. Dollard, Comm. Math. Phys. **12**, 193–203 (1969)
9. W. Amrein, J. Zuleta, Helv. Phys. Acta **70**, 1–38 (1997)

Lecture 11: Time Independent Formalisms. Flux-Across Surfaces. Enss Method. Inverse Scattering

At the beginning of Lecture 10 we have remarked that scattering Theory in Quantum as in Classical Mechanics, describes those effects of the interaction of a system of N particles which can be measured when the components of the system have become spatially separated so that the mutual interactions have become negligible.

As in Lecture 10, we limit ourselves here to a system of two quantum particles which interact through potential forces which are invariant under translation. In this case the problem can be reduced to the problem of one particle in interaction with a potential force. We remarked that scattering theory in the one-body problem with forces due a potential V is essentially the comparison of the asymptotic behavior in time of the system under two dynamics given by two self-adjoint operators H_1 and H_2.

We shall treat in some detail the case in which the ambient space is R^3, both systems are described in cartesian coordinates, and the Hamiltonians describing the free (asymptotic) motion and the motion during interaction are respectively

$$H_1 = -\frac{\hbar^2}{2m}\Delta \qquad H_2 = -\frac{\hbar^2}{2m}\Delta + V(x) \tag{1}$$

where m is the mass of the particle and $V(x)$ is the interaction potential. In general we shall choose units in which $2m = \hbar = 1$.

We shall make stringent assumptions on the potential $V(x)$, and in particular that it be Kato-small with respect to the laplacian so that the operator $-\frac{\hbar^2}{2m}\Delta + V$ is (essentially) self-adjoint. As in Lecture 10 we will assume also that $V(x)$ vanishes sufficiently fast at infinity (e.g. $lim_{|x|\to\infty}|x|^p\ V(x) = 0$ for a suitable value of $p > 1$).

The theory can also be applied when H_1 is periodic in space; this is the case if one describes scattering of a particle by a crystal.

In Lecture 10 we have formulated scattering theory as the comparison between the asymptotic behavior for $t \to \pm\infty$ of a generic element in $\mathcal{H}$ that evolves according the dynamics given H_2, $\phi(t) \equiv e^{-itH_2}\phi$, and the behavior of two elements $\phi_\pm(t)$ which evolve according to H_I and differ very little from $\phi(t)$ when $t \to \pm\infty$.

G. Dell'Antonio, *Lectures on the Mathematics of Quantum Mechanics II: Selected Topics*, Atlantis Studies in Mathematical Physics: Theory and Applications 2, DOI 10.2991/978-94-6239-115-4_11

The theory presented in Lecture 10 is the *time dependent formulation of scattering theory* because all definitions and theorems refer explicitly to temporal evolution. In this lecture we shall analyze a formulation called *time-independent (or stationary) scattering theory* which is centered on the analysis of the generalized eigenfunctions of the operators H_2 and H_1.

This formulation predates the time-dependent one and, although less intuitive, in the case $H_2 = -\Delta + V, \quad H_1 = -\Delta$ provides existence and completeness of the wave operators (or rather of their generalization) under *weaker conditions* on the potential V. Since the time-independent version is less intuitive, it is convenient to give first the connection between the two approaches. This will also clarify the role of the resolvents of $(H_k - \lambda)^{-1}, \quad k = 1, 2$ in the proof of existence of the wave operators.

In time-independent scattering theory the wave operators are found *as solutions of suitable functional equations*. To find these equations it is convenient to go back to the time-dependent formulation.

We now extend the previous definition of wave operator $W_+(H_2, H_1)$ by requiring convergence of $e^{-itH}e^{itH_0}$ for $t \to \pm\infty$ only in the sense of Abel. We shall define therefore

$$W'_+ \equiv lim_{\epsilon\to 0} 2\epsilon \; lim_{T\to\infty} \int_0^T e^{-2\epsilon t} e^{itH_2} \, e^{-itH_1} \, \Pi_1 \, dt$$
$$= lim_{\epsilon\to 0} 2\epsilon \; lim_{T\to\infty} \int_0^T e^{-\epsilon t + itH_2} \, [e^{-\epsilon t - itH_1}] \Pi_1 dt \qquad (2)$$

where the limit is understood in an abelian sense.

If W_+ exists, also W'_+ exists (and the two operators coincide). *The converse is not true*. It is convenient to recall the relation between the group of unitary operators e^{itH} and the *resolvent* of the self-adjoint operator H.

Under the assumption that H be bounded below by mI one has, for λ real and strictly less than m

$$i(H - \lambda + i\epsilon)^{-1} = \int_0^\infty e^{-\epsilon t} e^{it(H-\lambda)} dt \qquad (3)$$

for any $\epsilon > 0$ (make use of the spectral representation of H). Parseval's relation between Fourier transforms leads to

$$W'_+ = lim_{\epsilon\to 0} \frac{2\epsilon}{2\pi} \int_{-\infty}^0 (H_2 - \lambda - i\epsilon)^{-1} \, (H_1 - \lambda + i\epsilon)^{-1} \Pi_1 d\lambda \qquad (4)$$

It is convenient to write (4) in a different form *before taking the limit* $\epsilon \to 0$. Let $R(z) \equiv (H - z)^{-1}$ be the resolvent of the operator H and $E(\lambda)$ be its spectral family. By definition with $z = \lambda + i\epsilon, \quad \lambda \in R$

$$R(\bar{z})R(z) = \int_{-\infty}^{\infty} \frac{dE(\mu)}{(\mu - \bar{z})(\mu - z)}$$

$$= \int_{-\infty}^{\infty} \frac{dE(\mu)}{(\mu - \lambda)^2 + \epsilon^2} = \int_{-\infty}^{\infty} \frac{1}{\epsilon} \delta_\epsilon(\mu - \lambda) dE(\mu) \tag{5}$$

with

$$\delta_\epsilon(\mu - \lambda) \equiv \frac{\epsilon}{(\lambda - \mu)^2 + \epsilon^2} \tag{6}$$

The difficulty in taking the limit $\epsilon \to 0$ in (4) lies in the fact that the limits must be taken from different half-planes in the resolvent of H_2 and in that of H_1. To overcome this problem one proceeds as follows. On suitable domains one has

$$(H_2 - \lambda - i\epsilon)^{-1}(H_2 - \lambda + i\epsilon)^{-1}(H_2 - \lambda + i\epsilon)(H_1 - \lambda + i\epsilon)^{-1}$$

$$= (H_2 - \lambda - i\epsilon)^{-1}(H_1 - \lambda + i\epsilon)^{-1} \tag{7}$$

From (4) and (5) one has then

$$W'_+ = lim_{\epsilon \to 0} \int_{-\infty}^{\infty} \delta_\epsilon(H_2 - \lambda) \; G(\lambda + i\epsilon) d\lambda \; \Pi_1 \tag{8}$$

where we have defined for $Imz \neq 0$

$$G(z) = (H_2 - z)(H_1 - z)^{-1} \tag{9}$$

When $\epsilon \to 0$ the function δ_ϵ convergence (in the sense of measures) to the distribution δ at the origin. Therefore, in the weak sense

$$W'_+ = \int_{-\infty}^{\infty} \frac{dE_2(\lambda)}{d\lambda} G(\lambda + i0) \; d\lambda \; \Pi_1 \tag{10}$$

In the corresponding formula for W'_- the factor $G(\lambda + i0)$ is replaced by $G(\lambda - i0)$. Hence, at least formally,

$$W'_\pm = \int_{-\infty}^{\infty} \frac{dE_2(\lambda)}{d\lambda} \; G(\lambda \pm i0) \; d\lambda \; \Pi_1 \tag{11}$$

Remark that the boundary value $G(z)$ *may not be a continuous function,* and the derivative of the spectral measure *may only exist in distributional sense.* Therefore without further assumptions the definition of $W'_\pm$ is ill-posed.

We will prove that under suitable assumption on the potential V one can prove that the limit exists as a continuous map *between different function spaces*. This result goes under the name of *limit absorption principle*.

In stationary scattering theory whenever (11) is well posed it is *the definition* of (generalized) wave operator. One proves then that the operator so defined has all the properties of the wave operator defined in the time-dependent theory [1, 2].

Indeed under general assumptions one the pair H_2, H_1 one proves that $W'_\pm$ are isometries with domain $\mathcal{H}_{1,ac}$ and range $\mathcal{H}_{2,ac}$ and that $W'_\pm$ intertwine the groups e^{itH_2} and e^{itH_1}.

Under more restrictive assumptions one proves that $W_\pm(H_2, H_1) = W'_\pm(H_2, H_1)$ (without these further assumptions one proves only existence of $W'_\pm(H_2, H_1)$). Let $H_2 = H_1 + A$. One has

$$W(\tau) - W(t) = i \int_t^\tau e^{isH_2}\ A\ e^{-isH_1}\ ds \tag{12}$$

Similarly exchanging H_1 and H_2

$$W(\tau)^{-1} - W(t)^{-1} = -i \int_t^\tau e^{isH_1}\ A\ e^{-isH_2}\ ds \tag{13}$$

If A is unbounded (12) and (13) are valid in a suitable domain.

Let us assume that $W_+(H_2, H_1) \equiv s - lim\ W(t)\ \Pi_1$ exists. Multiply (12) to the left by $-W_+$, choose $t = 0$, take the limit $\tau \to \infty$ and use $e^{-itH_2} W_+ = W_+ e^{-itH_1}$ to obtain

$$W_+ - \Pi_1 = i\ lim_{\tau\to\infty} \int_0^\tau e^{itH_1}\ A\ W_+ e^{-itH_1} ds \tag{14}$$

where the limit is understood in the strong sense if A is bounded, in the weak sense otherwise.

To simplify notations it is convenient to introduce the following map $\Gamma^\pm_{H_1}(A)$, $A \in \mathcal{B}(\mathcal{H})$

$$\Gamma^\pm_{H_1}(A) = i\ lim_{\tau\to\infty} \int_0^{\pm\tau} e^{itH_1}\ A\ e^{-itH_1} dt \qquad A \in \mathcal{B}(\mathcal{H}) \tag{15}$$

if the limit exists in a weak or strong sense. With this notation (14) reads (for the sake of simplicity we omit the dependence on H_2 and H_1 and we write Γ_1 for Γ_{H_1}).

$$W_+ = \Pi_1\ +\ \Gamma_1^+\ (A\ W_+) \tag{16}$$

and similarly

$$W_- = \Pi_1\ +\ \Gamma_1^-\ (A\ W_-) \tag{17}$$

1 Functional Equations

Operators $W'_{\pm}(H_2, H_1)$ which satisfy (8) are found as solutions of *functional equations* (16) and (17). This construction has the virtue to allow iterative and approximate methods of solutions. In this scheme, the operator $W_{\pm}$ corresponds to a *strong solution* while $W'_{\pm}$ corresponds to a *weak solution*. If the solution $W_{\pm}$ exists and is unique, then $W'_{\pm} = W_{\pm}$.

As remarked above, the stationary formulation of scattering theory takes (16) and (17) as fundamental equations and determines $W_{\pm}$ as their (weak or strong) solutions. We must now show that the solutions have all the properties of the wave operators introduced in the time-dependent formulation.

In the case $\mathcal{H} = L^2(R^3)$, $H_1 = -\Delta$ and A multiplication by a function $V(x)$ the Eqs. (16) and (17) are an operator-theoretical version of the *Lippmann–Schwinger equation* for the generalized eigenfunctions of $-\Delta + V$.

Notice that while in the time dependent formalism the definition of Wave operator is based on the large time behavior of the solutions of the Schroedinger equation with hamiltonian H, in the time-independent formalism it is based on the *properties of the resolvent operator* $(H - z)^{-1}$ for $Imz \to 0$.

The relation between the two strategies is given by Paley-Wiener type theorems.

We return now to the time-independent approach. We will show that the solutions $W'_{\pm}$ of (16) and (17) coincides with the wave $W_{\pm}$ *when both are defined*. Remark that $B \in \mathcal{B}(\mathcal{H})$ commutes with H then $A \in D(\Gamma^{\pm})$ implies that both BA and AB belong to $D(\Gamma^{\pm})$ and

$$\Gamma^{\pm}(B\ A) = B\ (\Gamma^{\pm}(A)), \qquad \Gamma^{\pm}(A\ B) = (\Gamma^{\pm} A)\ B \tag{18}$$

We will consider only Γ^{+}: analogous results are valid for Γ^{-}.

Lemma 1 *Let $A \in D(\Gamma^{+})$ and define $R \equiv \Gamma^{+}(A)$. Then $R\ D(H) \subset D(H)$ and for every $u \in D(H)$ the following identity holds*

$$A\ u = R\ H\ u - H\ R\ u \tag{19}$$

Moreover $s - lim_{t\to\infty} R\ e^{-itH} = 0$. ◇

Proof Multiplying (16) from the left by e^{itH} and form the right by e^{-itH}

$$R(t) \equiv e^{itH}\ R\ e^{-itH} = i \int_t^{\infty} e^{isH}\ A\ e^{-isH} ds \tag{20}$$

Moreover $\frac{dR(t)}{dt} = -ie^{itH}\ A\ e^{-itH}$. Therefore if $u \in D(H)$ then

$$\frac{d}{dt} e^{itH}\ R\ u = -ie^{itH} Au + iR(t) e^{itH} Hu \tag{21}$$

This shows that $e^{itH} Ru$ is strongly differentiable in t; therefore $Ru \in D(H)$ and

$$\frac{d}{dt} e^{itH} \; R\, u = i e^{itH} \; H \; R \; u \tag{22}$$

For $t = 0$ on obtains

$$H \; R\, u = -A\, u + R \; H \; u \tag{23}$$

and the first part of the Lemma is proved. The second part follows from (15). ♡

Using Lemma 1 we now prove that the solution W'_+ of (16) coincides with the wave operator W_+ if the latter exists. In the proof we limit ourselves to the case in which the perturbation is a bounded operator. In this case both operators are defined.

Theorem 1 *Let H_1 be self-adjoint and A bounded and symmetric. Assume that $W_\pm \in \mathcal{B}(\mathcal{H})$ is a solution of (16) Then the generalized wave operators exist and $W'_\pm \equiv W_\pm(H_1 + A, H_1)$ where $W_+(H_2, H_1)$ is defined in time-dependent scattering theory.* ◇

Proof Since $W'_+ - \Pi_1 = \Gamma_1^+(A\; W'_+)$ it follows form Lemma 1 that

$$(W'_+ - \Pi_1)\; H_1\; u = -H_1\; (W'_+ - \Pi_1)\; u = W'_+\; H_1\; u \tag{24}$$

and therefore $W'_+ \; H_1 \subset H_2 \; W'_+$ and for any $z \notin R$

$$(H_2 - z)^{-1}\; W'_+ = W' + \; (H_1 - z)^{-1} \quad e^{itH_2}\; W'_+ = W'_+\; e^{itH_1} \quad t \in R \tag{25}$$

From Lemma (1) one derives $s - lim_{t\to\infty}(W'_+ - \Pi_1)e^{-itH_1} = 0$ and therefore, multiplying to the left by e^{itH_2}

$$W'_+ = s - lime^{itH_2}\; e^{-itH_1}\; \Pi_1 \tag{26}$$

An analogous result holds for W'_-. This concludes the proof of Theorem 1. ♡

We have seen that for scattering by a potential $V(x)$ in stationary scattering theory the wave operators are the solutions of the equation

$$W'_\pm = I + \Gamma^\pm(V W'_\pm) \tag{27}$$

where $\Gamma^\pm$ is defined on a suitable class of functions as

$$\Gamma^{\pm}(A) \equiv \int_0^{\pm\infty} e^{itH_0} A e^{-itH_0} dt \ , \qquad H_0 = -\Delta \tag{28}$$

These equations can be solved using different strategies. One can e.g. iterate equation $X = I - \epsilon\Gamma^{\pm}(VX)$ for sufficiently small values of the parameter ϵ and prove that the resulting solution can be continued to $\epsilon = 1$. Alternatively one can use fixed point techniques, either by contraction or by compactness (in the latter case one must prove uniqueness by other means).

2 Friedrich's Approach. Comparison of Generalized Eigenfunctions

We shall give some details of still another technique [2], which makes use of the properties of the operators $\Gamma^{\pm}$. This approach, often employed in the textbooks in Theoretical Physics, goes back to K. Friedrichs and consists in a comparison between the generalized eigenfunctions of $H = -\frac{1}{2m}\Delta + V$ and the ones of $H_0 = -\frac{1}{2m}\Delta$. The starting point is again (27), which must be satisfied by W'_+; in our case it reads

$$W'_+ = I + i\int_0^{\infty} e^{-itH_0} V W'_+ e^{itH_0} dt \tag{29}$$

The same holds for W'_-. Since the operator Γ must have in its domain the generalized eigenfunctions of H_0 it is convenient to interpret (28) in distributional sense, or equivalently to consider the limit as $\epsilon \to 0$ of the solutions of equation

$$W'_+ = I + i\int_0^{\infty} e^{-itH_0} V W'_+ e^{itH_0 - \epsilon t} dt \tag{30}$$

The functions $\phi^0_k(x) \equiv \frac{1}{(2\pi)^{3/2}} e^{ik\cdot x}$ are the *generalized eigenfunctions* of H_0 relative to the eigenvalue $\frac{k^2}{2m}$. The corresponding generalized eigenfunctions of H are then

$$\phi_k = W'_+ \phi^0_k \tag{31}$$

The map $\phi^0_k \to \phi_k$ given by the solution of (31) (with W'_+ solution of (24)) for $\epsilon > 0$ can be extended to a map between bounded differentiable functions. This extended map can be continued to $\epsilon \to 0$ under suitable regularity assumptions on the potential V. From (30)

$$\phi_k(x) = \phi^0_k(x) + lim_{\epsilon\to 0} i \int_0^{\infty} (e^{-itH_0 + i\frac{k^2}{2m}t - \epsilon t} V \phi_k)(x) dt \tag{32}$$

and therefore

$$\phi_k(x) = \phi_k^0(x) - lim_{\epsilon \to 0} \left(H_0 - \frac{k^2}{2m} - i\epsilon \right)^{-1} V\phi_k(x) \tag{33}$$

Equation (33) takes the name of *Lippmann–Schwinger equation*. If the integral on the right-hand side exists one can write as an integral equation

$$\phi_k(x) = \frac{1}{(2\pi)^3/2} e^{ik \cdot x} - \frac{m}{2\pi} \int \frac{e^{i|k||x-y|}}{|x-y|} V(y)\phi_k(y) d^3y \tag{34}$$

If the potential is of short range (e.g. $|V(x)| \leq C|1 + |x|)^{-\alpha}$ where $2\alpha > d + 1$ (d is space dimension) one verifies that the solution $\phi_\lambda(x)$ of the stationary equation

$$-\Delta\phi(x) + V(x)\phi(x) = \lambda\phi(x) \tag{35}$$

has the following asymptotic form when $|x| \to \infty$

$$\phi_k(|x|, \omega) = \frac{1}{(2\pi)^{3/2}} e^{i\lambda^{\frac{1}{2}}(x,\omega)} + \frac{1}{(2\pi^{3/2}} a(\phi, \omega, ; \lambda)|x|^{-\frac{(d-1)}{2}} \frac{e^{i\lambda^{\frac{1}{2}}|x|}}{|x|} + o(|x|^{-\frac{(d-1)}{2}}) \tag{36}$$

with $\omega = \frac{x}{|x|}$.

3 Scattering Amplitude

Notice that the right hand side of (36) is, modulo higher order terms, the sum of a plane wave and of a spherical wave multiplied by a factor $a(\phi, \omega; \lambda)$ that depends on ω (the direction of the incoming wave) and on the direction of $\hat{x}$. This factor takes the name of *scattering amplitude*.

In the Physical Literature the scattering amplitude is defined decomposing the solution of the Schroedinger equation in *incoming and outgoing* spherical waves

$$\phi(x) = r^{-\frac{d-1}{d}} [\gamma b_+(\omega) e^{i\lambda^{\frac{1}{2}}|x|} - \bar{\gamma} b_-(-\omega) e^{-i\lambda^{\frac{1}{2}}|x|}] + o(|x|^{-\frac{d-1}{2}}) \tag{37}$$

where $\gamma = e^{i\pi\frac{d-3}{4}}$. Notice that the notation *incoming* and *outgoing* comes from a time-dependent analysis. This decomposition can be proven by stationary phase techniques under suitable assumptions, e.g. the existence of a constant ρ such that $\int_{|x|<\rho} |\phi(x)|^2 dx < C\rho$. In this notation the S-matrix $\mathcal{S}$ is defined as the operator that satisfies $b_+(\omega) = (\mathcal{S}b_-)(\omega)$. Notice that the S-matrix is for $d \geq 2$ a unitary operator on $L^2(S^{d-1})$.

From stationary scattering theory one derives

$$\mathcal{S}(\lambda) = I - 2\pi i \Gamma_0(\lambda)(V - VR(\lambda + i0)V\Gamma_0^* \tag{38}$$

with $R(\lambda + i0) = (H - \lambda - i0)^{-1}$ and

$$(\Gamma_0(\lambda)\phi)(\omega) = \frac{1}{\sqrt{2}}\lambda^{\frac{d-2}{4}}(2\pi)^{-2}\int_{R^d} e^{-i\lambda^{\frac{1}{2}}(x,\omega)}\phi(x)dx \tag{39}$$

We shall see in the next lectures that the *Limit Absorption Principle*, valid for short range potentials, guarantees that $(H - \lambda - z)^{-1}$, $Imz \neq 0$ can be continued, for $Im\, z \to \pm 0$ to a bounded continuous operator $R(z)$ on $\mathcal{H}_\beta$, $\beta > \frac{1}{2}$ with values in $\mathcal{H}_{-\beta}$ where

$$\mathcal{H}_\beta \equiv \{f \ : \ \int_{R^d}(x^2 + 1)^\beta |f(x)|^2 dx \equiv \|f\|_\beta < \infty\} \tag{40}$$

In time-independent (sometime called *stationary*) scattering theory the S matrix $\mathcal{S}$ is defined by

$$\mathcal{S}(\lambda) = I - 2\pi i \Gamma_0(\lambda)(V - VR(\lambda + i0)V)\Gamma_0^*(\lambda) \tag{41}$$

Remark that the product can be regarded to be the product of bounded operators between different spaces and that, using the resolvent identity, the operator S can be rewritten as

$$\mathcal{S} = I - 2\pi i \Gamma_0(\lambda)V\Gamma_0^*(\lambda) \tag{42}$$

From this one sees that the two definition of S-matrix coincide. We don't give here the details of the proof and refer to [1] see also [2, 8].

4 Total and Differential Cross Sections; Flux Across Surfaces

Starting with this definition of S-matrix, with partly heuristic considerations one defines the *total cross section* and the *differential cross section*. The latter determines, for a beam of particles of momentum approximately equal to k_0 which cross the region where the gradient of the potential is localized, the percentage of those outgoing particles which have momentum approximately equal to k.

To concude this brief description of the *time-independent* method in Scattering Theory we mention the *flux across surfaces theorem* that connects the more mathematical aspect of time-independent scattering theory with the presentation on textbooks more oriented to Theoretical Physics.

In these textbooks in discussing quantum scattering theory from a potential V one considers the probability density of the following event: a particle *enters with momentum* $k_0 \neq 0$ the region Ω in which the force ∇V is different from zero and exits from Ω with momentum contained in a solid angle Σ.

Of course since the incoming particle is represented by a function in $L^2(R^3)$, it cannot have momentum precisely equal to k_0. In this formulation of the scattering process a limiting process is implied implicitly.

One can imagine a beam of N particles which do not interacting among themselves and are scattered by a potential. Each particle in a remote (but not too remote) time and at very large distance from the support of the potential has distribution in momentum space *approximately* equal to $\delta(k_0)$ and distribution *almost uniform* on a plane perpendicular to $\hat{k}_0$. Only a fraction of these particles reaches the region Ω and the probability to exit in the solid angle Σ refers *only to this fraction of the particles* (i.e. it is a conditional probability).

In most text of Theoretical Physics this leads to substitute the wave function of the incoming particles by the plane wave e^{ik_0x} and let the number of incoming particles go to infinity. This balances the fact that the percentage of particles which reach the interaction region goes to zero if one takes a uniform distribution in a plane perpendicular to k_0.) We are interested only in the particles that have interacted.

If S is the scattering matrix, one considers therefore the operator $T \equiv S - I$. A heuristic argument shows that the operator T has integral kernel (in Fourier transform)

$$\pi i \delta(k^2 - p^2) T(k, p) \tag{43}$$

where $T(k, p)$ is a smooth function.

The presence of the delta function reflects the conservation of energy for the asymptotic motion, due to the intertwining property of the wave operators.

By formal manipulations one shows that the probability density that a particle which enters with momentum k_0 and undergoes scattering is emitted in a solid angle Σ is

$$\sigma_{diff}^{k_0}(\Sigma) = 16\pi^4 \int_\Sigma |T(\omega\, |k|_0,\; k_0)^2 d\omega \tag{44}$$

The function $\sigma_{diff}^{k_0}$ is called the *differential cross section*. To find a heuristic connection between (43) and the scattering operator a defined in this lecture recall that in the time independent scattering theory the generalized eigenfunction corresponding to momentum k is obtained as solution of the Lippmann–Schwinger equation

$$\phi(x, k) = e^{-ik.x} - \frac{1}{2\pi} \int \frac{e^{-i|k||x-y|}}{|x_y|} V(y)\phi(y, k) d^3y \tag{45}$$

and its asymptotic behavior for large $|x|$ is

$$\phi(x,k) \simeq e^{-ik_0.x} + f^{k_0}(\omega)\frac{e^{-i|k||x|}}{|x|} \tag{46}$$

From (45) the integral kernel of T can be expressed as function of $\phi(x,t)$ as follows

$$T(k,p) = \frac{1}{2\pi}\int e^{-ikx}V(x)\phi(x,p)d^3x \tag{47}$$

Comparing terms of order $|x|^{-1}$ in (45) and (46) one sees that

$$f^{k_0}(\omega) = (2\pi)^{-1}\int e^{i|k_0|\omega.y}V(y)\phi(y,k_0)d^3y \tag{48}$$

and therefore $f^{k_0}(\omega) = 4\pi^2 T(\omega|k_0|, k_0)$. One arrives in this way to (44). This connection of (44) with the formalism of scattering theory does not clarify the connection with the measurements that one performs to measure the *cross section*.

We shall therefore mention briefly the relation between (44) and the scattering process based on the *theorem of flux across surfaces*.

A description of the scattering process closer to the experimental realization is the following.

In a scattering experiment the particles, after interaction, are recorded when they cross an array of counters situated at a large distance R from the region in which the scattering takes place. The distance must be large enough to consider the outcoming particles as free particles.

What is measured is the number of particles exiting *in a given direction*. In general one measures quantities that integrated over time, i.e. one does not determine the precise exit time. In other words, the scattering process is quantified by measuring the flux of particles which cross, between time T and T' a portion Σ of the area of a sphere placed at distance R from the origin.

If the radius R is large enough this quantity can be considered as independent of the precise localization of the interaction region. Recall that in Quantum Mechanics the *flux* is defined as follows

$$j^{\phi_t} \equiv Im\ \phi_t^* \nabla \phi_t \tag{49}$$

It satisfies the *continuity equation*

$$\frac{\partial \rho_t}{\partial t} + div j^{\phi_t}, \quad \rho_t(x) = |\phi_t(x)|^2 \tag{50}$$

One is tempted to assume that the probability for the particle to cross the portion Σ of spherical surface in the interval of time $T \leq t \leq T + \Delta$ is

$$\int_{\Sigma} d\sigma \int_{T}^{T+\Delta} (n \cdot j^{\phi_t})(\sigma, t)\, dt \tag{51}$$

where $n(\sigma, t)$ is the outward normal to the surface of the sphere in the point of coordinates σ.

This cannot be true in a strict sense, since $(n \cdot j^{\phi_t})(\sigma, t)$ may be negative (and even not well defined since the function may be non-differentiable). But we expect that it becomes non negative when $R \to \infty$ since we expect that the incoming portion of the wave vanish in that limit.

A more appropriate definition of cross section may be then

$$\sigma^{\phi}_{flux}(\Sigma) = lim_{R\to\infty} \int_{T}^{\infty} dt \int_{R\Sigma} (n \cdot j^{\phi_t})\, d\omega \tag{52}$$

where $R\Sigma$ is the intersection of the sphere of radius R with the cone generated by Σ and a point P in the support of ∇V. When $R \to \infty$ this quantity is independent from P.

Remark that the definition (52) does not depend on T since we have assumed that ϕ be a scattering state. Therefore we expect that the following theorem holds

Flux-across-surfaces theorem
One has

$$lim_{R\to\infty} \int_{T}^{\infty} \int_{R\,\Sigma} j^{\phi_t} d\Sigma = \int_{C_\Sigma} |\Omega_{+}^{-1}\, \phi(k)|^2 d^3k \tag{53}$$

◇

This theorem has been proved under various assumptions on the potential. One can consult e.g. [1, 2, 4]. It is worth noticing that in the course of the proof it also shown that in the limit $R \to \infty$ the measure $(n \cdot j^{\phi_t})\, d\omega$ converges to a positive measure and one has

$$lim_{R\to\infty} \int_{T}^{\infty} dt \int_{R\Sigma} (n \cdot j^{\phi_t}) d\omega = lim_{R\to\infty} \int_{T}^{\infty} dt \int_{R\Sigma} |(n.j^{\phi_t})| d\omega \tag{54}$$

Condition for this to be true are given by the *limit absorption principle* that we shall discuss in the next lectures. The physical intuition which suggest the analysis of the flux across surfaces is also at the basis of the alternative approach to Quantum Scattering Theory initiated to V. Enss, based on *a geometric analysis* of the behavior for $t \to \pm\infty$ of the solutions of Schroedinger's equation for initial data in the subspace of absolute continuity for the hamiltonian H.

5 The Approach of Enss [3, 5]

We have seen in Book I that the structure of free propagation is such that the behavior for $t \to \pm\infty$ of the solutions of the free Schroedinger equation differs little from free propagation along the direction of momentum. We recall briefly this analysis. Define for $t \neq 0$ the operators $M(t)$ and $D(t)$ through

$$M(t)(\phi x) = e^{-\frac{x^2}{2t}}\phi(x) \qquad D(t)f(x) = |t|^{-\frac{d}{2}}\phi\left(\frac{x}{t}\right) \tag{55}$$

One has
(a) For $|t| \neq 0$ $M(t)$ and $D(t)$ are isomorphisms of $\mathcal{S}'$ and of $\mathcal{S}$ and are unitary in $L^2(R^d)$.
(b)

$$U_0(t) = e^{\mp i\frac{d\pi}{4}} M(t)D(t)\mathcal{F}M(t) \tag{56}$$

($\mathcal{F}$ denotes Fourier transform). Defining for $t > 0$

$$(T(t)\phi)(x) = e^{\mp\gamma(d)}e^{i\frac{x^2}{2t}}\left(\frac{1}{t}\right)^{\frac{d}{2}}\hat{\phi}\left(\frac{x}{t}\right) \tag{57}$$

the operators $T(t)$ are unitary in $L^2(R^d)$ and one has, for every $\phi \in L^2(R^d)$

$$lim_{t\to\infty}\|[U_0(t) - T(t))]\phi\|_2 = 0 \tag{58}$$

The probability distribution in configuration space tends asymptotically to

$$\frac{1}{t^d}|\tilde{\phi}\left(\frac{x}{t}\right)|^2 dx = |\hat{\phi}(\xi)|^2 d\xi, \qquad \xi = \frac{x}{t} \tag{59}$$

Remark that this is the distribution in position of a classical free particle which is at the origin at time zero with $|\hat{\phi}(\xi)|^2$ as distribution if momentum. If the initial state is a gaussian $\psi(0,x) = Ce^{i\frac{|x-x_0|^2}{2}+i(x,p_0)}$ (which has as Fourier transform a gaussian centered in p_0) the solution of the free equation at time t is still a gaussian centered in tp_0 and with variance in x of order $t^{\frac{1}{2}}$.

If we choose a new (time dependent) coordinate system in which the space variables are scaled by a factor t^α, $0 < \alpha < \frac{1}{2}$ (and therefore momenta are scaled by $t^{-\alpha}$) in the new variables the variance tends to zero for $t \to \infty$ while the distance between the centers of two gaussians corresponding to different values of the momenta grows like $t^{\frac{1}{2}-\alpha}$.

On this scale the two wave packets are far apart in the far future. At the same time the range of the potential increases under dilation. The generator of this change of variables is the (dilation) operator $D = \frac{1}{2}(x.p + p.x)$. This suggests that the comparison with free motion will be successful only if the potential decays sufficiently rapidly at infinity.

We shall see later that a sufficient decay is $lim_{|x|\to\infty}|x|^{\frac{3}{2}}V(x) = 0$ (as suggested by dimensional analysis) and we shall give a more precise definition of *short range potentials*. Under free motion the observable x^2 satisfies

$$\frac{d^2}{dt^2}\hat{x}^2 = -[H_0, [H_0, \hat{x}^2]] \qquad H^0 = -\frac{1}{2}\Delta \tag{60}$$

Let $D = \frac{1}{2}(\hat{x}.\hat{p} + \hat{p}.\hat{x})$ be the generator of the group f space dilation. Then

$$[H_0, \hat{x}^2] = 2D, \qquad [H_0, D] = H_0, \qquad [H_0, [H_0, \hat{x}^2] = 2H_0 > 0 \tag{61}$$

Therefore for every ϕ setting $\phi(t) = e^{itH_0}\phi$ one derives

$$\frac{d^2}{dt^2}(\phi(t), x^2\phi) = 2(\phi(t), H\phi(t)) = 2(\phi, H\phi) \tag{62}$$

As a consequence if $(\phi, H\phi) > 0$

$$\frac{\bar{x}^2(t)_\phi}{t^2} \simeq Ct^2 \tag{63}$$

Of course in the free case we can obtain more detailed information from the explicit knowledge of the solution. From this brief analysis of the case $V = 0$ we draw the following simple conclusions: the dilation group plays an important role, the asymptotic motion is linear in time (ballistic) and the double commutator $[H_0, [H_0, X]]$ is positive and strictly positive above the onset of the continuum spectrum.

6 Geometrical Scattering Theory

The considerations, trivial if referred to free motion, have inspired a method elaborated by V. Enss [5–7] *Geometric Scattering Theory* a procedure that defines the wave operators placing emphasis *on the asymptotic properties of the solutions*. This method provides relevant information for potential scattering and can extended to the N body problem [6].

Later the method was generalized and put in more abstract form by Mourre [8] and it has acquired a central role in the modern scattering theory in Quantum Mechanics. The method of Mourre has been further generalized and applied to the N-body problem in [9].

We introduce now briefly Geometric scattering theory; it will be discussed more in detail in the next lecture.

Definition 1 *(space of scattering states)* Let ξ_{B_R} the indicator function of the ball of radius R centered at the origin. Define *space of scattering states* relative to the hamiltonian H the set

$$\mathcal{M}_\infty(H) \equiv \{\phi \in \mathcal{H} : lim_{t\to\pm\infty} |\xi_{B_R} e^{-itH}\phi|_2 = 0 \quad \forall R > 0\} \tag{64}$$

♢

This definition captures our expectation that if a particle is in a scattering state the probability to find it in a bounded region of space tends to zero as $t \to +\infty$.

Definition 2 *(space of bound states)* Define *space of bound states* the set

$$\mathcal{M}_0(H) \equiv \{\phi \in \mathcal{H} : lim_{R\to\infty} sup_t |(I - \xi_{B_R}) e^{-itH}\phi|_2 = 0\} \tag{65}$$

♢

This definition captures our expectation if a particle is in a bound state the probability to find it outside a ball of radius R vanishes when $R \to \infty$.

With these definitions existence and completeness of the wave operators $W_\pm(H, H_0)$ (with $H_0 = -\Delta$ and $H = H_0 + V$) may be stated in the following way

Proposition 1 (Enss [5]) *Let* $V \in L^2(R^3) + L^\infty(R^3)$ *and assume that* $\mathcal{H}_{sing} = \emptyset$. *Then*

$$\mathcal{M}_\infty(H) = \mathcal{H}_{a.c.} \qquad \mathcal{M}_0(H) = \mathcal{H}_p \tag{66}$$

♢

Notice that the spectrum of the hamiltonian H is continuous but not absolutely continuous. For every element $\phi \in \mathcal{H}_{con}$ the following weaker property holds

$$lim_{T\to\infty} \frac{1}{2T} \int_{-T}^{T} \|\xi_{B_R} e^{-itH}\phi\| dt = 0 \tag{67}$$

Moreover for every $\phi \in \mathcal{H}$

$$\frac{1}{T} \int_0^T \|\xi_{B_R} e^{-itH}\phi\| dt \le f_R(T) \|(H + i\ I)\phi\| \qquad lim_{T\to\infty} f_R(T) = 0 \tag{68}$$

This is an *ergodicity property*.

An important role in Geometric Scattering Theory is taken by the RAGE theorem (from the names Ruelle, Amrein, Georgescu, Enss) which illustrates the geometrical method we will describe presently.

We begin with a theorem of Wiener which has an independent interest. Recall that a *Baire measure* is finite and charges at most a denumerable collection of points.

Theorem (Wiener) *Let μ be a finite Baire measure on R and define $F(t) = \int e^{-ixt} d\mu(x)$. Then*

$$lim_{T\to\infty}\frac{1}{2T}\int_{-T}^{T}|F(t)|^2dt = \sum_{x\in R}|\mu(\{x\})|^2 \tag{69}$$

◇

Proof One has

$$\frac{1}{2T}\int_{-T}^{T}|F(t)|^2dt = \int d\mu(x)h(T,x) \tag{70}$$

where $h(T,x) \equiv \int d\mu(y)(T(x-y))^{-1}\sin((T(x-y))$. The integrand is uniformly bounded and when $T \to \infty$ the integral converges to zero if $y \neq x$ and to one if $y = x$. Therefore by the dominated convergence theorem

$$lim_{T\to\infty}\frac{1}{2T}\int_{-T}^{T}|F(t)|^2dt = \sum_{x\in R}|\mu(\{x\})|^2 \tag{71}$$

♡

We now state and prove the RAGE theorem.

Theorem (RAGE) *Let H be a self-adjoint operator and C a bounded operator such that $C(H+iI)^{-1}$ be compact. Denote by $\Pi_{cont}(H)$ the orthogonal projection on the continuous spectrum of H. Then*
(a) There exists a function $\epsilon(T)$ such that $lim_{T\to\infty}\epsilon(T) \to 0$ and for every $\phi \in D(H)$

$$\frac{1}{2T}\int_{-T}^{T}|Ce^{-itH}\Pi_{cont}\phi|_2^2dt \leq \epsilon(T)|(H+i)\phi|_2^2 \tag{72}$$

(b)

$$lim_{T\to\infty}\frac{1}{2T}\int_{-T}^{T}|Ce^{-itH}\Pi_{cont}\phi|_2^sdt = 0, \quad s = 1,2 \tag{73}$$

(c)

$$\frac{1}{2T}\int_{-T}^{T}|Ce^{-itA}\Pi_{cont}\phi|_2dt \leq \epsilon(T)^{1/2}|(H+i)\phi|_2 \tag{74}$$

◇

Proof Remark that (b) follows from (a) for a simple density argument and that (c) follows form (a) and (b) by Schwartz's inequality. Setting $\psi = (H+i\ I)\phi$ one can assume that C is compact and substitute $(H+i)\phi$ with ϕ. Let

$$\epsilon_C \equiv sup_{\phi \neq 0} \|\phi\|_2^{-2} \frac{1}{2T} \int_{-T}^{T} |Ce^{-itH} \Pi_{cont}(H)\phi|^2 dt \tag{75}$$

Since $\|\epsilon_C(T)\| \leq \|C\|$ it is sufficient to consider the case when C has rank one. $\Pi_{cont}(H)$ commutes with H and therefore it suffices to prove that if $\Pi_{cont}(H)\psi = \psi$ then

$$\frac{1}{2T} \int_{-T}^{T} |(\psi, e^{-itH}\phi)|^2 dt \leq \epsilon(T)\|\phi\|^2 \tag{76}$$

where $lim_{T\to\infty}\epsilon(T) = 0$.

By the spectral representation of H we have $(\psi, e^{-itH}\phi) = \int e^{-itx} h(x) d\mu(x)$. Making use once more of Schwartz's inequality

$$\frac{1}{2T} \int_{-T}^{T} |(\psi, e^{-itH}\phi)|_2^2 dt \leq \|\phi\|_2^2 \delta(T) \tag{77}$$

where

$$\delta(T) = \left[\int d\mu(x) d\mu(y) \left| \frac{sen((x-y)T)}{(x-y)T} \right|^2 \right]^{\frac{1}{2}} \tag{78}$$

The thesis of the RAGE theorem follows now from Wiener theorem. ♡

The RAGE theorem provides convergence *in the mean*; for the existence of the wave operator *strong convergence* is required, and for this the essential spectrum of H must be absolutely continuous.

7 Inverse Scattering Problem

The *inverse scattering problem* is the possibility to determine uniquely the potential from the knowledge of the S matrix. We shall use a geometric method, proposed also in this context by Enss. We shall study only the case of short range potentials which are Kato small with respect to the Laplacian and such that

$$G_V(R) \equiv \|\xi(|y| \geq R)V(y)(-\Delta + I)^{-1}\| \in L^1(R), \qquad y \in R^d \tag{79}$$

We denote by $\mathcal{V}_S$ the collection of these potentials. For them the wave operators

$$\Omega_{\pm,V} = s - lim_{t\to\pm\infty} e^{it(-\Delta+V} e^{-itH_0} \tag{80}$$

exist and are complete and the operator $S(V) \equiv (\Omega_{+,V})^* \Omega_{-,V}$ is unitary.

Define the *scattering map*

$$\mathcal{V}_S \ni V \to S(V) \tag{81}$$

We shall prove that this map in injective: the knowledge of the the S matrix determine the potential uniquely. Define *long range* the class of $\mathcal{V}_L$ of such that for a positive constant C

$$V^L \in C^4(R^d), \quad |D^\alpha V^L(y)| \le C(1+|y|)^{-1-\alpha(\epsilon+\frac{1}{2})}, \quad 1 \le \alpha \le 4, \quad 0 < \epsilon < \frac{1}{2} \tag{82}$$

Then the wave operators are complete if the reference hamiltonian is chosen to be

$$H^D = H_0 + V^L\left(t\frac{p}{m}\right) \tag{83}$$

Also in this case the scattering map is injective, but the proof of this statement is more elaborated. It should also be noted that for *short range potentials* the potential is completely determined by the knowledge of scattering data *at on fixed energy.*

The proof of injectivity of the scattering map is based on some a-priori estimates that we will state; for some of them we give complete proofs. More details can be found in [7]. We shall make use of the following Lemma, which is proved in the next lecture.

Lemma 2 [9] *For each function $f \in C_0^\infty(R^d)$ that has support in the ball B_η for each choice of the integer k it is possible to find a positive constant C_k such that*

$$\|\xi(x \in \mathcal{M}')\, e^{itH_0}\, \xi(p - mv)\xi(x \in \mathcal{M})\| \le C_k(1 + r + |t|)^{-k} \tag{84}$$

for every $v \in R^d$, $t \in R$ and every pair of measurable sets $\mathcal{M}$, $\mathcal{M}'$ for which

$$r \equiv dist\{\mathcal{M}', \mathcal{M}\} > 0 \tag{85}$$

◇

To show injectivity we need *separation estimates.*

Lemma 3 [9] *If the potential V satisfies for some $\rho \in [0, 1]$ and ever function $g \in C_0^\infty$ the estimate*

$$(1 + R)^\rho \|v(x)g(p)\xi(|x| > R)\| \in L^1((0, \infty), dR) \tag{86}$$

then for every function $f \in C_0^\infty(B_\eta)$ it is possible to find function h such that $(1 + \tau)h(\tau) \in L^1((0, \infty))$ and, for every $v \in R^d$, $|v| \ge 4\eta$ the following inequality holds

$$\|V(x+tv)e^{-itH_0}f(p)(1+x^2)^{-\frac{3}{2}}\| \le h(|vt|) \tag{87}$$

◇

Proof (*outline*) From Lemma 3

$$\mathcal{M} = \{|x| \le c|vt|\| \;\; \mathcal{M}' = \{|x| \ge C|vt|\} \qquad c < C \tag{88}$$

If C is chosen appropriately, for r large enough one has

$$\|Vg(p-mv)\|\|\xi(|x-vt|) > C|vt|))\, e^{-itH_0}\, f(p-mv)\left(\frac{1}{(1+|x|^2)}\right)^{\frac{3}{2}} \xi(|x| < c|vt|c)\|$$
$$\le k(1+c|vt|)^{-3} \tag{89}$$

moreover

$$\|Vg(p-mv)\|\|\xi(|x-vt|) > C|vt|)\, e^{-itH_0} f(p-mv)\left(\frac{1}{(1+|x|^2)}\right)^{\frac{3}{2}} \xi(|x| < c|vt|)$$
$$\| \in h_1(|vt|) \tag{90}$$

where $(1+y)^{\rho}h_1(y) \in L^1((0,\infty))$ under our assumption on the potential. Lemma 2 follows from (89) to (90). ♡

Notice the following corollary.

Corollary *If $\hat{\phi}_0 \in C_0^\infty(B(\eta)$ then, uniformly in $t \in R$*

$$\|(\Omega_\pm - I)e^{-itH_0}\phi_v\| = O(v^{-1}), \qquad \hat{\phi}_v(p) = \hat{\phi}_0(p-mv) \tag{91}$$

◇

Proof Let ϕ_0 be a wave function such that $\hat{\phi}$ has support in the ball of radius η and let ϕ_v be defined by $\hat{\phi}_v(p) = \hat{\phi}(p-mv)$.

From Duhamel's formula one derives

$$(\Omega_+ - I)e^{-itH_0}\psi = i\int_0^\infty d\tau e^{i\tau H_0} V e^{-i(t+\tau)H_0} \tag{92}$$

Using (87) one obtains (92). ♡

To conclude we give a *reconstruction formula* that gives the potential once the scattering matrix is known. This formula gives the potential by giving in each point $x \in R^d$ the integral of the potential along rays that originate from x (tomography); a theorem of Radon guarantees existence and uniqueness.

Theorem 2 (reconstruction formula [9]) *If (86) holds, then for each pair of functions which satisfy (84) one has*

$$((S - I)\phi_v, \psi_v) = \frac{i}{v} \int_{-\infty}^{\infty} d\tau V(x + \tau v)\phi_0, \psi_0) + o(v^{-\rho+1}) \tag{93}$$

◇

Proof (*outline*) By definition $\mathcal{S} - I = (\Omega_+ - \Omega_-)\Omega_-$. From Duhamel's formula one derives

$$i(\mathcal{S} - I)\phi_v = \int_{-\infty}^{\infty} e^{itH_0} V \Omega_- e^{-itH_0}\phi_v, \quad \hat{\phi}_v(p) = \hat{\phi}(p - mv) \tag{94}$$

Since $\Omega_- D(H_0) \subset D(H)$ one has

$$(\psi_v, i(\mathcal{S} - I)\phi_v) = \int_{-\infty}^{\infty} P_v(vt)dt + R(v) \tag{95}$$

where the principal term P_v and the residual term $R(v)$ are given respectively by

$$P_v(vt) = (e^{-itH_0}\psi_v, V(x)e^{-itH_0}\phi) \quad R_v = \int_{-\infty}^{\infty} ((\Omega_- - I)e^{-itH_0}\psi_v, V(x)e^{-itH_0}\phi)dt \tag{96}$$

It follows from the preceding results that

$$|R_v| \leq C \int_{-\infty}^{\infty} |V e^{-itH_0}\phi_v|_2 l \leq C \int_{-\infty}^{\infty} h(|vt)dt \tag{97}$$

This term satisfies therefore the requirements of the theorem. The term P_v can be rewritten as

$$P_v(t) = (V(x + vt)e^{-itH_0}\psi_o, e^{-itH_0}\phi) \tag{98}$$

Setting $\tau = vt$ one has, pointwise in τ

$$lim_{|v| \to \infty} P_v(\tau) = (V(x + \tau\hat{v})\psi_0, \phi_0) \tag{99}$$

and from

$$|P_v(\tau)| \leq C\|V(x)e^{-i\frac{\tau}{v}H_0}\frac{f(p-mv)}{(1+x^2)^{\frac{3}{2}}}\| \leq C_1 h(|\tau|) \tag{100}$$

Write $P_v(\tau)-$ as $P_v^1 + P_v^2$ where

$$P_v^1 = (V(x+vt)e^{-itH_0}\psi_0, (e^{-itH_0} - I)\phi_0), \quad P_v^2 = ((e^{-itH_0} - I)\psi_0, V(x+\tau\hat{v})\phi_0) \tag{101}$$

Since $\hat{\phi}_0$ is normalized to one and has compact support

$$|(e^{-i\frac{\tau}{v}} - I)\phi_0 P_v^2| \leq |H_0\phi_0|_2|\frac{\tau}{|v|}|, \qquad |(e^{-i\frac{\tau}{v}} - I)\phi_0 P_v^2| \leq 2 \tag{102}$$

From Lemma 2 one derives then

$$|P_v^1(\tau)| \leq \frac{C}{v^\rho}|\tau|^\rho h(|\tau|) \tag{103}$$

Since $lim_{|v|\to\infty} P_v^1(|v|^\rho(\tau) = 0$ from the dominated convergence theorem follows

$$\int_{-\infty}^{\infty} P_v^1 = 0(v^{-\rho}) \qquad 0 \leq \rho \leq 1 \tag{104}$$

For $\rho = 1$ one obtains $O(|v|^{-1})$.

As for the term P_v^2 one obtains analogous estimates by making use of $(1-\tau)^\rho|\xi(|x| > \frac{\tau}{2})\phi_0|_2 \in L^1((0,\infty))$. ♡

From Theorem 2 one derives

Corollary *The* scattering map *is injective.* ◇

Proof Suppose that V_1 and V_2 are short range potentials with the same scattering matrix. Denote by V their difference. In what follows we consider only vectors z which belong to a prefixed plane, which we choose to be $\{1,\ 2\}$.

Let ϕ and ψ be elements of $L^2(R^d)$, $d \geq 2$, such that $\hat{\phi},\ \hat{\psi} \in C_0^\infty(R^d)$. Define

$$\phi_z = e^{-ipz}\phi, \quad \psi_z = e^{-ipz}\psi \quad f(z) \equiv (V\phi_z, \psi_z) \tag{105}$$

This function is bounded and continuous. Under the assumption stated, we can choose $g \in C_0^\infty$ and such that $g(p)\phi = \phi$. We have then $g \in L^2(R^2, dz)$; indeed

$$|f(z)| \leq |Vg(p)\phi_z|_2 \leq \|Vg(p)\xi(|x| > \frac{|z|}{2})\| + \|Vg(p)\||\xi(|x| \leq \frac{|z|}{2})\phi_z|_2 \tag{106}$$

Choosing v in the $\{1,\ 2\}$ *plane*, the Radon transform of f is by defnition

$$\tilde{f}(v, x) = \int_{-\infty}^{\infty} f(z + \tau v)d\tau = \int_{-\infty}^{\infty} (V(x + \tau v)\phi_z, \psi_z)d\tau \qquad (107)$$

and by Theorem 2 this function is zero. Since $f \in L^2(R^2, dz)$ it follows $f(z) = 0$ due to the properties of the Radon transform. In particular $f(0) = 0$ and therefore $(V\phi, \psi) = 0$ if $\hat{\phi},\ \hat{\psi} \in C_0^\infty$, a dense set. It follows $V = 0$ as an operator, and therefore also as a function. ♡

References

1. S.Th. Yafaev, *Mathematical Scattering Theory*, Translations of Mathematical Monographs (American Mathematical Society, Providence, 1992)
2. W. Amrein, J. Jauch, K. Sinha, *Scattering Theory in Quantum Mechanics: Physical Principles and Mathematical Methods* (Reading, W.A. Benjamin, 1977)
3. V. Enss, in *Rigorous Atomic and Molecular Physics*, ed. by G. Velo, A. Wightman. Geometric Methods in Spectral and Scattering Theory (Plenum, New York, 1981)
4. V. Enss, Comm. Math. Phys. **89**, 245–268 (1983)
5. V. Enss, *Lecture Notes in Mathematics*, vol. 1159. p. 39–178 (Springer New York, 1985)
6. W. Hunziker, Math. Physics **41**, 3448–3510 (2000)
7. A. Mourre, Comm. Math. Phys. **78**, 391–408 (1981)
8. M. Reed, B. Simon, *Methods of Modern Mathematical Physics*, vol II (Academic Press, Cambridg, 1975)
9. V. Enss, R. Weder Journ. of Mathematical. Physics **36**, 3902–3921 (1995)

Lecture 12: The Method of Enss. Propagation Estimates. Mourre Method. Kato Smoothness, Elements of Algebraic Scattering Theory

In this lecture we give more details of an alternative approach to quantum scattering theory, initiated by V. Enss, This approach is based on *a geometric analysis* of the behavior for $t \to \pm\infty$ of the solutions of Schroedinger's equation for initial data in the subspace of absolute continuity for the hamiltonian H.

As we have seen in the previous lecture, by proving that the spectrum of H has not a singular continuous part one gains a complete control of the asymptotic properties for any initial data, and this corresponds to asymptotic completeness.

We have seen in Book I that an interesting property of free propagation is that the behavior for $t \to \pm\infty$ of the solutions of the free Schroedinger equation differs little from free propagation along the direction of momentum. We recall briefly this analysis.

Define for $t \neq 0$ the operators $M(t)$ and $D(t)$ by $M(t)(\phi x) = e^{-\frac{x^2}{2t}}\phi(x)$ and $D(t)f(x) = |t|^{-\frac{d}{2}}\phi(\frac{x}{t})$. One has (*Lemma 3.10* in vol.I).

(a) For $|t| \neq 0$ $M(t)$ and $D(t)$ are isomorphisms of $\mathcal{S}'$ and of $\mathcal{S}$ and are unitary in $L^2(R^d)$.
(b)

$$U_0(t) = e^{\mp i\frac{d\pi}{4}} M(t)D(t)\mathcal{F}M(t) \tag{1}$$

($\mathcal{F}$ denotes Fourier transform). Recall (*Theorem 3.10* in Book I) that, defining for $t > 0$

$$(T(t)\phi)(x) = e^{\mp\gamma(d)} e^{i\frac{x^2}{2t}} \left(\frac{1}{t}\right)^{\frac{d}{2}} \hat{\phi}\left(\frac{x}{t}\right) \tag{2}$$

the operators $T(t)$ are unitary in $L^2(R^d)$ and one has, for every $\phi \in L^2(R^d)$

$$lim_{t\to\infty} \|[U_0(t) - T(t))]\phi\|_2 = 0 \tag{3}$$

This theorem states that the probability distribution in configuration space tends asymptotically to

G. Dell'Antonio, *Lectures on the Mathematics of Quantum Mechanics II: Selected Topics*, Atlantis Studies in Mathematical Physics: Theory and Applications 2, DOI 10.2991/978-94-6239-115-4_12

$$\frac{1}{t^d}|\tilde{\phi}\left(\frac{x}{t}\right)|^2 dx = |\hat{\phi}(\xi)|^2 d\xi, \qquad \xi = \frac{x}{t} \tag{4}$$

Remark that this is the distribution in position of a classical free particle which is at the origin at time zero with $|\hat{\phi}(\xi)|^2$ as distribution if momentum. If the initial state is a gaussian $\psi(0,x) = Ce^{i\frac{|x-x_0|^2}{2}+i(x,p_0)}$ (which has as Fourier transform a gaussian centered in p_0) the solution at time t of the free equation is still a gaussian centered in tp_0 and with variance in x of order $t^{\frac{1}{2}}$.

Since the equation of motion are linear from the knowledge of the Gaussian case one derives the asymptotic structure of any (smooth) initial datum.

The method of Enss is a comparison, for a given initial datum, of the asymptotic structure of the wave function the interaction dynamics with the asymptotic structure corresponding to the free dynamics.

The geometric properties of these asymptotic propagations show that, for a dense subset in the support of the absolutely continuous spectrum of $-\Delta + V$ and for a suitable class of potentials V, the asymptotic (in time) spacial behavior of the wave function with the potential V differs little from that the free case.

In particular, in the remote future and at large spatial distances most of the states *in the absolutely continuous part of the spectrum* of H are represented by *outgoing waves* from a sphere of radius sufficiently large so that that outside the sphere the potential is very small. At the same time the component that describes *incoming waves* becomes negligible $t \to +\infty$.

We have seen that free propagation can approximated by a family of maps which, a part for a phase factor, are isometric dilations $\phi(x) \to (\frac{1}{t})^{\frac{d}{2}}\hat{\phi}(\frac{x}{t})$.

One can expect that, at least for short-range potential, the same be true for a quantum particle interaction though a potential V. If this is the case, it is useful to use a system of coordinates which dilate in time.

It is natural therefore to study the group generated by time translations and dilations. The generators of these subgroups do not commute.Therefore it is natural to study their commutator. In the free case one has $[D, H_0] = 2H_0$.

In the free case the method of stationary phase shows that the part of the wave function that corresponds to the negative part of the spectrum of D (which corresponds roughly speaking to incoming waves) has the property to become negligible for large enough times.

One can expect that these considerations can be extended to the interacting case and that also in that case the spectral properties of $[D, H] = 2H_0 + [D, V]$ be important for the proof.

Notice that $e^{i\lambda D}V(x)e^{i\lambda D} = V(\frac{x}{\lambda})$ and therefore $i[D, V] = \frac{d}{d\lambda}V(\frac{x}{\lambda})$. The property of having a negligible incoming part must hold for *scattering states*, that correspond to the positive part of the spectrum of D. On the contrary for *bound states* we expect that the outgoing part be negligible.

To turn these semi-heuristic remarks into a rigorous proof it is necessary to have convenient *a-priori estimates*.

In its original form Enss' method makes use of a decomposition of the Hilbert space that follows as closely as possible the behavior of classical trajectories in phase space (we have seen that for free motion this is possible).

The purpose is to prove that any state that belongs to the *continuum spectrum* of H can be approximated, in the far future and on a suitable scale of space, by and *outgoing state* and in the remote past by an *incoming state*. And to prove that this implies that *on the states of the continuous spectrum* of $H = H_0 + V$ there is unitary equivalence between the dynamics due to the hamiltonian $H_0 + V$ and to H_0.

But then the continuous spectrum of H is absolutely continuous and this implies asymptotic completeness.

1 Enss' Method

We give some details of the method of Enss. Choose a new (time dependent) coordinate system in which the space variables are scaled by a factor t^α, $0 < \alpha < \frac{1}{2}$ (and therefore momenta are scaled by $t^{-\alpha}$).

In the new variables under free motion *the variance* of the wave function tends to zero for $t \to \infty$ while the distance between the centers of two gaussians corresponding to different values of the momenta grows like $t^{\frac{1}{2}-\alpha}$. On this scale two wave packets are *far apart* in the far future.

In the presence of an interaction potential, one should keep in mind that the range of the potential increases under dilation.

This suggests that the comparison with free motion will be effective only if the potential decays sufficiently rapidly at infinity to compensate for this increase. The role of the parameter α will be to quantify this compensation.

We shall see later that a sufficient decay is $lim_{|x|\to\infty}|x|^{\frac{3}{2}}V(x) = 0$ (as suggested by dimensional analysis) and we shall give a more precise definition of *short range potentials*.

Under free motion the observable $\hat{x}$ satisfies

$$\frac{d^2}{dt^2}\hat{x}^2 = -[H_0, [H_0, \hat{x}^2]] \qquad H^0 = -\frac{1}{2}\Delta \tag{5}$$

Recall that $D = \frac{1}{2}(\hat{x} \cdot \hat{p} + \hat{p} \cdot \hat{x})$ and

$$[H_0, \hat{x}^2] = 2D, \qquad [H_0, D] = H_0, \qquad [H_0, [H_0, \hat{x}^2] = 2H_0 > 0 \tag{6}$$

Therefore for every ϕ, denoting by $\phi(t) = e^{itH_0}\phi$ the unitary propagation, it follows

$$\frac{d^2}{dt^2}(\phi(t), x^2\phi) = 2(\phi, H\phi) > 0 \tag{7}$$

For the average $< x^2 >_\phi (t)$ of $|x|^2$ over the state described by $\phi(t)$ one has, asymptotically in t

$$\frac{< x^2 >_\phi (t)}{t^2} \simeq C \tag{8}$$

Of course in the free case we can obtain much more detailed information from the explicit knowledge of the solution. Our purpose here is to find a method that provides information also in the case V different from 0 and can extended to the N body problem.

From this brief analysis of the case $V = 0$ we can draw the following simple conclusions: the dilation group plays an important role, the asymptotic motion is linear in time (ballistic) and the double commutator $[H_0, [H_0, X]]$ is positive and strictly positive above the onset of the continuum spectrum. The considerations, trivial if referred to free motion, have inspired the method elaborated by V. Enss [1–3].

Later the method was generalized and put in more abstract form by Mourre [4] and it has acquired a central role in the modern scattering theory in Quantum Mechanics. In this lecture will describe also Mourre's method. The method has been further generalized and applied to the N-body problem in [5].

The method of Enss relies on the intuitive nature of scattering theory by comparing, for a given initial datum, the *asymptotic structure* of the wave function for the free and for the interacting dynamics.

The geometric properties of these propagations show that, for a dense subset in the support of the absolutely continuous spectrum of $-\Delta + V$ and for a suitable class of potentials V, the asymptotic (in time) spacial behavior of the wave function with the potential V differs little from the free case.

As remarked, in its original form Enss' method makes use of a decomposition of the Hilbert space that follows as closely as possible the behavior of classical trajectories in phase space (we have seen that for free motion this is possible). In this sense it may be considered as a *semiclassical method*.

This decomposition makes use of free motion and dilation group: neglecting dispersion the support of the outgoing states is obtained by dilating the initial support. We give here only an outline of the method of Enss; for a detailed and clear exposition we refer to [3, 5–7].

Compared with the time-dependent and time-independent methods described before the strength of Enss' method is on the physical intuition that for a system of two particles once the effect of the interaction has (almost) disappeared the particles separate from each other and the vector that describes their separation grows linearly in time and becomes parallel to the relative velocity.

This can be seen as a *localization of the state in phase space*. The localization becomes weaker in the course of time (due to dispersive effects) but still sufficient to separate asymptotically states that correspond to different momenta.

The separation will be *less than in the classical case* (classically these states are asymptotically separated by a distance proportional to t).

One of the advantage of Enss' method is that it is close to the phenomenological description of the scattering process. This approach provides a closer connection with the terminology employed in a large part of the Theoretical Physics books in scattering theory, in particular in the definition of *total cross section* and *differential cross section.* It leads therefore precise estimates (or bounds) on the physically relevant quantities.

2 Estimates

We now provide some details. In order to turn these heuristic remarks into a rigorous proof it is necessary to have convenient *a-priori estimates.*

In Lecture 10 we have studied the limit $e^{-itH}e^{itH_0}\phi$. Denoting with Π_{cont} the orthogonal projection onto the continuous part of the spectrum of H we consider wave function such that $\Pi_{cont}\phi = \phi$ and we want to prove

$$lim_{T\to\infty}sup_{t\geq 0}|(e^{-itH} - e^{-itH_0})e^{-i\tau H}\Pi_{cont}\phi|_2 = 0 \tag{9}$$

This relation indicates that on the continuum part of the spectrum *the free dynamics and the interacting roughly coincide in the remote future.*

On the potential, in addition to be Kato-small, we make the following assumption

$$\|V(H_0 + I)^{-1}\eta(|x| > R)\| \in L^1(R^+) \tag{10}$$

($\eta(A)$ is the indicator function of the set A.) From (10) we derive

$$lim_{R\to\infty}(1 + R)\|\eta(|x| \geq R)V(H - z)^{-1}\| = 0 \tag{11}$$

Notice that condition (10) is weaker than

$$\exists\epsilon > 0 \ : \ \|V(H_0 + I)^{-1}\eta(|x| > R)\| \leq c(1 + R)^{-1-\epsilon} \tag{12}$$

Condition (10) implies that the difference between the resolvents is a compact operator; indeed for $Imz \neq 0$ on has

$$\frac{1}{H_0 - z} - \frac{1}{H - z} = \frac{1}{H_0 - z}(1 + |x|)^{-\frac{1}{2}}(1 + |x|^{\frac{1}{2}})V\frac{1}{H - z} \tag{13}$$

This is the product of a bounded operator times the operator $\frac{1}{H_0-z}(1 + |x|)^{-\frac{1}{2}}$ which is compact since

$$(H_0 - z)^{-\frac{1}{2}}(1 + |x|)V(H - z)^{-1} = A_< + A_> \tag{14}$$

$$A_{>R} = (H_0 - z)^{-\frac{1}{2}}\eta(|x| > R)(1 + |x|)V(H - z)^{-1} \qquad (15)$$

The operator $A_{<R}$ has a corresponding definition. The operator $A_{<R}$ is compact and the norm of the operator $A_{>R}$ tends to zero when $R \to \infty$. Therefore A is compact.

From Ruelle theorem (see Lecture 10) we know that a wave function that belongs to the continuous part of the spectrum of H exits *in the mean* in the far future from any bounded domain of configuration space.

For very large times we introduce a partition in outgoing and incoming states by means of the spectral decomposition of the generator D of the group of dilations.

The *outgoing part* belongs to the positive part of the spectrum of D modulo a term which vanish when $t \to \infty$.

We shall prove that the outgoing part for large enough times *does not any longer interact with the potential* (because the potential is short-range). Therefore on this states the operator Ω_- differs little from the identity.

The remaining part (incoming) becomes asymptotically orthogonal to the entire state space.

Therefore the state cannot be orthogonal to the range of Ω_- and the range of Ω_- is the entire subspace of $\mathcal{H}$ corresponding to the continuous part of the spectrum. It follows that the singular continuous spectrum of H is empty and asymptotic completeness holds.

3 Asymptotic Completeness

We shall now give some details of the proof of asymptotic completeness with Enss's method. Recall that, by Ruelle' theorem, if the operator $\xi(|x| < R)(H + iI)^{-1}$ is compact for every R and if ϕ is in the continuous spectrum of H then one has

$$lim_{T\to\infty}\frac{1}{T}\int_0^T dt\|\xi(|x| < R)e^{itH}\phi\|_2 = 0 \quad \forall R < \infty \qquad (16)$$

We shall prove that if (16) holds then ϕ is a scattering state, i.e. it belongs to the range of Ω_-. We make the crucial observation that using the definition of resolvent and by a diagonal procedure one can derive from (16) that the integral

$$\| \int_{-R}^{R} dt|\eta(|x| < R)e^{-i(t+\tau)H}(H + iI)\phi\|_2 \qquad (17)$$

goes to zero *in the mean* when $\tau \to \infty$, $R \to \infty$. It follows that it is possible to find a sequence of times τ_n in such a way that the quantity

$$\phi_n = e^{-i\tau_n H}\phi \tag{18}$$

represents a sequence of states localized further and further away form the essential support of the potential.

$$lim_{n\to\infty}|\eta(|x| < n)e^{-i\tau_n H}\phi|_2 = 0 \tag{19}$$

$$lim_{n\to\infty}\int_{-n}^{n} dt|\eta(|x| < n)e^{-i(t+\tau_n)H}(H + iI)^{-1}\phi|_2 = 0 \tag{20}$$

To convert these observations into a proof one needs accurate estimates of the convergence in space and in time of the outgoing part of the wave function.The method of Enss shows that for a dense set of initial states (roughly speaking those for which the absolute value of the velocity is bounded and separated away from zero) the wave function decays rapidly outside the classical permitted domain.

The necessary estimates are given for for free motion, with methods akin to those we have used in Lecture 3 of Book 1. These estimates are valid for short range potentials; the extension to long range potential requires more elaborated techniques and a modification in the construction of the scattering matrix.

All estimates exploit the fact that the states one consider have finite energy support and that on functions localized far away from the origin the operator H differs little from the free hamiltonian, which is a function of momenta only. A typical estimate is the following

$$lim_{R\to\infty}\int_0^\infty dt\|\eta(|x| > (1+a)(R+vt))e^{-itH}g(H)\eta(|x| < Rt)\| = 0 \tag{21}$$

where $a > 0$ and the function $g \in C_0^\infty$ has support in $(-\infty, \frac{mv^2}{2})$ (m is the mass of the particle) and $v \in R^d$ is arbitrary.

This estimate is obtained from a similar one valid for $V = 0$ by proving that a suitable class of functions of the total energy can be well approximated by the corresponding functions of the kinetic energy in domains where the potential is small. If $V = 0$ the estimate (21) can be sharpened. It is sufficient the consider the case of hyperplanes, e.g. the hyperplane orthogonal to the axis x_1.

If $g \in C_0^\infty(R)$ with $supp\ g \in [0, \infty]$, for each $\delta > \frac{1}{2}$ and each $n \in \mathcal{N}$ there exists a constant $C_{n,g,\delta}$ such that, for $r,\ t > 0$ one has

$$\|\xi(x_1 < -(t+r))e^{-itH_0}g(p_1)\eta(x_1 > 0)\| \leq C(1+t+r)^{-k} \tag{22}$$

One proves (22) taking Fourier transform and noticing by integration by parts, that a function which is in the domain of the pth power of the Laplacian tends to zero at infinity with a power $q(p)$ where q grows with p.

4 Time-Dependent Decomposition

The main part of Enss' method is the introduction of a suitable time-dependent decomposition of R^d as the union of a spherical region around the origin (of increasing size) and in a finite number of truncated cones.

Consider the set X of wave functions that belong to the continuous subspace of H and which have energy bounded and separated from zero. This set is dense.

Our purpose is to show that no member of this set can be orthogonal to a state which belongs to the absolutely continuous subspace of H and which has energy bounded and separated away from zero. This shows that the subspace of absolute continuity for H coincides with the subspace of continuity and provides a proof of asymptotic completeness.

It is necessary to consider states with energy *strictly larger than zero* because otherwise "the speed of separation" of the parts which belong to the truncated cones may become zero.

We may notice that asymptotically the partition that is used by Enss *can be considered as a partition of classical phase space.*

Choose a smooth function of the energy. Remark that for any function $f \in L^1(R)$ (and in particular in $\mathcal{S}$) one has, for each $\phi \in \Omega$

$$lim_{n\to\infty}\|(\hat{\phi}(H) - \hat{\phi}(H_0))\phi_n\|_2 = 0 \tag{23}$$

where ϕ_n is defined in (18). Indeed

$$\|\int_{-\infty}^{\infty} f(t)(e^{-itH} - e^{-itH_0})\phi_n\|_2 \le \int_{-\infty}^{\infty} dt|f(t)|\|(e^{-itH}e^{itH_0} - I)\|_2 + 2\int_{|t|>n} |f(t)|dt \tag{24}$$

Under the hypothesis on f, the second term to the left converges to zero when $n \to \infty$. The first term converges to zero since, by Duhamel's formula, it is bounded by

$$\int_{-n}^{n} dt|Ve^{-itH}\phi_n|_2 = \int_{-n}^{n} dt|V(H+iI)^{-1}e^{-itH}(H+iI)\phi_n|_2 \tag{25}$$

Decomposing the function ι (identically equal to one) as follows $\iota = \eta(|x| < n) + \eta(|x| \ge n)$ one obtains two terms each of which goes to zero when $n \to \infty$, one as a consequence of (21) and one due to the assumptions on the potential. It follows that

$$\psi_n \equiv \hat{f}(H_0)\phi_n \tag{26}$$

is a good approximation to ϕ_n.

Decompose now R^d in a ball at the origin B_n of radius n and in a finite number M of truncated cones C_m^n with axes $e_m \in R^d$ $m = 1, \ldots, M$ and defined by $|x| > n$ $x \cdot e_m \ge \frac{|x|}{2}$. It is convenient to smoothen the corresponding projection operators using convolution with a fixed $\zeta \in \mathcal{S}$ chosen in such a way that the support of the Fourier transform $\hat{\zeta}(p)$ be contained in small ball at the origin and $\hat{\zeta}(0) = 1$.

In this way we obtain a regular partition $F_0(B_n), \cup_m F_0(C^n_m)$ of R^d which takes into account our requirement that the bound states have energy away from zero. It follows from (20)

$$lim_{n\to\infty}|F_0(B_n)\psi_n|_2 = 0 \qquad lim_{n\to\infty}|\phi_n - \sum_m F_0(C^n_m)\psi_n|_2 = 0 \tag{27}$$

The states $F_0(C^n_m)$ are localized away from the origin. We decompose each in an *outgoing* part and in an *incoming* one. If the support in energy of the state ϕ is included in the interval $[a, b]$, $a > 0$ $b < \infty$ we choose $\zeta \in \mathcal{D}$ such that $\zeta_m(p) = 0$ for $p - e_m < -a$ and $\zeta(p) + \zeta(-p) = 1$ for $|p| < b$.

We define *outgoing* and *incoming* states in the mth sector

$$\psi^{out}_n(m) = F_0(C^m_n)\zeta(p) \qquad \psi^{in}_n(m) = F_0(C^m_n)\zeta(-p) \tag{28}$$

(remark that $\psi_n = [\zeta(p) + \zeta(-p)]\psi_n$.)

We want to prove that the states $\psi^{out}_n(m)$ evolve *almost* freely in the future and the states $\psi^{in}_n(m)$ evolve *almost* freely in the past. The following estimates are useful

$$lim_{n\to\infty}\int_0^\infty dt|\xi(|x| \le n + at)e^{-itH_0}(H_0 + I)\psi^{out}_n(m)|_2 = 0 \tag{29}$$

$$lim_{n\to\infty}\int_{-\infty}^0 dt|\xi(|x| \le n - at)e^{-itH_0}(H_0 + I)\psi^{in}_n(m)|_2 = 0 \tag{30}$$

Recall that a is the lower bound, arbitrary but finite, we have chosen for the energy (and therefore to the velocity). The speed which with the centers of the sectors separate from each other will decrease with $|a|$.

Schrödinger's equation is dispersive but in the low-energy region a greater part of the wave function will be supported near the barycenter and this gives sufficient separation between the wave function which belong to different energy shells.

Since the range of the potential is short this will lead to asymptotic independence. We will take advantage from the fact that all operators which enter the estimates are bounded and therefore it is sufficient to give estimates for a dense subspace.

Estimates (29) and (30) are easy to interpret but have rather elaborated proofs [6, 7].

We shall in the following, give some elements of their proofs and use them for the conclusion of the proof of asymptotic completeness. Let us remark that, *if one assumes the existence of the wave operators* $\Omega_\pm$, from (29), (30) follows

Lemma 1 *For every value of* m

$$lim_{n\to\infty}\|(\Omega_- - I)\psi^{out}(m)\|_2 = 0, \qquad lim_{n\to\infty}\|(\Omega_+ - I)\psi^{in}(m)\|_2 = 0, \tag{31}$$

◇

Proof We prove only the first relation; the second is proved in the same way. On each sector one has

$$\begin{aligned}|(\Omega_- - I)\psi_n^{out}(m)|_2 &\le \int_0^\infty dt|Ve^{-itH_0}\psi_n^{out}(m)|_2\\ &\le \|V(H_0+I)^{-1}\| \int_0^\infty |\xi(|x| \le n+at)e^{-itH_0}(H_0+I)\psi_n^{out}(m)\\ &+ \|(H_0+I)\psi_n^{out}(m)\| \int_0^\infty dt\|V(H_0+I)^{-1}\xi(|x|>n+at)\|\end{aligned} \tag{32}$$

Estimate (29) implies that the first term goes to zero if $n \to \infty$; the second term vanishes in this limit due to the fact that the potential is short range. ♡

Lemma 1 implies that $\psi_n^{in}(m)$ and ϕ_n tend to become orthogonal in the limit $n \to \infty$. In fact

$$|(\psi_n^{in}(m), \phi_n)| \le |(I - \Omega_+)\psi_n^{in}(m)|_2 + |(e^{itH_0\tau_n}\psi_n^{in}(m), \Omega_+^*\phi)| \tag{33}$$

The second summand to the right is bounded by

$$\|\xi(|x| \le n + a\tau_n)e^{i\tau_n H_0}\psi_n^{in}(m)\|_2 + \|\xi(|x| > n + a\tau)\Omega_+^*\phi\|_2 \tag{34}$$

The second term decreases to zero, and so does the first as can be seen using estimates analogous to those that lead to the proof of (29), (30).

We complete now the proof of asymptotic completeness by proving that there are no states that belong to the continuous spectrum of H and are orthogonal to the range of Ω_-.

Since the range of Ω_- contains every state that belongs to the absolutely continuous spectrum of H this shows that the singular continuous spectrum of H is empty.

Assume then that there exists ϕ which is in the continuous spectrum of H and orthogonal to the range of Ω_-. Then this is true for every ϕ_n.

On the one hand, every one of the $\phi_n^{in}(m)$ belongs to the range of Ω_-. On the other hand, ϕ_n is well approximated by the sum of $\psi_n^{out}(m)$, $m = 1 \dots M$.

Since these states belong to the range of Ω_- we get a contradiction. In the same way one shows that ϕ belongs to the range of Ω_+.

We now give an outline of the proof of (29), (30). We shall reduce the problem to the one-dimensional case and then make use of the explicit form of the free propagator. Remark that the ball $|x| \le n + at$ is contained in the half-plane $(u, x) \le (n + at)$ for each unit vector u.

We write any function $\zeta_m(p)$ as sum of a finite number of functions $\xi_{m,k}(p) \in \mathcal{D}$ each with support in a cone with axis $w_{m,k}$ and we choose the axes in such a way that

$$supp\ \xi_{m,k}(p) \in \{p \in R^d,\ (p, w_{m,k}) \geq 2a\} \tag{35}$$

A simple but laborious geometric analysis shows that this can be achieved. Estimate (29) follows then form the following simple estimate valid for each value of the indices m and k

$$lim_{n\to\infty} \int_0^\infty dt \eta(x < n + at) e^{itH_0}(H_0 + I) F_0(C_m^n) \xi_{m,k}(p) \psi_n|_2 = 0 \tag{36}$$

To simplify notation, in each sector we call axis 1 the axis $w_{m.k}$. This procedure allows us to do the estimate for a state well localized in a neighborhood of the $x_1 = 0$ plane and with Fourier transform supported in $[a, b]$. Remark that $\sum_{s>1} p_s^2$ commutes with $\xi(|x_1| < n + at)$; therefore we are reduced to an estimate in one dimension. In this case we have the explicit form of the free propagator

$$(e^{-itH_0})\phi(x) = (2\pi i t)^{-\frac{1}{2}} e^{i\frac{x^2}{2t}} \int \left[1 + i\frac{y^2}{2t} - \frac{1}{2}\left(\frac{y^2}{2t}\right)^2\right] e^{-i\frac{xy}{t}} \phi(y) dy \tag{37}$$

Using this information and other of similar nature (see, e.g. [8]), recalling that by assumption the energy spectrum belongs to $[a, b]$ and making separate estimates for the regions corresponding to $2n + m < x_1 < 2n + m + 1$ it is possible to prove

$$|\xi(x_1 < n + at) e^{-itH_0}(H_0 + I) \eta_{m,k}(p) \psi_n|_2 \leq C[(1 + t)(n - 1 + at)]^{-1} \tag{38}$$

This completes the proof of (29). The proof of (30) is analogous.

5 The Method of Mourre

We now outline a procedure followed by E. Mourre [4] to prove asymptotic completeness for potential scattering. The origins of this methods are in Enss' method and in the smoothness and dispersive estimates of T. Kato.

The method has been generalized [5] in particular to cover asymptotic completeness and spectral structure in the quantum mechanical N-body problem. The generalizations have various names (double commutator method [9, 10], subordinate operators, weakly conjugate operators, ...).

Mourre's metod aims at providing estimates through which one can derive the absence of singular continuous spectrum and the asymptotic behavior (in time) of the states which belong to the absolutely continuous part of the spectrum of the Hamiltonian.

The method of Mourre and its generalizations are now the standard tools in the recent mathematical literature on scattering theory in Quantum Mechanics.

Mourre's method is similar to Enss' method, but it uses more effectively the generator of dilations to produce a partition of the Hilbert space $L^2(R^3)$ that depends on two parameters: time and a dilation factor.

This provides a convenient partition in *outgoing* and *incoming* states and gives a link between geometric scattering theory and the more traditional approach of time dependent scattering theory.

The aim is, as in Enss' method, to prove that every state that belongs to the *continuous spectrum* of the hamiltonian H is well approximated by an *outgoing state* at times sufficiently remote in the future and by an *ingoing state* at times sufficiently remote in the past.

In Mourre's method the partition is given by the spectral decomposition of the dilation operator $D \equiv \frac{1}{2}(x \cdot \hat{p} + \hat{p} \cdot x)$. One can notice that on a dense subset of $\mathcal{H}$ (the domain of the operator $ln|\hat{p}|$) the following relation holds

$$e^{i\lambda D} ln\ |\hat{p}| e^{-i\lambda D} = ln\ |\hat{p}| + \lambda\ I \tag{39}$$

so that the operators $ln\ |\hat{p}|$ and D are *a pair of canonical variables* in the sense Weyl. This simplifies the estimates.

Moreover, noting that $e^{i\lambda\, ln\, |p|} = |p|^{i\lambda}$ one is led to introduce the Mellin transform and therefore to describe the wave function ϕ as a function of $|p|$ and a direction $\omega \in S^3$ as follows

$$\tilde{\phi}(\lambda, \omega) = \frac{1}{\sqrt{2\pi}} \int \frac{d|p|}{|p|} |p|^{\frac{3}{2}} |p|^{i\lambda} \hat{\phi}(|p|, \omega) \tag{40}$$

Remark that for any measurable function F on R

$$F(D)\tilde{\phi}(\lambda, \omega) = F(\lambda)\tilde{\phi}(\lambda, \omega) \tag{41}$$

With this notation it is easy to construct the projection operators P_+ and P_- one the positive (resp.negative) part of the spectrum of D.

One can see that this definition *is not equivalent* to the one in Enss' method, Elements of the form $\xi_j(x)\eta_j(\hat{p})\phi$ are localized (in the spectral representation of D) near the point $(x_i.p_j)$ but their localization becomes weaker when $|p_j|$ and $|x_i|$ increase.

The fact that D and H do not commute will imply that the flow of H will conserve *only approximately* the decomposition of the Hilbert space in incoming and outgoing states. A crucial role in this respect is played by the commutators $[H_0, D]$ and $[H.D]$.

6 Propagation Estimates

Definition 1 (*propagation estimates*) Let A be a self-adjoint operator in the Hilbert space $\mathcal{H}$. We shall say that H_0 satisfies *propagation estimates* (or dispersive estimates) with respect to a self-adjoint operator A if there exists constants $s > s' > 1$ such that for every function $g \in C_0^\infty(R)$ the following estimates hold

$$|(1+A^2)^{-s/2} e^{-itH_0}\ g(H_0)\ (1+A^2)^{-s/2}| \leq c(1+|t|)^{-s'} \quad \forall t \in R \tag{42}$$

$$|(1+A^2)^{-s/2} e^{-itH_0}\ g(H_0)\ P_A^{\pm}| \leq c(1+|t|)^{-s'} \quad \forall \pm t > 0 \tag{43}$$

where we have denoted by P_A^+ the projection on the positive part of the spectrum of A and we have used the notation $P_A^- \equiv I - P_A^+$. ◇

Often it is convenient to use *a local version.* In the local version one requires only that the estimate be satisfied for all functions $g \in C_0^\infty(I_0)$ where I_0 is an open interval. In this lecture we will use always the global version (42) and (43).

Definition 2 (*short range*) Let A be a self-adjoint operator. The potential V is said to be *a short range perturbation* of H_0 *with respect to* A if for $H = H_0 + V$ one has
(i) The operator

$$(H+i)^{-1} - (H_0+i)^{-1} \tag{44}$$

is compact.
(ii) There exist a real number $\mu > 1$ and integers $k,\ j \geq 0$ such that the operator

$$(H+i)^{-j}\ V\ (H+i)^{-k}\ (1+A^2)^{\mu/2} \tag{45}$$

extends to a bounded operator in $\mathcal{H}$. ◇

The abstract theorem we will use is

Theorem 1 ([4, 8, 10]) *Assume that there exists a self-adjoint operator A such that H_0 satisfies the propagation estimate with respect to A and suppose that V is a short range perturbation of H_0 with respect to A. Let $H = H_0 + V$. Then the wave operators $W_\pm(H, H_0)$ exist and are asymptotically complete.* ◇

Often in the application the operator A is the generator of the group of dilations. This leads to identify the range of A_+ with the outgoing states. In other cases a different choice of A is useful. For example in the case of the hamiltonian

$$H = -\Delta + f \cdot x_1 \qquad f \neq 0$$

which is used to discuss the Stark effect, a useful choice is $A = \frac{i}{f}\frac{\partial}{\partial x_1}$. This Hamiltonian has an absolutely continuous spectrum which covers the entire real axis. In this case one has $i(HA - AH) = I$ on a dense set of vectors which are analytic and invariant for both operators.

Proof of Theorem 1 We begin proving the existence of $W_+(H, H_0)$. For $W_-(H, H_0)$ the procedure is similar. For the standard Cook-Kuroda argument it suffices to prove

$$\int_0^\infty |(H+i)^{-j}\ V\ e^{-itH_0}\ g(H_0)\ \psi|_2 dt < \infty \tag{46}$$

Making use of (42) and (43) one has

$$|(H+i)^{-j}\ V\ e^{-itH_0}\ g(H_0)\ \psi|_2$$
$$\leq \|(H+i)^{-j}\ V\ (H+i)^{-k}(1+A^2)^{\mu/2}\|\|(1+A^2)^{-\mu/2}(H+i)^k e^{-itH_0}\ g(H_0)\ (1+A^2)^{-s/2}\psi\|_2 \tag{47}$$

Remark that $(H+i)^k\ (H_0+i)^{-k}$ is a bounded operator which differs from the identity by a compact operator and that one can substitute $g(H_0)$ with $f(H_0) \equiv (H_0+i)^j g(H_0)$ since both belong to C_0^∞.

The operator $e^{itH_0}\ g'(H_0)\psi$ tends weakly to zero when $t \to \infty$ and this convergence is preserved under the action of a compact operator; therefore

$$|(H+i)^{-j}\ V\ e^{-itH_0}\ g(H_0)\ \psi|_2$$
$$\leq \|(H+i)^{-j}\ V\ (H+i)^{-k}(1+A^2)^{\mu/2}\|\|(1+A^2)^{-\mu/2}e^{-itH_0}\ f(H_0)\ (1+A^2)^{-s/2}\psi|_2$$
$$\leq (1+|t|)^{-s'} \tag{48}$$

This proves existence of $W_+(H, H_0)$.

We begin the proof of asymptotic completeness by proving that the operators

$$g_1(H)\ (W_\pm - I)\ g_2(H_0)P_A^\pm \tag{49}$$

are compact if $g_1,\ g_2 \in C_0^\infty$. This follows from

$$g_1(H)\ (e^{itH}\ e^{-itH_0} - I)\ g_2(H_0)P_A^\pm = \int_0^t g_1(H)\ e^{i\tau H}\ V\ e^{-i\tau H_0}\ g_2(H_0)P_A^\pm d\tau \tag{50}$$

where the integrand is norm continuous and compact. Therefore also the integral is a compact operator. For $\tau > 0$ we have the estimates

$$\|g_1(H)\ e^{itH}\ V\ e^{-itH_0}\ g_2(H_0)P_A^\pm\|$$
$$\leq \|g_1(h)\ V\ (H_0+i)^{-k}(1+A^2)^{s/2}\|\ \|(1+A^2)^{-s/2}e^{-itH_0}\ g_2'(H_0)P_A^+\| \leq (1+|t|)^{-s'} \tag{51}$$

It follows that also the limit $t \to \infty$ exists and defines a compact operator. Compactness of $g(H) - g(H_0)$ and the intertwining properties of the wave operators imply that from the compactness of

$$g_1(H)\ (W_\pm - I)\ g_2(H_0)P_A^\pm \tag{52}$$

one can derive the compactness of

$$(W_\pm - I)\, g(H_0)P_A^\pm, \quad g(H)\,(W_\pm - I)\, P_A^\pm \tag{53}$$

To prove asymptotic completeness we first prove that $\sigma_s(H) \cap I_0$ is a discrete set in every bounded open interval $I_0 \subset R$. This implies the singular continuous spectrum is empty and that there at most denumerably many eigenvalues and they have finite multiplicity.

Let $J \subset I_0$ be relatively compact and let $g \in C_0^\infty, \quad g(\lambda) = 1, \lambda \in J$. Since always $Range(W_\pm) \subset \mathcal{H}_s^\perp$, one has always $P_s(H)\; W_\pm = 0$. Therefore

$$\begin{aligned} P_s(H)E_H(J) &= P_s(H)E_H(J)g(H) = P_s(H)E_H(J)g(H)(P_A^+ + P_A^-) \\ &= P_s(H)E_H(J)g(H)(I - W_+)P^+(A) + P_s(H)E_H(J)g(H)(I - W_-)P^-(A) \end{aligned} \tag{54}$$

From (54) it follows that $P_s(H)E_H(J)$ is compact, and then, being a projection operator, it is of finite rank. We can now prove

$$Range(W_\pm) = \mathcal{H}_{a.c.}(H) \tag{55}$$

For every open bounded interval we have shown that $I_0/\sigma_p(H)$ is an open set. Let now $g \in C_0^\infty(I_0/\sigma_p(H))$. We must prove

$$s - lim_{t\to\pm\infty} e^{itH_0}e^{-itH}\phi = W_\pm^*\, \phi \qquad \phi \in \mathcal{H}a.c. \tag{56}$$

The procedure we follow is a typical *localization procedure* in the spectrum of H. Choose $\phi \in \mathcal{H}_{a.c.}$ such that $\phi = g(H)\phi$ and compute

$$\begin{aligned} &\|e^{tH_0}e^{-itH}g(H)\phi - W_+^*g(H)\phi\|_2 \\ &\quad = \|((P_A^+ + P_A^-)e^{itH_0}e^{-itH}g(H)\phi - W_+^*g(H)\phi)\|_2 \le A_+(t) + A_-(t) \end{aligned} \tag{57}$$

where $A_\pm(t) \equiv |P_A^\pm(I - W_+^*)g(H)e^{-itH}\phi|_2$. (we have made use of the intertwining properties of $W_\pm$).

The operator $P_A^+(I - W_+^*)$ is compact and $e^{-itH}\phi$ converges weakly to zero. Therefore $lim_{t\to\infty}A_+(t) = 0$. On the other hand one has

$$A_-(t) \le \|P_A^- e^{-itH}g(H)\phi\|_2 + \|P_A^- e^{-itH_0}W_+^*g(H)\phi\|_2 \tag{58}$$

and the propagation estimates (43) and (44) imply

$$s - lim_{t\to\infty}P_A^- e^{-itH_0}g(H_0) = 0 \tag{59}$$

From $W_+^*g(H)\phi = g(H_0)W_+^*\phi$ and from (54) we deduce that the second term in (57) converges to zero in the limit $t \to \infty$. But

$$|P_A^- e^{-itH} g(H)\phi| \le P_A^- W_-^* e^{-itH} g(H)\phi| + |P_A^- e^{-itH}(I - W_-^*) g(H)\phi| \qquad (60)$$

and by (56) both terms converge to zero in the limit $t \to \infty$. This completes the proof of Theorem 1. ♡

We give now an indication of the procedure one may follow to prove the propagation estimates (43) and (44) that we have used in the proof of asymptotic completeness. Consider first the case $H_0 = -\Delta$ on $\mathcal{H} \equiv L^2(R^n)$ and choose for A the dilation operator $A = \frac{i}{2}(\nabla \cdot x + x \cdot \nabla)$.

Lemma 2 *The operators H_0 and A satisfy for every $s > s' > 0$ the estimates (43) and (44).* ◇

Proof By means of functional calculus define $K_0 \equiv log H_0$. Making use of Fourier transform it is easy to prove that on a common domain of essential self-adjointness which is invariant under the action of both operators one has $i(K_0 A - A K_0) = 2I$ and therefore

$$e^{itH_0} A e^{-itH_0} = A + 2tI \qquad (61)$$

This implies the desired propagation estimates (43), (44). Moreover that $P_A^{\mp} e^{-itK_0} P_A^{\pm} = 0 \;\; \forall \pm t > 0$. To see this, apply the uni-dimensional Mellin transform that we now briefly recall.

In momentum space the term $e^{-itH_0} g(H_0)$ reads

$$e^{-itp^2} g(p^2) \equiv (p^2)^{-it} g(p^2) \qquad (62)$$

Let $g \in C_0^\infty(R_+)$. An easy application of the non-stationary phase theorem (see Lecture 8) proves that the function

$$G_t(\lambda) = \frac{1}{2\pi} \int_0^\infty e^{-it\rho} g(\rho) \rho^{-i\lambda-1} d\rho \qquad (63)$$

satisfies the following estimates, where C_N are suitable constants

$$|G_t(\lambda)| \le C_N |t|^{-N} (1 + |\lambda|)^N \qquad \forall t \in R, \quad \forall N \ge 1 \qquad (64)$$

$$|G_t(\lambda)| \le C_N (1 + |t + \lambda|)^N \qquad \forall t, \lambda > 0 \;\; \forall N \ge 1 \qquad (65)$$

From (64), (65) follows for $s > 1$

$$|(I + A^2)^{-s/2} e^{-itK_0} (I + A^2)^{-s/2}| \le \int_{-\infty}^{\infty} G_t(\lambda)(1 + |t|)^{-s} d\lambda \le C_{N,s} |t|^{-N} \quad \forall t \in R \qquad (66)$$

if $N < s - 1$. Moreover one has

$$(I + A^2)^{-s/2} e^{-itH_0} g(H_0) P_A^+ = G_t(\lambda)(I + A^2)^{-s/2} H_0^{i\lambda} P_A^+ d\lambda \tag{67}$$

The contribution to the integral of the region $\lambda < 0$ is estimated with (67) and provides the bound

$$\| \int_{-\infty}^{0} G_t(\lambda)(I + A^2)^{-s/2} H_0^{i\lambda} P_A^+ d\lambda \| \leq C_{N,s} |t|^{-N} \tag{68}$$

The contribution to the integral for positive values of λ is estimated for every $m > 1$ making use of (68)

$$\| \int_{0}^{\infty} G_t(\lambda)((1 + A^2)^{-s/2} H_0^{i\lambda} P_A^+ d\lambda \| \leq c \int_{0}^{\infty} (1 + t + \lambda)^{-N} d\lambda \leq c t^{-N+1} \tag{69}$$

The proof of Lemma 2 is then completed for any value of $1 < s' < s$ by interpolation using the estimates (68) and (69). ♡

7 Conjugate Operator; Kato-Smooth Perturbations

The procedure we have followed to prove asymptotic completeness in the case of short range potentials is a particular case of the *method of* conjugate operator [4, 10, 14].

The *conjugate operator method* is used to deduce the spectral properties on an open part $\Omega \in R$ of the spectrum of a self-adjoint operator H from the existence of another self-adjoint operator A with suitable properties. In the applications to scattering theory the operator A is usually the generator of the dilation group.

The method has its roots in T. Kato's theory of *smooth perturbations*. We shall briefly review this theory following [8].

Definition 3 Let H be a self-adjoint operator on a Hilbert space $\mathcal{H}$ with resolvent $R(\mu) = \frac{1}{H - \mu I}$. Let A be a closed operator. The operator A is called H*-smooth* iff for every $\psi \in \mathcal{H}$ and for every $\epsilon \neq 0$ the vector $R(\lambda + i\epsilon\psi$ belongs to $D(A)$ and

$$\|A\|_H \equiv sup_{|\psi|_2=1,\ \epsilon>0} \frac{1}{4\pi} \int_{-\infty}^{\infty} \|AR(\lambda + i\epsilon)\psi_2 + AR(\lambda - i\epsilon)\psi_2\|^2 d\lambda < \infty \tag{70}$$

◇

It is convenient for what follows to formulate H-smoothness in different ways using the following generalization of Plancherel Lemma.

Lemma 3 ([8]) *Let ψ be a weakly measurable function from R to the separable Hilbert space $\mathcal{H}$ such that $\int |\phi|_2 dx < \infty$.*

Define $\hat{\psi} : R \to \mathcal{H}$ by

$$\hat{\psi} = \frac{1}{\sqrt{2\pi}} \int e^{-ipx} \psi(x) dx \tag{71}$$

Then

$$\int |A\hat{\psi}(p)|_2^2 dx = \int |A\psi(x)|_2^2 dx \tag{72}$$

where by convention the integrals are set to be ∞ if either $\hat{\psi}$ or ψ are not in the domain of H. ◇

Proof We give only an outline of the proof. Given a family $\psi(x) \in \mathcal{H}$ $x \in R$ let A be a bounded operator. For any $\phi \in \mathcal{H}$ we have that $(\phi, A\hat{\phi}(p)) \equiv (A^*\phi, \psi(p))$ is the Fourier transform of the function $(A^*\phi, \psi(x))$. Therefore by Plancherel lemma

$$\int |(\phi, A\hat{\psi}(p))|^2 dp = \int |(\phi, A\psi(x))|^2 dx \tag{73}$$

if either integral is finite. Summing over an orthonormal basis gives (72).

If A is self-adjoint consider first the operator $E_{[-N,N]}A$ where E_I is the spectral projection on the interval I.

Then (72) applies to the bounded operator $E_{[-N,N]}A$ and if both $\hat{\psi}(p)$ and $\psi(x)$ belong to the domain of A for all x, $p \in R^1$. An easy limit procedure gives (72) for A.

Finally, if A is unbounded and not self-adjoint, there is a self-adjoint operator $|A|$ (formally $|A| = \sqrt{A^*A}$ such that $D(|A| = D(A)$ and $||A|\phi|_2 = |A\phi|_2$. Thus (72) follows from the self-adjoint case. ♡

We can now reformulate H-smoothness in terms of the unitary group e^{itH}.

Lemma 4 *The operator A is H-smooth iff for all $\psi \in \mathcal{H}$ one has $e^{itH}\psi \in D(A)$ and for almost all $t \in R$*

$$\int_{-\infty}^{\infty} |Ae^{itH}\psi|_2^2 dt \leq (2\pi)\|A_H\|^2|\psi|_2^2 \tag{74}$$

◇

Proof Fix $\epsilon > 0$. One has

$$\int_0^\infty e^{-\epsilon t} e^{i\lambda t} e^{-itH} \phi dt = -i R(\lambda - i\epsilon)\psi \tag{75}$$

By Lemma 3

$$\int_{-\infty}^{\infty} |AR(\lambda + i\epsilon)\psi|_2^2 d\lambda = 2\pi \int_0^\infty e^{-2\epsilon t} |Ae^{-itH}\psi|_2^2 dt \tag{76}$$

Taking the limit $\epsilon \to 0$ proves Lemma 4. ♡

The connection between A-smoothness and the spectral properties of H is given by the following theorem.

Theorem 2 *If A is H-smooth then $Range A^* \subset \mathcal{H}_{ac}(H)$.* ◇

Proof Let $\psi \in D(A^*)$ set $\phi = A^*\psi$ and let $d\mu_\phi$ be the spectral measure for H associated to ϕ. Define

$$F(t) = \frac{1}{\sqrt{2\pi}} \int e^{-itx} d\mu_\phi(x) = \frac{1}{\sqrt{2\pi}} (A^*\psi, e^{-itH}\phi) \tag{77}$$

Then $|F(t)| \leq \frac{1}{\sqrt{2\pi}} |\psi|_2 |Ae^{-itH}\phi|_2$. By Lemma 3 $\hat{F}$ belongs to $L^2(R^3)$ and $d\mu_\phi$ is absolutely continuous with respect to Lebesgue measure. ♡

We describe now briefly the Kato-Putnam theorem, which links Kato smoothness with commutator estimates.

Theorem 3 (Kato-Putnam [4, 10, 15]) *Let A and H be self-adjoint operators. Suppose $C \equiv i[H, A]$ is positive. Then $C^{\frac{1}{2}}$ is H-smooth. If $Ker\ C = \{0\}$ then H has purely continuous spectrum.* ◇

Proof The second statement follows from the first and Theorem 2 by noting that

$$Ker\sqrt{C} = (Range\sqrt{C})^\perp = \{0\}. \tag{78}$$

For the first statement, compute $\frac{d}{dt}[e^{itH} A e^{-itH}] = e^{itH} C e^{-itH}$ and then use

$$\int_s^t (\phi, e^{i\tau H} C e^{-i\tau H} \phi) d\tau = (\phi, e^{itH} A e^{-itH} \phi) - (\phi, e^{isH} A e^{-isH} \phi) \tag{79}$$

Therefore

$$\int_s^t |\sqrt{C}e^{-i\tau H}\phi|_2^2 d\tau \leq 2\|A\||\phi|_2^2 \tag{80}$$

Since t and s are arbitrary it follows that $\sqrt{C}$ is H-smooth and $\|\sqrt{C}\|_H^2 \leq \frac{\|A\|}{\pi}$. ♡

A generalization of this Theorem has been given by Putnam; his method, that we shall call *positive commutator method*, allows to deduce various estimate for the resolvent of H from the positivity of a commutator

$$P_I(H)[H, iA]P_I(H) \geq aP_I(H) \qquad a > 01 \tag{81}$$

where I is an open finite set contained in the spectrum of H.

8 Limit Absorption Principle

Among the conclusion one can draw which have relevance in scattering theory is the *limit absorption principle*

$$sup_{z\in J^\pm}\|(1+A^2)^{-s/2}(H-z)^{-1}(1+A^2)^{-s/2}\| < \infty \tag{82}$$

for every closed interval $J \subset I$ and every $s > \frac{1}{2}$.

One makes the following assumptions on the operator A

(i) The map

$$s \to e^{-isA}f(H)e^{isA}\phi \tag{83}$$

is twice continuously differentiable for every $f \in C_0^\infty(I)$ and every $\phi \in \mathcal{H}$. We will use the notation $H \in C^k(A)$ when the map (83) is k-times differentiable.

(ii) For every $\lambda \in I$ there exist a neighborhood Δ strictly contained in I and a positive constant a such that

$$E_\Delta(H)[H, iA]E_\Delta(H) \geq aE_\Delta(H) \tag{84}$$

where E_Δ is the spectral projection of H relative to the interval Δ.

Remark that due to (i) the commutator $[H, A]$ is well defined as quadratic form on the union $\cup_{\mathcal{K}} E_K(H)\mathcal{H}$ where the union is taken over all compact set which are contained in Δ.

In [4, 10] the following results are obtained.

(a) For all $s > \frac{1}{2}$ and every $\phi,\ \psi \in \mathcal{H}$, uniformly for λ in every compact subset of I, the limit

$$lim_{\epsilon \to 0_+}(\psi, (I + A^2)^{-\frac{s}{2}} \frac{1}{H + \lambda \pm i\epsilon}(I + A^2)^{-\frac{s}{2}}\phi) \tag{85}$$

exists. This implies in particular that the spectrum of $H$is pure absolutely continuous in I.
(b) If $\frac{1}{2} < s < 1$ and $f \in C_0^\infty$ then

$$\| < A >^{-s} e^{-itH} f(H) < A >^{-s} \| = O(t^{\frac{1}{2}-s}. \tag{86}$$

These decay estimates play an important role in the proof of asymptotic completeness.
(c) Under the further assumption that $H \in C^4(A)$ for every closed interval $J \subset I$

$$sup_{z \in J^\pm} \| P_\pm(A)(H - z)^{-1} P_\mp(A) \| < \infty \tag{87}$$

where $P_\pm(A)$ is the spectral projection of A on its positive (resp. negative) part. In case A is the dilation operator $P_\pm(A)$ is interpreted as projection over the outgoing (resp.incoming) states.

For details and further results one can consult the references to this lecture.

9 Algebraic Scattering Theory

We end this lecture with a brief description of the analysis of scattering processes which can be performed in the Heisenberg representation. The role of the group of spacial dilations can be seen also in this representation by studying the asymptotic behavior *of the expectation values of the observables*.

This possibility has been emphasized by K. Hepp [11] and others, especially D. Ruelle [14] H. Araki [13], and has been given a central role by V. Enss [12]

Algebraic Scattering has a less ambitious program than the scattering theories we have analyzed so far. It does not aim at proving existence and completeness of the Wave operators but considers only the asymptotic behavior for $t \to \pm\infty$ of the *expectation values* of relevant observables in *scattering states* which *by definition* are the the state in the *continuous part* of the spectrum of the Hamiltonian H.

Recall the the Wave operators W_{H,H_0} exist only if there are no singular part in the continuous spectrum of H. One of the typical results of Algebraic Scattering Theory is the proof that

$$\psi \in \mathcal{H}_{cont}(H) \to lim_{t \to \infty} \frac{mx}{t} e^{-iHt}\psi = 0 \tag{88}$$

It is shown [11, 12, 14] that under mild conditions on the potential V this result holds; the conditions are not strong enough to prove the existence of the wave Operators.

The potential V may contain a *long range* part V_l and a *short range* part V_s The long range part must satisfy

$$lim_{|x|\to\infty} V_l(x) = 0, \qquad lim_{|x|\to\infty} x \cdot \nabla V_l(x) = 0 \tag{89}$$

The sort range part V_s of the potential may have a part $V_{s,1}$ that is Kato small with respect to H_0 but also another part $V_{s,2}$ which describes highly singular perturbations and that may be responsible for the presence of a singular continuous part in the spectral measure of H. The theory requires that

$$D(H_0) \cap D(V_s) \cap D(|x|^2) \tag{90}$$

be dense in $\mathcal{H}$. For the short range part $V_s(x)$ of the potential it is required that

$$(H_0 - z)^{-\frac{1}{2}}(1 + |x|^2)^{\frac{1}{2}} V_s(x)(H_0 - z)^{-1} \in \mathcal{K} \tag{91}$$

where z is a sufficiently negative negative number which is in the resolvent set of the three operators $H,\ H_0$, $H_0 + V_l$ and $\mathcal{K}$ is the class of compact operators.

Notice that the decay conditions on V_l are weaker that the integrability conditions under which the Wave operators exist. It allows e.g. the potential $V_l(x) = \xi(|x| \geq 1)(|x|log|x|)^{-1}$ where ξ is the indicator function.

The proofs are much simplified if one makes the stronger assumption

$$(1 + |x|)^{\frac{1}{2}} V_s(H_0 - z)^{-1} \in \mathcal{K} \tag{92}$$

Let P^{cont} be spectral projection on the continuous part of the spectrum of H. Denote by $D = \frac{1}{2}(p.x + x.p)$ the generator of space dilations. Under these assumptions one proves [11, 12].

Theorem 4 *Let $H = H_0 + V$ satisfy the assumptions above. Then in the sense of strong resolvent convergence one has*

$$lim_{|t|\to\infty} \frac{mx^2(t)}{2t^2} = H P^{cont} \tag{93}$$

$$lim_{|t|\to\infty} \frac{D(t)}{t} = 2H P^{cont} \tag{94}$$

◇

One has also

Theorem 5 *Let f be Fourier transform of an integrable function. Then*

$$lim_{|t|\to\infty} \|[f(m\frac{x}{t}) - f(p)]e^{-itH}\psi\| = 0 \tag{95}$$

if $P^{cont}\phi = \phi$. ◇

The evolution of the observables is given by the Heisenberg equation of motion This is the basis of the *algebraic scattering theory* [13, 14] extended later to Quantized Field Theory and to the Algebraic Theory of Local Observables.

Algebraic scattering theory gives less information then in the context of Quantum Mechanics because some important tools are not directly available.

On the other hand, in an infinite-dimensional context (Quantum Field Theory), in absence of a Schroedinger representation, it is the only instrument available. In this approach one studies *asymptotic fields* that acting on the vacuum generate states that evolve according to the free hamiltonian.

One can find in [15] a general outline of the study of asymptotic completeness in Quantum Mechanics through the study of the asymptotic behavior of suitable class of observables.

It can be proven that the temporal evolution in the Heisenberg representation of a suitable class of observables under H_{cont} for very long times differs little from the evolution under H_0.

In the infinite dimensional case the *observable fields* can be asymptotically described in term of free fields. By studying the asymptotic behavior of suitable observables one can show e.g.

Theorem 6 *If $D(H_0) = D(H)$ for every $f \in C^\infty(R)$ and every $\phi \in \mathcal{H}_{cont}(H)$ one has*
(i)

$$lim_{t\to\infty} f\left(\frac{x(t)^2}{t^2}\right)\phi = f(2H)\phi \tag{96}$$

(ii)

$$lim_{t\to\infty} f\left(\frac{A(t)}{t}\right)\phi = f(2H)\phi$$

$$lim_{t\to\infty} f\left(H_0(t)\right)\phi = f(H)\phi \tag{97}$$

◇

V. Enss uses a similar method but, working as he does in the Schroedinger representation, he obtains *accurate asymptotic estimates* for the asymptotic behavior of the solutions. For example one can prove the following theorem.

Theorem 7 ([12]) *Let $H = H_0 + V, \quad H_0 \equiv -\Delta$ with V Kato small with respect to H_0. If*

$$(H_0 - z)^{-\frac{1}{2}}(1 + |x|^2)^{\frac{1}{2}} V(H_z)^{-1} \in \mathcal{K} \tag{98}$$

one has, in strong resolvent sense

$$lim_{|t|\to\infty}\frac{x^2(t)}{t^2} = HP_{cont} \qquad lim_{|t|\to\infty}\frac{D(t)}{t} = 2HP_{cont} \tag{99}$$

◇

We do not here give the proof of this Theorem, but remark the following corollary:

Corollary *If ϕ belongs to the continuum subspace of H then*
(i)

$$lim_{|t|\to\infty}|(I - \xi(v_1 t < x < v_2 t)e^{-itH}\xi(v_1^2 < H < v_2^2)\phi|_2 = 0 \tag{100}$$

(ii)

$$\forall R \quad lim_{|t|\to\infty}|\xi(|x| < R)e^{-itH}\phi|_2 = 0 \tag{101}$$

♣

References

1. V. Enss, Geometric methods in spectral and scattering theory, in *Rigorous Atomic and Molecular Physics*, ed. by G. Velo, A. Wightman (New York Plenum, New York, 1981)
2. V. Enss, Commun. Math. Phys. **89**, 245–268 (1983)
3. V. Enss, *Chapter Notes in Mathematics* (Springer, New York, 1985), pp. 39–178
4. A. Mourre, Commun. Math. Phys. **78**, 391–408 (1981)
5. W. Hunziker, I. Segal, J. Math. Phys. **41**, 3448–3510 (2000)
6. V. Enss, B. Simon, Commun. Math. Phys. **76**, 177–209 (1980)
7. V. Enss, J. Funct. Anal. **52**, 219–251 (1983). Proposition 2.10
8. M. Reed, B. Simon, *Methods of Modern Mathematical Physics* (Academic Press, cambridge, 1975)
9. W. Amrein, J. Zuleta, Helv. Phys. Acta **70**, 1–38 (1997)
10. V. Amrein, J. Jauch, K. Sinha, *Scattering theory in Quantum Mechanics*, V. Benjamin, (Reading Mass, 1977)
11. K. Hepp, Helv. Phys. Acta **42**, 425–458 (1969)
12. V. Enss, *Chapter Notes in Mathematics* vol. 1159, 1984
13. H. Araki, R. Haag, Comm. Math. Phys. **4**, 77–91 (1967)
14. D. Ruelle, Helv. Phys. Acta **35**, 147–163 (1962)
15. H. Cycon, R. Froese, W. Kirsch, B. Simon, *Schroedinger Operators*, Text and Monographs in Physics (1988)

Lecture 13: The N-Body Quantum System: Spectral Structure and Scattering

We shall make use of the methods outlined above to study the quantum N-body problem in its general aspects and in the asymptotic behavior. For a more complete analysis and further references we refer to [1–4].

The quantum N-body system is a collection of N particles with masses $\{m_k\}, m_k > 0$ interacting among themselves through potential forces. The system is described by a wave function $\Phi(x), \quad x = x_1, \dots x_N \;\; x_k \in R^3$.

Introducing in R^{3N} the scalar product $< x, y > \equiv \sum_k m_k(x_k, y_k)$ the classic kinetic energy of the system is $T = \frac{1}{2} < \dot{x}, \dot{x} >$. With a suitable of units we write the Schroedinger operator as

$$H = -\frac{1}{2}\Delta + V(x), \quad x \in R^{3N} \tag{1}$$

and assume that V is invariant under translations of each argument in R^3. In this case the motion of the center of mass is free and the Hilbert space has a natural decomposition

$$\mathcal{H} \equiv L^2(R^3) \otimes L^2(X), \quad X \equiv \{x_1, \dots x_n\}, \; x_k \in R^3, \; \sum_k m_k x_k = 0 \tag{2}$$

This decomposition is invariant for the evolution generated by H. We shall assume that the potential has the structure

$$V(x) = \sum_{i<k} V_{i,k}(x_i - x_k), \qquad lim_{|y|\to\infty} V_{i,k}(y) = 0 \tag{3}$$

It important to notice that (3) *does not imply* $lim_{|x|\to\infty} V(x) = 0$ because there may exist directions in which $x_i - x_k, \; x_i \neq x_j$ remains bounded for some values of the indices.

G. Dell'Antonio, *Lectures on the Mathematics of Quantum Mechanics II: Selected Topics*, Atlantis Studies in Mathematical Physics: Theory and Applications 2, DOI 10.2991/978-94-6239-115-4_13

This corresponds to our intuition that a N-particle system can be studied mathematically only if one finds first a mechanism through which in correspondence to some initial data the systems can be subdivided, asymptotically in time, into isolated subsystems.

It follows that to study mathematically the asymptotic behavior in time of the system one must find directions which correspond to fragmentation in subsystems. This requires to a description of the system not only with functions on the configuration space $(R^3)^N$ but with functions on subspaces corresponding the possible fragments.

1 Partition in Channels

This task can be accomplished by the introduction of N unit vectors $a_k \in R^3$, $k = 1, \dots N$ and by an analysis of the fragmentations that correspond to the translation $x_k \to x_k + \lambda a_k$ for λ very large.

Of course complete fragmentation is obtained only in the limit $\lambda \to \infty$ but the assumptions we shall make on the potentials $V_{i,k}$ will guarantee that the error made is *negligible* if λ is taken *sufficiently large.*

Notice that if $a_k = a_h$ the difference $x_k - x_h$ is invariant under the given translation. The *clusters* of particles are therefore described by closed subspaces $\Lambda_{I_i,\dots I_s}$ of R^{3N} defined by

$$\Lambda_{I_1,\dots I_s} \equiv \{a \in R^{3N}, \quad k = 1, \dots s \ \ \{i,j\} \in I_k \leftrightarrow a_i = a_j\} \tag{4}$$

where I_k are disjoint collections of indices.

Within one of these subspaces the translation considered are rigid translations of the set of points which correspond to the given partition. This *partition in channels* will allow the study of the asymptotic behavior of the entire system considering *separately its projection* on the different channels.

The subsets $\Lambda_{\Sigma_1,\dots\Sigma_n}$ are a ortho-complemented lattice L' closed under intersection and such that $\emptyset \in L'$. We shall always use the reference system in which the center of mass is at rest in the origin and therefore we shall always refer to the lattice L obtained by intersecting L' with X. Notice that for every $P \in L$ one has a unique orthogonal decomposition

$$X = P \oplus P^{\perp} \tag{5}$$

and therefore every $x \in X$ can be decomposed it in a unique way as

$$x = x_P + x^P, \quad x_P \in P, \quad x^P \in P^{\perp} \tag{6}$$

The coordinates x^P are relative coordinates within each cluster, the coordinates x_P are the coordinates of the center of mass of each cluster. Therefore one will set $\sum_k M_k x_k^P = 0$ where we have denoted by M_k the total mass of the kth cluster. For example if $N = 4$, the cluster P is described by $\{1, 2\}$, $\{3, 4\}$ and the particles have equal mass m one has

$$x_P = (\eta, -\eta), \quad x^P = \left(x_i - \frac{1}{2}\eta, x_2 - \frac{1}{2}\eta, x_3 + \frac{1}{2}\eta, x_4 + \frac{1}{2}\eta\right) \tag{7}$$

where we have denoted by $\eta, \quad -\eta$ the coordinates of the centers of mass of the two clusters. Notice now that for each partition P one can write

$$V(x) = V^P(x^P) + R_P(x) \quad |R_P| \le f(|x_P|), \quad f(s)_{s\to\infty} \to 0 \tag{8}$$

The term R_P is the sum of the potentials between pairs of bodies *which do not belong to the same cluster* (and therefore by assumption $lim_{|x|\to\infty} R_P(x) = 0$) while the term V^P is the sum of potentials between pairs of bodies *which belong to the same cluster*. Define

$$H_P \equiv H - R_P = -\frac{1}{2}\Delta + V^P(x^P) \tag{9}$$

We expect that H_P describes with fair approximation the motion of the system *when the distances between the clusters defined by the partition P become very large.*

Therefore we expect that *almost every initial datum* ϕ one can associate functions $\phi_P, \ P \in L$ which depend only on the x^P and are such that for times t very large one has approximately

$$e^{-iHt}\phi \simeq \sum_P e^{-iH_P t}\phi_P \tag{10}$$

We therefore expect that *for almost all initial data in the remote future* (and past) the system can be described as decomposed into aggregates (may be not the same in the past as in the future) each of which describes the motion of the cluster of material points which interact among themselves and remain *approximately localized* in a finite region of space.

To prove that this is the case it will be necessary (see [3–5]).

(a) To provide *propagation estimates* in order to show that for each initial datum ϕ the decomposition (4) becomes more and more accurate when t increases.
(b) To provide *separation estimates* in order to show at times remote in the future the clusters at a large distance form each other.

In order to obtain these estimates one needs a *regular decomposition* (e.g. by C^∞ functions) of configuration space which at very large distance on a suitable scale tends to coincide with the partition in the elements P of L.

This decomposition is achieved through functions of class $F_P \in C^\infty$; these are functions which sum up to one everywhere and are *mollifiers* of the indicator functions associated to the given partition.

The functions F_P *may be time dependent* and *may converge* for $t \to \infty$ in a suitable sense to indicator functions.

The possibility to achieve these goals depends on the possibility to provide accurate estimates on the spatial behavior of $e^{-iHt}\phi$ for very large times This estimates are linked to compactness estimates for the integral kernel of the operator $(H - z)^{-1}$, $z \in C/R$ and are somewhat related to the *uncertainty principle* which provides, at each instant of time, a lower bound for the product of the dispersions of $e^{-iHt}\phi$ in position and momentum.

The partitions introduced above permit to extend to the N-body problem the estimates typical of the methods of Enss and of Mourre. We remark that the possibility to use these estimates makes the quantum N-body problem *much easier that the corresponding classical one*. Indeed in the classical case the decomposition along asymptotic directions *is too fine* and this makes a measurable decomposition impossible.

2 Asymptotic Analysis

In Quantum Mechanics the Hilbert space in which the system will be studied is

$$\mathcal{K} \equiv \oplus_{\alpha_1,\ldots\alpha_K} \otimes L^2((R^3)^{n_{\alpha_k}}) \tag{11}$$

where K is the number of *channels*, α_k denotes a generic channel and n_{α_k} is the number of particles in channel α_k.

It is clear that if at least two channels exist, the Hilbert space $\mathcal{K}$ *is not isomorphic in a natural way* to $L^2((R^3)^N)$ and rather contains this space as *proper subspace*. Therefore the analysis we will make will be an *asymptotic analysis* adapted to scattering theory.

For example in the case $N = 3$ the possible channels are labeled

$$\{1, 2, 3\}, \{1, 2\}\{3\}, \{1\}\{2, 3\}, \{1, 3\}\{2\}, \{1, \}\{2\}\{3\} \tag{12}$$

In this case one has

$$\mathcal{K} = L^2(R^3) \oplus [L^2(R^3) \otimes L^2((R^3)^2)]^3 \oplus L^2((R^3)^3) \tag{13}$$

The first channel correspond to bound states of the system, the three next channels correspond to the case in which two of the particles form a bound state and a third particle is asymptotically free and the remaining channel corresponds to asymptotic states in which the three particles do not interact with each other. Of course for some system one or more of these channels may not be present.

To fix ideas, we can think of the system composed of the Helium nucleus and of two electrons. In this case the first channel will be composed of the states the Helium atom, the second and third will be parametrized by the states of singly ionized Helium atom and a free electron, the fourth cannel will not be present (it would consist of a bound state of the two electron a free Helium nucleus, and the fifth channel will be composed of two free electrons and a free Helium nucleus.

These parametrizations (except for the first) refer to *scattering states*. As a consequence there are two *distinct parametrizations* which refer to the behavior in the remote past and in the distant future. They are both valid but, e.g., a state which belongs to a channel composed on the remote past of a free electron and a singly ionized *He* atom may in the remote future have a components in the same channel and a component in a channel described by two free electrons and a Helium nucleus.

This explains the greater difficulty in the treatment of the N-body problem, $N \geq 3$ as compared to the case $N = 2$ and to potential scattering.

Before entering, even briefly, into the study of the N-body problem let us recall some general properties of the Schroedinger operator.

3 Assumptions on the Potential

We shall make always the assumption that V be locally in $L^2(X)$ and belongs to Kato class, i.e. there are numbers $0 < \alpha < 1$ and $\beta > 0$ such that

$$|V\phi|_2 \leq \alpha|\Delta\phi|_2 + \beta|\phi|_2 \qquad \forall\phi \in C_0^\infty(X) \tag{14}$$

Let us recall that if V belongs to the Kato class then $H \equiv -\Delta + V$ is (essentially) self-adjoint, bounded below and has the same domain as Δ. Notice that $V \equiv \sum_{i<j} V_{i,j}(x_i - x_j)$ is in L^2_{loc} if $V_{i,j}$ are in $L^2(R^3)$ and that V is of Kato class if for all j, i the potentials $V_{i,j}(y)$ are of Kato class.

In this case Kato theorem assures then that the quantum dynamics of the N-body system is well posed. Through the spectral representation of H the energy distribution of the state ϕ is well defined.

The Hilbert space is the direct sum of the subspace $\mathcal{H}_B$ related to the point spectrum of H and of the subspace $\mathcal{H}_C$ in which the spectral measure is continuous. This can be repeated for each of the Hilbert spaces and Hamiltonians for the different channels. Notice that corresponding to each channel (partition) P one has

$$H = H_P + R_P, \qquad |R_P| \leq f(|x_P|), \quad lim_{s\to\infty} f(s) = 0 \tag{15}$$

where H_P is the Hamiltonian of a N-body system in which one neglect all forces between particles belonging to different clusters. Therefore the hamiltonian H_P is the sum of operators H_k which act *independently* on the direct product of the Hilbert spaces associated to each cluster.

Each operator H_k satisfied the condition for the applicability of Ruelle theorem. This theorem implies that, under the assumptions made on V, if $f \in L^\infty$, $lim_{|x|\to\infty}$ $f(|x|) = 0$ and for all $z \in \rho(H)$ the operator $f(x)(z - H)^{-1}$ is compact. Denoting by $\xi(R)$ the indicator function of the ball of radius R in R^d we have

$$(i) \;\; \phi \in \mathcal{H}_B \leftrightarrow lim_{R\to\infty}|(I - \xi(R))e^{-iHt}\phi| = 0 \tag{16}$$

$$(ii)\ \ \phi \in \mathcal{H}_C \leftrightarrow lim_{t\to\infty} t^{-1} \int_0^t ds|\xi(R)e^{-iHt}\phi|^2 = 0 \ \ \forall R < \infty \tag{17}$$

In order to apply Ruelle's theorem let us remark that if V is Kato-small with respect to H and if $lim_{t\to\infty} V(x) = 0$ then $V(H + iI)^{-1}$ is a compact operator. To prove this, notice that V can be approximated with a function V_R with compact support and in the same way one can replace the function $h(p) = (1 + p^2)^{-1}$ with its restriction $h_R(p)$ to a ball of radius R up to an error $f_R(p) \leq 2(R^2 + p^2)^{-1}$.

An explicit computation proves that $V_R \cdot h(i\nabla)$ is a Hilbert–Schmidt operator and therefore $V(H + iI)^{-1}$ differs from a Hilbert–Schmidt operator by an operator with norm bounded by $C_1|p|^2|f_R(p)| + |f_R(p)|$.

This term can be made arbitrary small by taking C_1 small and R large and therefore $V(H + iI)^{-1}$ is norm limit of Hilbert–Schmidt operator and hence compact. We conclude that for the N-body potentials we are considering when analyzing the behaviour of the system under hamiltonian H_P within each cluster we can make use of Ruelle's theorem.

The analysis of the different partition can be done by induction. Recall that in Quantum Mechanics when considering identical particles the Hilbert space is a subspace of $L^2(X)$ which corresponds to an irreducible representation of the permutation group.

The formalism that we are describing is adapted to these cases simply projection the estimates in this subspaces. It is necessary of course that the operators we are considering be invariant under permutations.

4 Zhislin's Theorem

We shall now study the spectral properties of the Schroedinger operator for the system we are discussing.

Recall that we denote by $\sigma_{disc}(H)$ the collection of the eigenvalues of finite multiplicity of a self-adjoint operator H and with $\sigma_{ess}(H)$ the complement of $\sigma_{disc}(H)$ in $\sigma(H)$. One has

$$H = H_P + H^P + R_P \quad H_P = -\frac{1}{2}\Delta_P \otimes I + I \otimes H^P, \quad H^P \equiv -\frac{1}{2}\Delta^P + V^P \tag{18}$$

where Δ_P (resp. Δ^P) are the Laplace operators in the coordinates x_P (resp. x^P). If P is not empty one has $\sigma(-\Delta_P) = [0, +\infty)$ and therefore

$$\sigma(H_P) = [\mu_P, +\infty), \qquad \mu_P \equiv inf\,\sigma(H_P) \tag{19}$$

(μ_P is the minimal energy for a system composed of the clusters described by P and not interacting among themselves). Notice that the lower bound of the spectrum can be lower if one takes into account this inter-cluster interaction. In fact we have

Lemma 1 *If* $P < Q$ *then* $\sigma(H_Q) \subset \sigma(H_P)$.

◇

Proof By definition $H_P = H_Q + R_{P,Q}$. Let T_s be the translation operator by $s \in Q^*$ where $Q^* \equiv Q - \cup_{C \subset Q}$. Since T_s commutes with H_Q one has

$$|(\lambda I - H_P)T_s\phi| \leq |(\lambda I - H_Q)\psi| + |R_{P,Q}T_s\phi| \tag{20}$$

Let $\lambda \in H_Q$. The first term to the right in (20) can be made arbitrary small with a suitable choice of ϕ and for the properties of $R_{P,Q}$. The second term can be made arbitrary small by choosing s suitably large. It follows that for a suitable choice of ϕ the left hand side can be made arbitrary small and this implies $\lambda \in \sigma(H_P)$. ♡

Notice that for every choice of clusters the Hilbert space is always $L^2((R^3)^N$ but the approximate Hamiltonians are different according to the structure of the clusters. It follows form Lemma 1 that $\sigma(H) \supset [\mu, +\infty) \quad \mu \equiv \min_{P > \emptyset} \mu_P$.

Theorem 1 (Zhislin [1])

$$\sigma_{ess}(H) = [\mu, +\infty). \tag{21}$$

◇

Proof We give this proof in detail, because it is the prototype of all the other proofs.

The strategy is to approximate the decomposition into clusters by means of a regular partition of unity in such a way that for large $|x|$ we can use the partition given by the lattice L with a good estimate of the error made. Passing to the limit one obtains the proof of (21).

Recall that a regular partition of the unity in X is a collection of positive and regular functions $j_\alpha \in C^\infty$ (we shall call *elements of the partition)* such that

$$\sum_\alpha j_\alpha^2 = 1 \tag{22}$$

(the choice of the square in (22) will be convenient in the following).

The partition in channels can instead be seen as choice of hyperplanes in X and in this sense it associates to every channel (except $\emptyset$) a product of distributions δ. In order to make partitions of unity adapted to L we shall take the partitions $A \in L$.

Roughly speaking a regular partition corresponds to smoothening the δ functions that describe P and substitute them with C^∞ functions with support in a conical neighborhood of the support of the corresponding distribution. The solid angle of the cone must be finite but it may be made arbitrary small if we are only interested in the asymptotic behavior for large times.

According to Ruelle's theorem in each channel the outgoing and incoming states can be seen as *localized at infinity*. This will lead to the asymptotic estimates we shall describe. The following identity holds in the domain of definition of all terms

$$H = \sum_\alpha j_\alpha H j_\alpha + \frac{1}{2} \sum_\alpha [j_\alpha, [j_\alpha, H]]. \tag{23}$$

Notice the *double commutator* in (23). The use of double commutators will be important in what follows. Notice also that if the functions that implement the partition were substituted by distributions, the error made would be a distribution with support in the intersection of hyper-planes.

Recalling that $\sum_\alpha j_\alpha^2 = 1$ one can write (23) as

$$H = \sum_\alpha j_\alpha H j_\alpha - \frac{1}{2} \sum_\alpha (|\nabla j_\alpha|^2. \tag{24}$$

We have made use of the fact that the term $[j_\alpha, H]$ depends only of the position coordinates in $P^\perp$. For every partition P we define the corresponding element $j_{\alpha(P)}$ as follows. If $P \equiv \{\emptyset\}$ (i.e. the partition considered is $\{x_1\}, \ldots \{x_N\}$) we set $j^2_{\alpha,\emptyset} \equiv 1 - \sum_{P \neq \emptyset} j_P^2$. If $P \neq \emptyset$ consider the open covering of the unit sphere $S^1 \subset X$ obtained as

$$S_P \equiv \{x, \, : \, |x| = 1, \; |x_P| \neq 0\} \tag{25}$$

and the corresponding partition of unity $J_{\alpha(P)}$, $supp(J_{\alpha(P)} \subset S_P$. Notice that since $J_{\alpha(P)}$ has compact support for every P one can find $\epsilon > 0$ such that if $x \in suppJ_{\alpha(P)}$ then $|x_P| > \epsilon$.

The functions J_α we have introduced are defined on the unit sphere. We shall extend them to X in the following way: for $|x| < 1$ choose any extension which satisfies (22), for $|x| > 1$ set $j_\alpha(x) \equiv J_\alpha(\frac{x}{|x|})$.

The function that we have chosen have the following properties

$$|x| > 1, \quad \lambda \geq 1 \rightarrow j_\alpha(\lambda x) = j_\alpha(x) \tag{26}$$

$$|x| \geq 1, \quad x \in supp \, j_{\alpha(P)} \rightarrow |x|_P \geq \epsilon |x| \tag{27}$$

In the case of two-body potential of Coulomb type it is easy to verify that (27) implies for every partition P.

$$|\nabla j_{\alpha(P)}| = O\left(\frac{1}{|x|}\right), \quad |x| \rightarrow \infty \tag{28}$$

Therefore the second term to the right in (24) is compact relative to H. In the first term set $H = H_P + R_P$ and notice that $j_{\alpha(P)} R_P j_{\alpha(P)}$ is a Kato class potential with respect to H_P which vanishes at infinity, and is therefore compact relative to H_P. Hence

$$H = j_{\alpha(P)} H_A j_{\alpha(P)} + K \tag{29}$$

where K is compact relative to H_P. From Weyl theorem one derives

$$\sigma_{ess}(H) = \sigma_{ess}\left[\sum_P j_{\alpha(P)} H_P j_{\alpha(P)}\right] \tag{30}$$

Remark now that for every partition P, $H_P \geq \mu\, I$ (by the definition of μ) and therefore

$$\sum_P j_{\alpha(P)} H_P j_{\alpha(P)} \geq \mu \sum_P j^2_{\alpha(P)} = \mu \tag{31}$$

From this one derives

$$\sigma_{ess}(H) \subset [\mu, +\infty) \tag{32}$$

Since we have already shown that the reverse inclusion holds, the proof of Zhislin theorem is complete. ♡

5 Structure of the Continuous Spectrum

In order to achieve the asymptotic decomposition we must now study the spectrum of H in $[\mu, +\infty)$ and in particular prove that in this region the spectrum is absolutely continuous, property that is needed to prove asymptotic completeness.

We shall begin providing a qualitative analysis with the purpose of introducing and justifying some a priori estimates that we will prove later.

This will permit us to focus on the new role played by the double commutators and by the dilation group (we will study the description of the system for λ large enough after the scaling $x_i \to \lambda\, x_i$ for *some* of the coordinates.

From the experience acquired in the study of potential scattering we expect that in each channel the asymptotic behavior of the system when $t \to \infty$ approaches free motion (the meaning of *free motion* is different in the different channels).

We expect also that if the wave function ϕ has a sufficiently localized momentum spectrum one should have roughly

$$(\phi_t, x^2\phi_t) = \frac{1}{2}\theta_E t^2(1 + O(t^{-1})), \qquad t \to \infty \tag{33}$$

where θ_E must be somehow linked with a group velocity.

Equation (33) can be written as

$$\frac{d^2}{dt^2} < x^2 >_t \simeq \theta_E, \qquad t \to \infty \tag{34}$$

One has

$$\frac{d^2}{dt^2} < x^2 >_t = < i[H, A] >_t, \quad A \equiv i[H, x^2] = \frac{1}{2}(x \cdot p + p \cdot x) \tag{35}$$

whence

$$i[H,A] = p^2 - x \cdot \nabla V(x), \quad p \equiv i\nabla \tag{36}$$

From (33) and (34) if E does not belong to the point spectrum of H (in particular if it belongs to the continuous part of the spectrum) denoting by E_Δ the spectral projection of H in the neighborhood Δ of E we expect that if Δ is sufficiently small the following inequality holds

$$B_\Delta(H) \equiv iE_\Delta(H)[H,A]E_\Delta(H) \geq (\theta_E - \epsilon_\Delta)E_\Delta(H) \tag{37}$$

for a suitable ϵ_Δ such that $lim_{\Delta\to 0}\epsilon_\Delta \Delta^{-1} = 0$. Remark that if ϕ and ψ are eigenvectors of H to the eigenvalue E the following equivalent relations hold (the second one is frequently called *virial theorem*).

$$(\phi, (x \cdot \nabla V)\psi) = 0, \qquad < \phi, \hat{p}^2\psi >=< \phi, (x \cdot \nabla V)\psi > \tag{38}$$

We will prove (Mourre's theorem) that (37) holds up to addition of a compact operator. We shall see also that θ_E is a function of the *thresholds* for H.

6 Thresholds

Definition 1 A *threshold* for H is an eigenvalue λ_i^P of H^P for some $P \neq \emptyset$. It is therefore a bound state in a non-trivial channel. ◇

To better understand the relation between θ_E and the threshold values remark that for energies greater than λ_k^P we expect to be able to construct states which are approximately the tensor product of a bound state of H^P with energy λ_k^P and a state of free particle with momentum P in a complementary cluster. The evolution of this state will be given approximately by

$$\phi(t) \simeq e^{-i\left(\frac{p_P^2}{2}+\lambda_P\right)t}\phi_P \otimes \phi^P, \quad \phi_P \in L^2(X_P), \quad H_P\phi_P = \lambda^P\phi_P \tag{39}$$

For this state one has

$$< x^2 >_t \simeq < x_P^2 >_t \simeq < p_P^2 > t^2 \simeq 2(E - \lambda^P)t^2, \qquad t \to \infty \tag{40}$$

From (40) we deduce that for this state one has $\frac{1}{2}\theta(E) \simeq E - \lambda^P$ if the energy is concentrated around E.

This *heuristic argument* reflects the fact that if a state has energy approximately equal to E, denoting by λ_0 the lower bound of the energy of the cluster P, the energy at disposal of the other clusters is $E - \lambda_0$. We expect therefore that (39) holds a part from terms which depend only on the properties of the system at finite distances.

The estimates suggested above on purely local properties are consequences of the following fact: if we denote by $\eta_R(y)$ the indicator function of the ball of radius R, the operator $\eta_{R_1}(x) \cdot \eta_{R_2}(\hat{p})\eta_{R_3}(x)$ is compact for any choice of finite values for R_1, R_2, R_3.

From (39) we shall conclude that the eigenvalues which do not correspond to thresholds can have only a threshold as limit point. Since the thresholds are eigenvalues of a cluster it follows that the set of thresholds is closed and denumerable. To give precise estimates we shall study more in detail the term $i[H, A]$; one has

$$i[H \cdot A] = -\Delta + x \cdot \nabla V \tag{41}$$

We shall assume that $W(x) \equiv x \cdot \nabla V$ (called the virial of V) satisfies all the assumptions we have made on V. In particular we assume that $V(x)$ be of Kato class. Setting

$$W(x) = \sum_m x_m \cdot \nabla_{x_m} \left(\sum_{i<k} V_{i,k}(x_i - x_k) \right) \equiv \sum_{i<k} W_{i,k}(x_i, x_k) \tag{42}$$

one has

$$lim_{|y| \to \infty} sup_x W_{i,k}(x, y) = 0 \tag{43}$$

Under these assumptions one can prove.

Lemma 2 (Virial lemma) *If (42) and (43) hold and if ϕ and ψ are eigenstates of H to the eigenvalue E then*

$$(\phi, [H, A]\psi) = 0 \tag{44}$$

◇

Proof For the proof it is convenient to introduce a *regularization* of the dilation operator A e.g.

$$A_\epsilon \equiv \frac{1}{2} \left[\hat{p} \cdot x e^{-\epsilon x^2} + e^{-\epsilon x^2} x \cdot \hat{p} \right], \qquad \epsilon > 0 \tag{45}$$

Since A_ϵ is bounded with respect to $\hat{p}^2$ (this is not true for A) it leaves invariant the domain of $\hat{p}^2$ and on this domain one has

$$[A_\epsilon, H]e^{-\epsilon x^2} = -\epsilon(\hat{p} \cdot x)^2 e^{-\epsilon x^2} - \epsilon(x \cdot \hat{p})^2 e^{-\epsilon x^2} - e^{-\epsilon x^2} x \cdot \nabla V(x) \tag{46}$$

Since $\phi, \ \psi \in D(A_\epsilon)$ one has by the standard virial theorem $(\phi, [H, A_\epsilon]\psi) = 0$. Passing to the limit $\epsilon \to 0$ one obtains (44). ♡

7 Mourre's Theorem

Denote by $\tau(H)$ the collection of all thresholds. Define

$$\Theta(E) = inf_{\lambda \in \tau(H),\ \lambda \leq E}\ 2(E-\lambda)\xi(E-\mu) \tag{47}$$

where $\mu \equiv inf_\lambda\{\lambda : \lambda \in \tau(H)\}$ and ξ is the indicator function of R^+. One has then the following theorem proved for the case $N = 3$ by Mourre [5] and then extended to the N-body case in [1].

Theorem 2 (Mourre [5]) *Let V and $x \cdot \nabla V$ satisfy (43). Let E_J be the spectral projection of H associated to the interval J. Then*
(i) $\forall E \in R,\ \epsilon > 0$ there exists a compact operator K such that

$$B_J(H) \equiv iE_J[H, A]E_J \geq (\Theta(E) - \epsilon)E_J + K \tag{48}$$

(ii) The eigenvalues of H which are not thresholds have finite multiplicity and can have only a threshold as limit point. Therefore the set $\tau(H)$ is closed and denumerable. ◇

We shall use the notation $J_n \to \{E_0\}$ to indicate a sequence of decreasing intervals which have $\{E_0\}$ as limit. Multiplying (48) from the right and from the left by per E_J and recalling that K is compact and that $E_J \to_s 0$ when $J_n \to \{E_0\}$ and E_0 is not an eigenvalue of H we conclude that if E is not an eigenvalue then

$$lim_{n\to\infty}\|KE_{J_n}\| = (E_{j_n}K^*KE_{j_n})^{1/2} = 0$$

and therefore if J is sufficiently small

$$B_J(H) \geq (\Theta(E_0) - \epsilon) \tag{49}$$

Proof of Mourre's theorem
We proceed by induction. The result holds if $P = \emptyset$. Suppose that it holds for H^P with $P > \emptyset$ (the symbol $>$ denotes the partial ordering in the lattice).

In this case $E_n(H^P)$ are the eigenvalues of $H^Q\ \ \forall Q > P$ and the threshold $\Theta^P(E)$ is defined relative to $E(H^P)$. Therefore (48) reads

$$B_J(H^P) \geq (\Theta(E) - \epsilon)E_J(H^P) + K \quad on\ \ L^2(H^P) \tag{50}$$
$$B_J(H^P) \equiv iE_J(H^P)[H^P, A^P] \quad i[H^P, A^P] = -\Delta^P - (x^P \cdot \nabla V^P(x^P))$$

$$V^P = \sum_{i,j \in \alpha(P)} V_{i,j} \tag{51}$$

We have denoted A^P the generator of the dilation group on the variables x^P). If the theorem holds for H^P it follows from that if E is not an eigenvalue of H^P then $B_J(H^P) \geq (\Theta^P(E) - \epsilon)E_J(H^P)$ and therefore since $\Theta^P(E) \geq \Theta(E)$

$$B_J(H^P) \geq (\Theta(E) - \epsilon)E_J(H^P) \tag{52}$$

Let now E be an eigenvalue of H^P with projection operator Π_E^P. We must prove (52) with $\Theta(E) = 0$. The dimension of Π_E^P may be infinite.

Choose an increasing sequence of projection operators $P_n < \Pi_E^P$ that converge strongly to Π_E^P. Since E is an eigenvalue, from the virial theorem one derives

$$B_J = (I - \Pi_n)B_J(I - \Pi_n) + (\Pi_n B_J(I - \Pi_N^P) + (I - \Pi_n^P)B_J\Pi_N \tag{53}$$

From (52) and (53)

$$\begin{aligned} B_J &\geq -\epsilon E_J + (1 - \Pi_n)K(I - \Pi_n) - 2||\Pi_n B_J E_J(I - \Pi_E^P)||I \\ &\geq -\epsilon E_J - ||K(\Pi_E^P - \Pi_n||I - ||KE_J(I - \Pi_E^P)||I \\ &\quad - 2||\Pi_n E_J(I - \Pi_E^P||I(I - \Pi_n)B_J(I - \Pi_N) \\ &\quad + (\Pi_n B_J(I - \Pi_N^P) + (I - \Pi_n^P)B_J\Pi_N \end{aligned} \tag{54}$$

Since K and $\Pi_n B_J$ are compact and since $E_J(I - \Pi_E^P)$ converges strongly to zero when $J \to \{E\}$ one can choose first n sufficiently large and then J sufficiently small in such a way to obtain (52) also when $\Theta(E) = 0$.

We want now to improve on this estimate and prove that for any open set $S \subset R$, $E \in S$ and for any given $\epsilon > 0$ one can choose $\delta > 0$ in such a way that for all $E \in S$ and $|J| < \delta$ one has

$$B_J(H^P) \geq (\Theta(E + \epsilon) - 2\epsilon)E_J(H^P) \tag{55}$$

Indeed of this were not true, the inequality would not hold for a sequence

$$E_n \to E, \quad E_n \in S, \quad E_n \in J_n, \quad lim_{n\to\infty}|J_n| = 0 \tag{56}$$

Choose n so large that $|E_n - E| \leq \epsilon/2$. It follows from the definition that $\Theta(E + x) \leq \Theta(E) + x$ for all $x \geq 0$ and therefore

$$\Theta(E) \geq \Theta(E_n + \epsilon) - \epsilon + E - EN \geq \Theta(E_n - \epsilon) - \frac{3\epsilon}{2} \tag{57}$$

Keeping into account that $|J_n| < |J|$ $B_J(H^P) \geq (\Theta(E) - \epsilon/2)E_J(H^P)$ we derive

$$B_J(H^P) \geq (\Theta(E_n + \epsilon) - 2\epsilon)E_{J_n}(H^P) \tag{58}$$

and this proves (57). In order to give an estimate for $B_J(H)$ we must now supplement (57) with an estimate $B_J(H^\alpha)$. To achieve this we prove that for every $E \in R$ and for every $\epsilon > 0$ there exist an interval J containing E and such that

$$B_J(H_P) \geq (\Theta(E + \epsilon) - 2\epsilon)E_J(H_P) \tag{59}$$

To prove (57) take Fourier transform with respect to x_α. In this representation vectors in $L^2(X)$ are represented by function in $L^2(\tilde{X}_\alpha)$ with values in $L^2(X^\alpha)$ and one has

$$(H_P\psi)(k) = (k^2 + H^P)\psi(k), \quad (E_j(H_P)\psi)(k) = E_{J-k^2}(H^P)\psi(k)$$

$$i([H_P, A]\psi)(k) = (k^2 + i[H^P, A^P])\psi(k) \tag{60}$$

Set $\phi = E_J(H_P)\psi$. Therefore

$$(\phi, B_J(H^P)\phi) = \int_{\tilde{X}_P} [(\phi(k), (k^2 + B_{J-k^2}(H^P))\phi(k)]dk \tag{61}$$

where we have denoted by (,) the scalar product in $L^2(\tilde{X}_P)$ and with [,] the scalar product in $L^2(X^P)$. Since H^P is bounded below the integrand vanishes outside a compact set. From (57) one derives

$$(k^2 + \Theta(E - k^2 + \epsilon) - 2\epsilon)||\phi(k)||^2 \geq (\Theta(E + \epsilon) - 2\epsilon)||\phi(k)||^2 \tag{62}$$

and this completes the proof of (61). To conclude the proof of Mourre's theorem we use now the localization formula we have discussed above

$$H = \sum_P j_{\alpha(P)} H j_{\alpha(P)} + \frac{1}{2}[j_{\alpha(P)}[j_{\alpha(P)}, H]] = \sum_P j_{\alpha(P)} H j_{\alpha(P)} - \frac{1}{2}\sum_P |\nabla j_{\alpha(P)}|^2 \tag{63}$$

where $\{j_{\alpha(P)}, \}$ is a partition of unity by means of C^∞ on X. Choose $f \in C^\infty$, real valued and such that $f \equiv 1$ in J, $E \in J$. Then

$$if(H)[H, A]f(H) = i\sum_\alpha f(H)j_\alpha[H_\alpha, A]j_\alpha f(H) + K \tag{64}$$

where K is compact. We shall prove

$$L \equiv f(H)j_{\alpha(P)} - j_{\alpha(P)}f(H) \in \mathcal{K} \tag{65}$$

Given (65) Eq. (61) reads

$$if(H)[H,A]f(H) \geq (\Theta(E+\epsilon)-2\epsilon)f^2(H)+K \tag{66}$$

Multiplying both terms by E_{J_1} one has $B_{J_1}(H) \geq (\Theta(E+\epsilon)-2\epsilon)E_{J_1}+K$. This inequality is equivalent to (48) if E is not an eigenvalue (indeed if E is not an eigenvalue one has $\Theta(E+\epsilon)=\Theta(E)$ if ϵ is sufficiently small).

To achieve the proof of Mourre's theorem we must therefore prove (65). Let $\hat{f}$ the Fourier transform of f and define $R_P \equiv (i+H_P)^{-1}$. Therefore

$$\begin{aligned} LR_P &= \int dt\,\hat{f}(t)e^{-itH}(j_{\alpha(P)}-e^{itH}j_{\alpha(P)}e^{-itH})R_{\alpha(P)} \\ &= i\int dt\,\hat{f}(t)\int_0^s e^{-i(t-s)H}Ke^{-isH_P} \end{aligned} \tag{67}$$

$$K \equiv (Hj_{\alpha(P)}-j_{\alpha(P)}H_{\alpha(P)}R_{\alpha(P)}) = ([p^2,j_{\alpha(P)}]-j_{\alpha(P)}I_{\alpha(P)})R_{\alpha(P)} \tag{68}$$

and we have proved that this operator is compact. Since $\|LR_P\| < C\|K\|$ it is sufficient to consider the case in which K has rank one, i.e. $K\psi=(u,\psi)v$. But then the integrand reads

$$\psi \to \hat{f}(t)(e^{isH_P u}\psi)e^{-i(t-s)H}v \tag{69}$$

which is norm continuous both in t and in s. Therefore LR_P is compact. Set $f(x)=(i+x)g(x)$. Then the operator

$$g(H)j_{\alpha(P)}-j_{\alpha(P)}g(H_p) = LR_\alpha + g(H)([p^2,j_{\alpha(P)}]R_P + j_{\alpha(P)}I_PR_P) \tag{70}$$

is compact. Since g was arbitrary (65) is proved. This concludes the proof of Mourre's theorem. ♡

Mourre's theorem is useful both for giving a priori estimates for the exponential decay of the eigenfunctions of H and for proving the absence of singular continuous spectrum and asymptotic completeness in the N-body problem. A typical estimate of the asymptotic behavior of the eigenfucntions is given in the following theorem, which we will state without proof.

Theorem 3 (Froese–Herbst I [4]) *Under the hypothesis of Mourre's theorem, let $H\phi = E\phi$ and let $a \equiv \sup\{b \in R,\ e^{bx}\phi \in L^2(X)\}$. If $E+\frac{1}{2}a^2$ is finite, then it is a threshold for H.*

Remark that both Froese–Herbst's theorem (and the ones that we will state later) can be proved along the lines of Mourre's theorem under the following assumptions on V and on its virial.

(i) V belongs to Kato's class.

(ii) For every non-trivial partition P when x is sufficiently large one has a decomposition.

$$V(x) = V^P(x^P) + I_P(x), \qquad |I_P(x)| < f(|x_P|) \tag{71}$$

with $\lim_{s\to\infty} f(s) = 0$. *In the case* $V = \sum_{i<j} V_{i,j}(x_i - x_j)$ *these conditions are satisfied if each term in the sum is of Kato class and vanishes at infinity.* ♣

8 Absence of Positive Eigenvalues

From the theorem Froese–Herbst I one derives an important result.

Theorem 4 (Froese–Herbst II [7]) *Under the assumptions of Mourre's theorem H* has no positive eigenvalue. ◇

Proof From Theorem 3 we know that if H has no positive thresholds and $H\phi = E\phi, \quad \phi \in L^2$ then

$$e^{a|x|}\phi(x) \in L^2(X) \quad \forall a > 0 \tag{72}$$

By induction starting with $\alpha \equiv X$ it is enough to prove that if (74) holds, then there are no positive eigenvalues. Choose ρ_0 such that

$$\int_{r<\rho_0} |\phi(x)|^2 dx \leq \int_{r>2\rho} |\phi|^2(x) dx \tag{73}$$

and choose $F(r) \in C^\infty$ with the property

$$F(r) \leq r, \qquad F(r) \geq 0 \qquad r \geq r_0 \to \quad F(r) = r \tag{74}$$

Set $\phi_a(|x|) \equiv e^{aF(|x|)}\phi|e^{aF(|x|)}|\phi^{-1}$. From (74) one derives $\int_{|x|<\rho_0} dx|\phi_a(x)|^2 \leq e^{-2\rho_0}$. Notice that there exists $c_1 > 0$ such that for every $a > 0$

$$(\phi_a, H\phi_a) \geq E + a^2/2 - c_1 a^2 e^{-2a\rho_0} \tag{75}$$

Indeed set $H_a \equiv e^{aF} H e^{-aF} = H - \frac{a^2}{2}|\nabla F|^2 + i\frac{a}{2}(\nabla F \cdot \hat{p} + \hat{p}\nabla F)$. One has

$$H_a\phi_a = E\phi_a, \qquad (\phi_a, H\phi_a) = E + \frac{a^2}{2}(\phi_a, |\nabla F|^2\phi_a) \tag{76}$$

and $|\nabla F| = 1$ for $|x| \geq \rho_0$. Therefore

$$|(\phi_a, |\nabla f|^2\phi_a) - 1| \leq c_1 e^{-2a\rho_0} \tag{77}$$

from this one derives (76).

In the same way we can estimate $i(\phi_a, [H, A]\phi_a)$. One obtains

$$i(\phi_a, [H, A]\phi_a) = \frac{ia^2}{2}(\phi_a, [|\nabla F|^2, A]\phi_a) - 2aRe(\phi_a, \gamma A\phi_a) \quad \gamma = \frac{1}{2}(\nabla F \cdot \hat{p} + \hat{p} \cdot \nabla F) \tag{78}$$

The first term in (79) is bounded by $c^2 a^2 e^{-2a\rho_0}$. For the second term

$$2Re(\phi_a, \gamma A\phi_a) = \hat{p}_k(x_k F_{,l} + F_{,k} x_l) - \frac{d}{2}F_{,ll} - \frac{x_k}{2}F_{,llk} \tag{79}$$

The first term in (79) is positive and the remaining two are bounded. Therefore there are positive constants c_2, c_3 such that

$$i(\phi_a, [H, A]\phi_a) = < \hat{p}^2 >_a - < x, \nabla V >_a \leq c_2 a^2 e^{-2a\rho} + ac_3 \tag{80}$$

where $<,.>_a \equiv < \phi_a, .\phi_a >$. Subtracting (75) from (80)

$$\frac{1}{2} < \hat{p}^2 >_a - < V >_a - < x \cdot \nabla V >_a \leq -E - \frac{a^2}{2} + (c_1 + c_2)a^2 e^{-2\rho_0} + ac_3 \tag{81}$$

Inequality (81) leads to a contradiction. Indeed the term to the left is bounded below for every value of the parameter a because both V and $x \cdot \nabla V$ are small relative to $\hat{p}^2$). The term to the right diverges to $-\infty$ when $a \to \infty$. Since the only assumption we have made is $H\phi = E\phi, \ \phi \in L^2(X), \ E > 0$ we conclude that there are no positive eigenvalues. ♡

A second important consequence of Mourre's theorem is an accurate estimate of the rate at which the essential support of the states in the continuous part of the spectrum of H leaves any compact in X (dispersive estimates).

As a corollary of the estimates we shall prove that there is *no singular continuous part of the spectrum.*

From the local compactness (expressed in Mourre's theorem) it follows that if ϕ belongs to the continuous part of the spectrum then

$$lim_{t \to \infty} |\xi_R e^{-iHt}\phi| = 0 \tag{82}$$

Remark that Ruelle's theorem implies only *convergence in the mean.*

Under further assumptions on the potential it will also be possible to estimate the rate of convergence. Equation (82) can also be proved if one makes the assumption that the *second virial* (i.e. $[[V, A]A]$) satisfies the assumption made for the potential and its first virial. For potentials of the form $V = \sum_{i<j} V_{i,j}(x_i - x_j)$ this new condition means that for each pair $i \neq j$ the function $(x_i - x_j)^2 \nabla^2 V_{i,j}(x_i - x_j)$ be a Kato class potential.

The following theorem is useful to prove that in the N-body problem under the stated conditions on the potential the singular continuous part of the spectrum is empty.

Theorem 5 *([2, 5]) Assume that $V(x)$ and $x \cdot \nabla V$ satisfy the hypothesis (i) and (ii) of Mourre's theorem and assume also that the second virial is bounded. Denote with S the collection of the thresholds and eigenvalues of H. Then if $E_J \phi = \phi$ for every $a > 1/2$ and compact J in $R - S$ one has, for a suitable constant $c_\phi(J, a)$*

$$\int_{-\infty}^{\infty} dt |(1 + x^2)^{-a/2} e^{-itH} \phi|^2 < c_\phi(J, a) |\phi|^2 \tag{83}$$

◇

We shall not prove this theorem but we shall state and prove a corollary.

Corollary *If the conditions in Theorem 5 are satisfied, then the singular continuous spectrum of H is empty.*

◇

Proof Let $f \in C_0^\infty$ and $\phi = E_j \phi$. From (82) one derives

$$|(1 + x^2)^{\frac{-a}{2}} f(H) \phi| \le \frac{1}{2\pi} \int dt \hat{f}(t) |(1 + |x|^2)^{\frac{-a}{2}} e^{itH} \phi| \le |f|_2 |\phi| \tag{84}$$

Taking point-wise limits this inequality extends to the characteristic functions of bounded Borel sets. Therefore $\phi \in \mathcal{H}_{a.c.}$. Since $R - S$ us open the states which satisfy $\phi = E_J \phi$ for some compact $J \in S^\perp$ span the range of E_{R-S}. As a consequence the range of E_{R-S} is contained in $\mathcal{H}_{a,c}$.

On the other hand the range of E_S contains all bound states since S is denumerable and contains all eigenvalues. ♡

We shall now use Theorem 5 to derive inequalities which will be useful in the proof of asymptotic completeness.

Lemma 3 *Set $\mathcal{R} \equiv (i + H)^{-1}$. If $A \mathcal{R}^m (1 + x^2)^{\frac{a}{2}}$ is a bounded operator for some $a > 1/2$ $m \ge 1$ then for every compact $J \subset R - S$ one has*

$$\int_0^\infty |A e^{-itH} E_J(H) \psi|^2 < c |\phi|^2 \tag{85}$$

◇

Proof One has

$$|A e^{-itH} E_J(H) \phi| \le |A \mathcal{R}^m (1 + x^2)^{\frac{a}{2}}| \cdot |(1 + x^2)^{\frac{a}{2}} e^{-itH} E_J (i + H)^m E_J \phi|^2 \tag{86}$$

If $a > \frac{1}{2}$ the second factor is less than $c_1 |\phi|$. ♡

In the following it will be convenient to make use of the following notation

$$A = O_m(|x|^{-a}) \leftrightarrow A\mathcal{R}^m(1+|x|^2)^{\frac{a}{2}} \in \mathcal{B}(\mathcal{H}) \tag{87}$$

and also of the following inequality

$$A = O_0(|x|^{-b}) \leftrightarrow A\hat{p}_k = O_1(|x|^{-a}) \;\; \forall a \equiv \max(1, b) \tag{88}$$

To prove (88) set $< x >^a \equiv (1+|x|^2)^{\frac{a}{2}}$. Making use of $[< x >^a, \mathcal{R}] = \mathcal{R}[< x >^a, H]\mathcal{R}$ one has

$$A\hat{p}_k\mathcal{R} < x >^a = A < x >^a \hat{p}_k\mathcal{R} + A[p_k, < x >^a]\mathcal{R} + \hat{p}_k\mathcal{R}[< x >^a, H]\mathcal{R} \tag{89}$$

If $a \leq 1$ all the terms in the right-hand side are bounded. They remain bounded $a \leq b$ and $A = O_0(|x|^{-b})$. This proves (88) A further important consequence is contained in the following lemma.

Lemma 4 *Suppose that the operator A can be written as $A = B \cdot C$ where B and C are $O_m(|x|^{-a})$ for some $a > \frac{1}{2}$. Then the following limit exists*

$$lim_{T\to\infty} \int_0^T dt E_J(H) e^{itH} A e^{-itH} E_J(H)\phi \tag{90}$$

◇

Proof Denote by $\phi(t)$ the integrand in (92). Then

$$\begin{aligned} |\int_T^{\mathcal{T}} dt|\phi(t)|^2 &= sup_{|\psi|=1}|\int_T^{\mathcal{T}} dt(\psi, \phi(t)|^2 \\ &\leq sup_{|\psi|=1}|\int_T^{\mathcal{T}} dt|BE_J(H)e^{-itH}\psi|^2 \int_T^{\mathcal{T}} dt|Ce^{-itH}E_J(H)\psi|^2 \end{aligned} \tag{91}$$

By assumption J is separated from the eigenvalues of H and also from the thresholds. Therefore the first factor is bounded and the second converges to zero as $t, \tau \to \infty$. ♡

9 Asymptotic Operator, Asymptotic Completeness

We want now to use these estimates derived from Mourre's theorem to prove asymptotic completeness in the N-body problem if the potentials $V_{i,j}(x_i - x_j)$ are of short range and satisfy a further regularity property that we will state presently.

Definition 2 (*short range*) The potential $V(x)$ is *short range* if, for every partition α one has for $|x_\alpha|$ sufficiently large

$$V(x) = V^{\alpha}(x_{\alpha}) + J(x^{\alpha}), \qquad |J(x^{\alpha})| \le |x^{\alpha}|^{-\mu}, \quad \mu > 1 \tag{92}$$

We have set $|x_{\alpha}| \equiv \min_{i \perp k} (m_i m_k)^{1/2} (m_i + m_k)^{-1/2} |x_i - x_k|$ where with the symbol $x_i \perp x_k$ we mean that x_i and x_k *belong to different clusters.* ◇

Under the conditions for the applicability of Mourre's theorem and if all potentials are of short range one can prove asymptotic completeness. An important role has the following theorem of Segal and Soffer; we only outline the proof (see [6, 8]).

Theorem 6 (Sigal–Soffer) *Assume that $V(x)$ is short range, satisfies the conditions in Mourre's theorem and moreover that*

$$|\nabla I_P(x)| \le c\, |x_P|^{-\mu}, \quad \mu > 1 \tag{93}$$

Then $\mathcal{H}^+ = \mathcal{H}_c = \mathcal{H}_n$ and each orbit in these spaces has the asymptotic behavior

$$\phi_t \equiv e^{-itH} \phi \simeq \sum_{P \neq \{\emptyset\}} e^{-itH_P} \left(I \otimes \Pi_B(H^P) \right) \phi_P \tag{94}$$

where $\Pi_B(H_P)$ is the projection operator on the bound states of channel P and we have used the notation

$$u(t) \simeq v(t) \leftrightarrow lim_{t \to \infty} |u(t) - v(t)| = 0 \tag{95}$$

◇

If $V(x) = \sum_{i<j} V_{i,j}(x_i - x_j)$ the conditions for the validity of the theorem of Sigal–Soffer are that each $V_{i,j}$ be small in the sense of Kato with respect to the Laplacian and for every term of the sum one has

$$|V_{i,j}(y)|, |\nabla V_{i,j}(y)| \le c|y|^{-\mu}, \quad \mu > 1 \tag{96}$$

We shall give only a brief outline of the proof of Theorem 6. The proof uses iteration starting with the partition which has no bound states. An important role is taken by the generators of *partial dilations* in which only part of the coordinates are dilated.

More precisely, if one wants to analyze the asymptotic behavior in time of a given decomposition P in clusters one makes use of the generator of dilations of the *center-of-mass coordinates of the clusters.*

This has the effect that, roughly speaking, the evolution of the cluster P according to the full hamiltonian and that according to H^P tend to coincide. The method of Mourre is efficient because of this property.

The proof of the Sigal–Soffer theorem is therefore based on the construction of a collection of observables which commute *locally* with the hamiltonian H and have the property that their evolution gives a control of the asymptotic behavior of the system in the various channels.

An important role is played by the asymptotic behavior in time of the operator $\gamma_P \equiv i[H, g_P]$ where g_P are smooth functions that characterize a regular partition *asymptotically linear* (so that on a large scale it is similar to the partition according to hyper-planes).

One has $\gamma = \gamma_0 + \sum_P \gamma_P$ where for each P we have denoted by $\gamma_P^\pm$ an approximate dilation operator that is used to describe the asymptotic behavior of the solution of the Schroedinger equation in the P sector.

Correspondingly we introduce the *asymptotic operator*

$$\gamma_P^\pm = s - lim_{t\to\pm\infty} e^{itH} \gamma_P e^{-itH_P} E_\Delta(H_P) \tag{97}$$

Is easy to prove that γ maps $\mathcal{H}_\Delta$ into itself and on $\mathcal{H}_\Delta$ the relation $\gamma^+ = \sum_P \gamma_P^+$ holds. Every vector $\psi \in \mathcal{H}_\Delta$ can be written

$$\psi = \sum_P \gamma_P^+ \phi, \quad \phi \in \mathcal{H}_\Delta \tag{98}$$

and therefore

$$\psi_t = e^{-itH}\psi \simeq \sum_P \gamma_P e^{-itH}\phi = \sum_\alpha e^{-itH_P} e^{itH_P} \gamma_P e^{-itH}\phi \simeq e^{-itH_P}\psi_P \tag{99}$$

where $\psi_P = W_P^+\phi$ and W_P^+ is the wave operator in channel P

$$W_P^+ \equiv s - lim_{t\to\infty} e^{itH_P} \gamma_P e^{-itH} E_\Delta(H) \tag{100}$$

From this one develops an iteration procedure that leads to the proof of the Sigal–Soffer theorem and asymptotic completeness.

References

1. W. Hunziker, I. Sigal, The general theory of N-body quantum systems, CRM Proceedings and Lecture Notes, Am. Math. Soc. **8**, 35–72 (1995)
2. M. Reed, B. Simon, *Methods of Modern Mathematical Physics*, vol. 4 (Academic Press, New York, 1978)
3. W. Hunziker, I.M. Sigal, J. Math. Phys. **41**, 3448–3510 (2000)
4. R.G. Froese, I. Herbst, Duke Math. J. **49**, 1075–1085 (1982)
5. E. Mourre, Comm. Math. Phys. **91**(83), 279–300
6. I. Segal, I. Soffer, Comm. Math. Phys. **132**, 73–101 (1990)
7. R.G. Froese, I. Herbst II, Comm. Math. Phys. **87**, 429–447 (1982)
8. I. Segal, I. Soffer, Ann. Math. **1126**, 35–108 (1987)

Lecture 14: Positivity Preserving Maps. Markov Semigropus. Contractive Dirichlet Forms

In Volume I we have remarked that in order that the operator $U = -\Delta + V$ be self-adjoint the conditions on the positive part V_+ of V are much weaker than the conditions on its negative part. In particular it not required that V_+ be small with respect to the laplacian.

Notice that, as multiplication operators, the positive function *preserve positivity*.

This trivial remark admits a non trivial extension, since the multiplication operators are not left invariant, as a set, by a generic transformation in the Hilbert space $\mathcal{H}$ while the property to be small with respect to another operator (e.g. the Laplacian) does not depend on the representation.

In the case $\mathcal{H} = L^2(X, d\mu)$ and $\mathcal{V}$ is the cone of positive functions, by using properties of the Laplacian (e.g. to have a resolvent that is described by a positive kernel), it is possible to associate to e^{-tH} a stochastic process, a modification of Brownian motion.

We are led therefore to consider the case in which in the Hilbert space there exists a convex cone $\mathcal{V}$ that is left invariant by a suitable class of transformations.

1 Positive Cones

Let Y be a linear topological space and consider in Y a strict convex cone generating cone K (Y is spanned by the convex combinations of the element in $-K \cup K$). Let K_0 be the interior of K. We shall call *positive* the elements of K, *strictly positive* the elements of K_0.

Definition 1 (*preservation of positivity*) We say that a map T of Y into itself is
(i) *positivity preserving* if $T(x) \in K$ for every $x \in K$.
(ii) *positivity improving* if $T(x) \in K_0$ for every $x \in K$. ◇

We shall study in some detail only the case $Y \equiv L^2(X, d\mu)$ where X is a measure space, and we often specialize to the case in which $X = R^d, d < +\infty$ and μ is

G. Dell'Antonio, *Lectures on the Mathematics of Quantum Mechanics II: Selected Topics*, Atlantis Studies in Mathematical Physics: Theory and Applications 2, DOI 10.2991/978-94-6239-115-4_14

Lebesgue measure. In this case K will be the cone of positive-valued functions and K_0 will be the cone of functions that are strictly positive in very compact.

Analogous results are obtained in the case Y is a C^* algebra and K is the cone of its positive elements.

We shall consider only linear maps. In this case Definition 1 takes the form

Definition 2 The operator T on $L^2(X, \mu)$ is
(i) *positivity preserving* if $f \geq 0$ implies $(Tf)(x) \geq 0$
(ii) *positivity improving* if $f \geq 0$ implies $Tf(x) > 0$ *on compact sets.* ◇

Definition 3 (*ergodic*) The operator T on $L^2(X, \mu)$ is *ergodic* if it is positivity preserving and for any positive function g and *strictly positive function* f there exist an integer n such that $(f, T^n g) > 0$. ◇

Note that if T is positivity improving, it is ergodic since the relation is satisfied for every integer n. If $x \to \phi(t, x)$ is a dynamical system in X, the evolution $f \to T_t f(x) \equiv f(\phi(t, x))$ is positivity preserving *but not improving.*

One can prove that the semigroup T_t is ergodic iff the dynamical system is ergodic in the traditional sense (the only invariant sets are the empty set and X).

The evolution described by the semigroup $e^{t\Delta}$ on $L^2(R^d, dx)$ is *positivity improving*. For every $t > 0$ one has

$$(e^{t\Delta} f)(x) = C_n \int e^{-\frac{|x-y|^2}{2t}} f(y)dy > 0 \quad \forall x \tag{1}$$

and the integral kernel of $e^{t\Delta}$ is strictly positive. If H is self-adjoint and positive, the semigroup e^{-tH} is positivity preserving (resp. improving) iff $(H + \lambda)^{-1}$, $\lambda > 0$ has the same property.

This is a consequence of the following identities

$$(H + \lambda)^{-1} = \int_0^\infty e^{-t(H+\lambda)} dt \qquad e^{-t(H+\lambda)} = lim_{n\to\infty} \left(1 - \frac{t}{n}(H + \lambda)\right)^{-n} \tag{2}$$

Lemma 1 *If μ is absolutely continuous with respect to the Lebesgue measure and T is positivity preserving, then $|Tf|(x) \leq T(|f|)(x) \quad \forall f \in L^2(R^n, dx)$ (the inequality is understood to hold a.e.).* ◇

Proof By density it is sufficient to give a proof when f is continuous. If f is real valued $|f| \pm f \geq 0$. Since T is linear and positivity preserving, $T|f(x)| \pm |Tf|(x) \geq 0$. Therefore $T|f(x)| \geq |Tf|(x)$.

If f is not real valued, $|f(x)| = sup_{\theta \in Q} Re(e^{i\theta} f(x))$ (Q is the set of rational numbers). Since Q is denumerable and f is continuous

$$sup_{\theta\in Q}Re(e^{i\theta}Tf(x)) = Tsup_{\theta\in Q}Re(e^{i\theta}f(x)) \leq T|f|(x)$$

♡

2 Doubly Markov

Definition 4 Assume μ finite. A bounded positivity preserving operator T which satisfies

$$T\iota = \iota, \quad T^*\iota = \iota, \quad \iota(x) = 1 \ \forall x \tag{3}$$

is said to be *doubly Markov*. This notation is due to the fact that ι is an eigenfunction to the eigenvalue one *for both* T and T^*. ◇

Lemma 2 *If T is doubly Markov then*

$$||Tf||_p \leq ||f||_p \quad 1 \leq p \leq +\infty \tag{4}$$

(T is a contraction on L^p for $1 \leq p \leq +\infty$) ◇

Proof By interpolation it suffices to give the proof when $p = 1$ and $p = +\infty$. If $f \geq 0$ $||Tf||_1 = (\iota, Tf) = (T^*\iota, f) = (\iota, f) = ||f||_1$. If f is not positive, from the preceding lemma $||Tf||_1 \leq ||T|f|||_1| = ||f||_1$.

If $f,\ g \in L^2, f \geq 0,\ g \geq 0$ then $(f, Tg) \geq 0$. It follows that also T^* is positivity preserving, and therefore $||T^*f||_1 \leq ||f||_1$. By definition $||g||_\infty = sup_{f,||f||_1=1}|(g,f)|$ and therefore

$$\begin{aligned}||Tg||_\infty &= sup_{f,||f||_1=1}|(Tg,f)| = sup_{f,||f||_1=1}|(g,T^*f)| \leq sup_{f,||T^*f||_1=1}|(g,f)| \\ &= ||g||_\infty\end{aligned} \tag{5}$$

♡

We remark that the integral kernel $T(x, x')$ of a doubly Markov operator can be used to define the *transition probability* of a stochastic process, in analogy to what we have seen in the cases of Brownian motion and of the Ornstein–Uhlenbeck process.

We shall see in this Lecture (Beurling–Deny Theorem) that if the quadratic form associated to H has suitable contraction properties then $e^{-tH}(x, x')$ defines a doubly Markov semigroup. We shall describe now the relevant properties of the positivity preserving operators.

Theorem 1 *Let the operator T be bounded, closed and positive on $L^2(X, d\mu)$. Let T be positivity preserving and assume that $||T||$ be an eigenvalue (and then the largest eigenvalue). The following statements are equivalent to each other*
(a) $||T||$ is a simple eigenvalue and the corresponding eigenfunction ϕ_0 can be chosen to be positive.

(b) T is ergodic.
(c) $L^\infty \cup \{T\}$ *is irreducible i.e. if a bounded operator A commutes with T and with the operator of multiplication by any essentially bounded function, then A is a multiple of the identity.* ◇

This theorem is an extension of the classic Frobenius theorem on matrices; L^∞ takes the place of the collection of matrices which are diagonal in a given basis.
Proof of Theorem 1
(a) implies (b)
Let $B \equiv \frac{T}{||T||}$, and let λ_n be the eigenvalues B in decreasing order. By assumption

$$\lambda_0 = 1, \quad \lambda_n < 1 \quad \forall n \geq 1 \tag{6}$$

It follows that $s - lim\ B^n = P_0$, the orthogonal projector onto ϕ_0. Therefore for $\phi \in L^2(X, d\mu)$ one has

$$\lim_{n\to\infty} (\phi, B^n\phi) = |(\phi, \phi_0)|^2 > 0 \tag{7}$$

(the last inequality follows because ϕ_0 is strictly positive on compact sets). Therefore there is at least one n_ϕ such that $(\phi, B^n\phi) > 0$.
(b) implies (c)
Let the closed subspace $\mathcal{S} \in L^\infty$ be left invariant by L^∞ and by T. If $f \in \mathcal{S}$ define $g(x) \equiv \frac{\bar{f}(x)}{|f(x)|}$ if $f(x) \neq 0$. Then $g \in L^\infty$ and $gf = |f| \in \mathcal{S}$. In the same way one proves that if $g \in \mathcal{S}^\perp$ then also $|g| \in \mathcal{S}^\perp$. But then $(|g|, T^n|f|) = 0\ \forall n$ and therefore $f \equiv 0$.
(c) implies (a)
Let ϕ_0 be eigenfunction of T to the eigenvalue $||T||$.
From Lemma 19.2 it follows that also $|\phi(x)|$ is an eigenfunction to the same eigenvalue, because $(\psi, T\phi_0)| \leq ||T||$ for any ψ. We must prove that for every compact $\mathcal{K}$ one has $inf_{\mathcal{K}}|\phi_0(x)| > 0$. Let $\Gamma \equiv \{f \in L^2, f\phi = 0 \quad a.e\}$.
By construction Γ is a closed subspace invariant under multiplication by L^∞ functions. Let $\Gamma = \Gamma_+ - \Gamma_- + i\Gamma_+ - i\Gamma_+$, $\Gamma_+ \equiv \{f \in \Gamma,\ \ f(x) \geq 0\}$. Then $T\Gamma_+ \subset \Gamma_+$ because if $f \in \Gamma_+$ one has $(Tf, |\phi|) = (f.T|\phi|) = ||T||(f, |\phi|) = 0$.
Analogous inclusions hold for the other three terms in the decomposition of Γ. Therefore $T\Gamma \subset \Gamma$.
From (c) one has the alternative $\Gamma = \{0\}$ or $\Gamma = L^2(x, d\mu)$. The second alternative is excluded because $\phi_0 \notin \Gamma$. Therefore $\Gamma = \{0\}$ and this implies that no function $f \in L^2$ such that a.e. $f(x)\phi_0(x) = 0$.
Uniqueness follows because it is not possible for two functions to be strictly positive and orthogonal to each other.

♡

3 Existence of the Ground State

We make now use of Theorem 1 to prove the following result which provides necessary and sufficient conditions in order that the ground state be simple. Later we shall give necessary and sufficient conditions for the existence of a ground state (here we *assume* existence).

Theorem 2 *Let H be self-adjoint and bounded below. Let $E \equiv inf\ \sigma(H)$. The following statements are equivalent to each other*
(a) E is a simple eigenvalue and the corresponding eigenfunction can be chosen to be strictly positive.
(b) There exists $\lambda < E$ such that $(H - \lambda I)^{-1}$ is ergodic.
(c) There exists $t > 0$ such that e^{-tH} is ergodic.
(d) $\forall\lambda < E$ the operator $(H + \lambda)^{-1}$ is positivity improving.
(e) $\forall t > 0$ the operator e^{-tH} is positivity improving. ◇

Proof From Theorem 1 we know that (a)–(c) are equivalent to each other, that (d) implies (b) and that (e) implies (c). We shall now prove the two remaining implications
(c) implies (d)

By assumption there are $s_0 > 0$ and non-negative functions $u,\ v$ which are not identically equal to zero such that $(u, e^{-s_0 H} v) > 0$. By continuity $(u, e^{-sH} v) > 0$ when s is in a neighborhood of s_0. Then

$$(u, (H + \lambda)^{-1} v) = \int_0^\infty e^{s\lambda}(u, e^{-sH} v) ds > 0$$

and therefore $((H + \lambda)^{-1} v)(x) > 0 \quad \forall x$.
(c) implies (e)

Let $u,\ v$ be non-negative functions not identically equal to zero. Define $\mathcal{N} \equiv \{t > 0\ (u, e^{-tH} v) > 0\}$. The function $(u, e^{-tH} v)$ is analytic in a neighborhood of R^+ therefore the set $((0, \infty) - \mathcal{N})$ cannot have 0 as accumulation point. It follows that $\mathcal{N}$ contains arbitrary small numbers.

In order to prove that $\mathcal{N} \equiv (0, +\infty)$ it suffices therefore to prove that $t > s, s \in \mathcal{N}$ implies $t \in \mathcal{N}$. Let $s_0 \in \mathcal{N}$. By assumption $(u, e^{-s_0 H} v) > 0$ and then $\bar{u}(x)(e^{-s_0 H} v)(x)$ is not identically equal to zero.

Let $w(x) = \min_s\{u(x), (e^{-sH} v)(x)\}$. Since the operator e^{-tH} is positivity preserving

$$(u, e^{-tH}(e^{-sH} v)) \geq (u, e^{-tH} w) = (u, e^{-tH} w) \geq (w, e^{-tH} w) = |e^{-\frac{tH}{2}} w| > 0 \quad (8)$$

It follows that if $t > 0$ and $s \in \mathcal{N}$ then $t + s \in \mathcal{N}$. This ends the proof of Theorem 2 ♡

Example Let $A \geq cI, c > 0$ an operator on $\mathcal{H}_1 \equiv L^2(R^d)$ and denote by $H = d\Gamma(A)$ on $\mathcal{H} = \Gamma(\mathcal{H}_1)$ its second quantization. Identify $\mathcal{H}$ with $L^2(X, d\mu)$ for a suitable measure space X, μ.

In Quantum Field Theory $\mathcal{H}_1$ is the *one-particle space* (e.g. $L^2(R^3)$), A is the one-particle hamiltonian, X is a space of distributions in R^3 and μ is a Gauss measure on X. Denote by Ω the *vacuum state* in Fock space. By construction

$$H\Omega = 0, \qquad H_{\Omega^\perp} \geq cI \tag{9}$$

Therefore H has a ground state which is simple and can be chosen positive. From Theorem 2 one derives that if e^{-tA} is positivity preserving, then $\Gamma(e^{-tA}) \equiv e^{-tH}$ is positivity improving in $L^2(X, d\mu)$. ♣

We apply now Theorem 2 to the N-body problem in Quantum Mechanics.

Theorem 3 *Let H be the hamiltonian of the N-body system in the frame in which the center of mass is at rest. If the infimum of the spectrum is an eigenvalue, then this eigenvalue is simple, and the corresponding eigenfunction can be chosen positive.* ♢

Proof According to Theorem 2 it is sufficient to prove that e^{-tH} is positivity preserving and that $\{e^{-tH} \cup L^\infty(R^{3N-3})\}$ is irreducible. We know that both statements hold for $H_0 \equiv -\sum_n \Delta_n$. Set $V_{i,j}^N(x) \equiv inf\{N, V_{i,j}(x) \text{ if } |V_{i,j}\}$. Then $e^{-tV_{i,j}^N(x)} \in L^\infty$ and is invertible.

Therefore the algebra $\mathcal{A}$ generated by $e^{-t(H_0+V_{i,j}^N)}$ together with the elements of $L^\infty(R^{3N-3})$ (considered as multiplication operators) is irreducible. Moreover $e^{-t(H_0+\sum_{i,j} V_{i,j}^N)}$ is positivity preserving (use Trotter–Kato formula and remark that each factor has this property).

It is easy to verify that $\sum_{i,j} V_{i,j}^N$ converges in L^2, when $N \to \infty$, to $\sum_{i,j} V_{i,j}$. Therefore when $N \to \infty$ $H_0 + \sum_{i,j} V_{i^N,j}$ converges in the strong resolvent sense (and therefore in the strong sense for the associated semigroups) to $H_0 + \sum_{i,j} V_{i,j}$.

Since the strong limit in L^2 preserves positivity and $\mathcal{A}$ is weakly closed , the proof of Theorem 3 is complete.

♡

Recall that if T is bounded and doubly markovian on $L^2(X, d\mu)$ with μ finite measure, then T is a contraction on all L^p, $1 \leq p \leq \infty$.

We now introduce a stronger condition on T, namely we require that T be a contraction from L^p to L^q where p and q are positive constants, with $p < q$.

Recall that , since the measure has finite total weight, one has always $||.||_q \geq ||.||_p$ if $p < q$ and the inequality is strict unless the measure is carried by a finite number of points.

4 Hypercontractivity

Definition 5 Let (X, μ) a measure with finite total weight. A bounded operator T is said to *hypercontrative* if there exist $q > 2$ such that T be a contraction from L^2 to L^q (i.e. $\forall f \in L^2 \quad |Tf|_q \leq |f|_2$) ♢

The importance of this notion is given by following theorem.

Theorem 4 (Gross [1]) *Let $H \geq 0$ be the generator of a positivity preserving semi-group and suppose that there exist $t_0 > 0$ such that $e^{-t_0 H}$ is hypercontractive between L^2 and L^4.*

Then
(1) inf $\sigma(H) \equiv E$ is an eigenvalue.
(2) The eigenvalue E is simple.
(3) The corresponding eigenfunction can be chosen positive. ◇

Proof It follows from Theorem 3 that it suffices to prove point (1) since the (2) and (3) follow. Consider a finite partition $\alpha \equiv \{S_1, \dots S_N\}$ of X, i.e. a finite collection of measurable sets such that

$$\cup_n S_n = X, \qquad S_i \cap S_k = \emptyset \quad i \neq k \tag{10}$$

Denote by $\xi(S_k)$ the indicator function of S_k and call P_α the operator

$$(P_\alpha f)(x) = \sum_i \frac{1}{\mu_{S_i}} \xi(S_i)(x) \int_{S_i} f(y)\mu(dy) \tag{11}$$

Then one easily verifies that
(1) $P_\alpha \leq P_\beta$ if the partition β is finer that α and P_{α_n} converges strongly to the identity when the partition is refined indefinitely.
(2) P_α is positivity preserving.
(3) P_α is a contraction on every L^p as one sees by interpolation: by construction $|P_\alpha f|_\infty \leq |f|_\infty$ and P_α contracts in L^1 because it is symmetric and $P_\alpha \iota = \iota$.

Set $A \equiv e^{-t_0 H}$ and $A_\alpha \equiv P_\alpha A P_\alpha$. From properties (1)–(3) one derives (the notation $lim_{\alpha \to \infty}$ indicates the limit when the partition if refined indefinitely)
(a) $s - lim_{\alpha \to \infty} A_\alpha = A$
(b) $||A|| = lim_{\alpha \to \infty} ||A_\alpha||$
(c) for every $\phi \in L^2$ there exists an integer K such that $|A_\alpha \phi|_4 < K\ ||A||_\alpha |\phi|_2$

Property (c) follows from (a) to (b) and the assumptions we have made on $e^{-t_0 H}$. For every finite partition we can identify the operator A with a $N \times N$ matrix that preserves positivity. From the Perron–Frobenius theorem follows the existence of $\phi_\alpha \in P_\alpha \mathcal{H}$ such that

$$A_\alpha \phi_\alpha = ||A||_\alpha \phi_\alpha \tag{12}$$

Normalizing this vector with $|\phi|_2 = 1$ it follows from (c) $|\phi_\alpha|_4 \leq K$. Hölder inequality gives

$$|\phi_\alpha|_2 \leq |\phi_\alpha|_1^{\frac{1}{3}} |\phi_\alpha|_4^{\frac{2}{3}} \qquad |\phi_\alpha|_1 = \int_X |\phi_\alpha|(x)|d\mu(x) \geq \frac{1}{K^2} \tag{13}$$

The unit ball in $L^2(X, d\mu)$ is compact for the weak topology, and we can extract a sequence ϕ_{α_n} with $\alpha_n < \alpha_{n+1}$ that converges weakly to ϕ and it follows

$$|\phi_\alpha|_1 \equiv \int_X \phi_\alpha d\mu \to \int_X \phi(x) d\mu = |\phi|_1 \tag{14}$$

Notice that $||\phi||_1 \geq \frac{1}{K^2}$ and therefore $\phi \neq 0$. On the other hand, (a) and (b) imply for every element $\psi \in L^2(X, d\mu)$

$$(\psi, A\phi) = (A\psi, \phi) = lim_\beta (A\psi, \phi_\beta) = lim_\beta ||A||_\beta (\psi, \phi_\beta) = ||A||(\psi, \phi) \tag{15}$$

Since this relation holds for every ψ one derives $A\phi = ||A||\phi$.

♡

Remark that if the measure space has total measure $\mu(X) > 1$ and

$$|A\phi|_q \leq M|\phi|_2, \quad M > 0 \tag{16}$$

an analysis similar to that presented above [1] proves that if $||A||$ is an eigenvalue of T its multiplicity m is bounded by

$$m \leq m_0 \equiv \left(\frac{M}{||T||}\right)^{\frac{2q}{q-2}} \mu(X) \tag{17}$$

The proof makes use of the fact that for every solution of $A\phi = ||A||\phi$ one has

$$(\phi, \phi) \geq \left(\frac{||T||}{M}\right)^{\frac{q}{q-2}} \tag{18}$$

From this one derives that the number of orthogonal solutions cannot be greater than M. ♡

We study now conditions on the operators A and B under which if A has the properties we are considering (preserve or improve positivity, being doubly Markov, be hypercontrative...) also the operator $A + B$ has the same property.

We are particularly interested to the case $A \equiv -\Delta$ and B is a multiplication operator by a function $V(x)$.

Theorem 5 *Let $\mathcal{H} = L^2(R^d, \mu)$, $H_0 \geq 0$ where μ is absolutely continuous with respect to Lebesgue measure, and assume $(H_0 + \lambda)^{-1}$ is positivity preserving for all $\lambda > 0$. Let $U(x)$, $-W(x)$ be real positive functions. Denote by $Q(H)$ the form-domain of the operator H.*

Let $Q(H_0) \cap Q(U)$ be dense in $\mathcal{H}$ and let W be small with respect to H_0 in the quadratic-form sense. Define

$$H \equiv H_0 + U + W \tag{19}$$

as quadratic forms. Let λ_0 *be the infimum of the spectrum of* $H_0 + W$. *Then for* $\lambda > \lambda_0$ *the operator* $(H - \lambda I)^{-1}$ *is positivity preserving.* ◇

Proof Denote by η_F the indicator function of the set F. Set

$$U_k(x) = \eta_{U(x)\le k}(x)U(x), \qquad W_h = \eta_{|W(x)|\le h}(x)W(x) \tag{20}$$

Consider

$$H_{k,h} \equiv H_0 + U_k + W_h \tag{21}$$

Since U_k and W_h are bounded the series

$$(H_{k,h} + \lambda I)^{-1} = (H_0 + \lambda I)^{-1} \sum_{n=0}^{\infty} [(U_k + V_h)(H_0 + \lambda I)^{-1}]^n \tag{22}$$

is absolutely convergent for λ sufficiently large and each term in the series preserves positivity. But if $H_0 + W > -b\, I$ for each value of the parameter h one has $H_0 + W_h > -b\, I$. It follows that $H_{k,h} + b$ is invertible and

$$(H_{h,k} + b\, I)^{-1} = (H_{h,k} + \lambda I)^{-1} \left(I + \sum_n [(\lambda - b)(H_{h,k} + b\, I)^{-1}]^n \right). \tag{23}$$

The series converges uniformly and each term preserves positivity. Therefore $(H_{h,k} + b\, I)^{-1}$ preserves positivity for $b > \lambda_0$ with λ_0 the infimum of the spectrum of $H + W$.

Since the cone of positive functions is weakly closed, to pass to the limit $h,\ k \to \infty$ it is enough to prove that $H_{h,k}$ converges to H in strong resolvent sense.

This has been proved, under the assumptions of Theorem 5, in Book I (convergence of operators). ♡

Remark that the (open) cone of *strictly positive* functions is not closed under weak convergence. Therefore even if each of the resolvent of each of the $H_{h,k}$ improves positivity *this needs not be true* for H.

Recall also that in the Feynman–Kac formula for the proof of the self-adjointness of $H_0 + V$ an important role is played by the requirement that $Q(H_0) \cap Q(V)$ be dense in $\mathcal{H}$.

Theorem 6 *Let* $\mathcal{H} = L^2(X, d\mu)$ $H_0 \ge 0$ $H_0\phi_0 = 0,$ $\phi_0 \in \mathcal{H}$ *and let* ϕ_0 *be strictly positive. Assume that* $(H_0 + \lambda I)^{-1}$ *be positivity preserving for each* $\lambda > 0$. *Let* $V(x) \ge 0,$ $(\phi_0, V\phi_0) < +\infty$. *Then* $D(H_0) \cap D(V_0)$ *is dense in* $\mathcal{H}$ *and therefore* $Q(H_0) \cap Q(V_0)$ *in dense in* $\mathcal{H}$. ◇

Proof Define $L^\infty_{\phi_0} \equiv \{f\ :\ \pm f \le t\phi_0\}$ for some $t > 0$. Notice that from $\phi_0(x) > 0\ \forall x$ it follows that $L^\infty_{\phi_0}$ is dense in $\mathcal{H}$ and also that

$$(H_0 + \lambda I)^{-1}\, L^\infty_{\phi_0} \subset L^\infty_{\phi_0} \tag{24}$$

Indeed $(H_0 + \lambda\ I)^{-1} f \leq t(H_0 + \lambda\ I)^{-1}\phi_0 = \frac{t}{\lambda}\phi_0$. By assumption $V \in L^1(X, \phi_0^2(x))dx)$ and therefore $L^\infty_{\phi_0} \subset Q(V)$. Therefore

$$(H_0 + \lambda\, I)^{-1} L^\infty_{\phi_0} \subset D(H_0) \cap Q(V) \tag{25}$$

Since $L^\infty_{\phi_0}$ is dense in $\mathcal{H}$ we conclude that $(H_0 + \lambda\ I)^{-1} L^\infty_{\phi_0}$ is dense in $(H_0 + \lambda\, I)^{-1}\mathcal{H} \equiv D(H_0)$. ◇

As a consequence of Theorem 5 and of the theorems we have proved for quadratic forms we know that $H + V$ is self-adjoint. We prove now that $D(H_0) \cap D(V)$ is a core for it.

Theorem 7 *Let $H_0 \geq 0$, $\phi_0 \in L^2$, $\phi_0(x) > 0$ and suppose that $(H_0 + \lambda\ I)^{-1}$ is positivity preserving. Let $V \in L^2(X, \phi_0^2 dx)$. Then $H_0 + V$ is essentially self-adjoint on $D(H_0) \cap D(V)$.* ◇

Proof We must show that $D(H_0) \cap D(V)$ is dense in $D(H)$ *in the graph topology*. We know that if λ is sufficiently large $(H - \lambda\ I)^{-1}$ preserves positivity and leaves $L^\infty_{\phi_0}$ invariant. Set $\mathcal{K} \equiv (H - \lambda\ I)^{-1}\ L^\infty_{\phi_0}$. Then

$$\mathcal{K} \subset D(H) \cap L^\infty_{\phi_0} = D(H) \cap D(V) \tag{26}$$

If $g \in \mathcal{H}$, $H_0 g \in \mathcal{H}$ it follows $g \in D(H)$ and therefore $\mathcal{K} \subset D(H_0) \cap D(V)$. But $L^\infty_{\phi_0}$ is dense in $\mathcal{H}$ and therefore $L^\infty_{\phi_0}$ is dense in $(H - \lambda\ I)^{-1}\mathcal{H}$. The closure of $(H - \lambda\, I)^{-1}\mathcal{H}$ in the graph topology of H coincides with $D(H)$; therefore $\mathcal{K}$ is dense in $D(H)$ in the graph topology of H.

Notice that $\mathcal{K} \subset D(H_0) \cap D(V)$ and therefore also this set is dense $\mathcal{H}$ in the same topology. ♡

5 Uniqueness of the Ground State

From Theorem 1 we know that for a bounded positivity preserving operator the smallest eigenvalue is simple if the operator is ergodic.

In the ergodic theory for classical ergodic systems it is known that ergodicity is equivalent to indecomposability (metric transitivity). An analogous definition can be introduced for operators on $L^2(X, d\mu)$; this definition coincides with the classical one if the operators are obtained by duality from maps $X \to X$.

We give here the definition in the operator case, prove that indecomposability implies uniqueness of the ground state and we give two conditions on V under which if H_0 satisfies indecomposability also $H + V$ satisfies this property.

Definition 6 (*indecomposable*) The bounded and closed operator T on $L^2(X, d\mu)$ is *indecomposable* if it does not commute with the projection on $L^2(Y, d\mu)$ where $Y \subset X$ is a measurable proper subset with $\mu(Y) \neq 0$. ◇

Theorem 8 *Let T be an operator on $L^2(X, d\mu)$ bounded self-adjoint and positivity preserving. Let $\|T\|$ be an eigenvalue. T is indecomposable iff the eigenvalue $\|T\|$ is simple and the corresponding eigenfunction can be chosen positive.* ♢

Proof Assume $Tu = \|T\|u$, $u \in L^2(X, d\mu)$. We can take u real because T preserves reality. Then

$$\|T\||u|^2 = (u, Tu) \leq (|u|, T|u|) \Rightarrow |u| = u \tag{27}$$

therefore $Tu_+ = \pm\|T\|u$. If $f \geq 0, f \in L^2(X, d\mu)$ one has $(Tf, u_\pm) = \|T\|(f, u_\pm)$. Since T is indecomposable either $u_- = 0$ or $u_+ = 0$. This implies uniqueness.

Conversely, assume that $\|T\|$ is a simple eigenvalue with eigenfunction $u > 0$ and that there exists a measurable set Y, such that the the orthogonal projection P_Y onto Y commutes with T.

Therefore $P_Y u = u$. But this is only possible if $\mu(Y) = 0$ or $\mu(X - Y) = 0$.

♡

In the case of unbounded self-adjoint operators the definition of irreducibility requires more attention.

Definition 7 Let A be self-adjoint unbounded on $L^2(x, d\mu)$. A is *indecomposable* if one cannot find a measurable subset Y of X with $0 < \mu(Y) < \mu(X)$ such that $f \in D(A)$ implies $P_Y f \in D(A), P_Y A - AP_Y = 0 \ \ on \ \ D(A)$.

If A is bounded below the condition is equivalent to $(A+\lambda I)^{-1}$ be indecomposable (in the sense of the previous definition) for λ sufficiently large. ♢

We consider now conditions under which if H_0 is indecomposable so is $H_0 + V$. If $H_0 + V$ is bounded below, this implies that if H_0 has a unique ground state, also $H_0 + V$ has this property.

Theorem 9 *Let $\mathcal{H} = L^2(X, \mu)$, and H_0 positive. Let U and $-W$ be measurable positive functions on X. Let $Q(U) \cap Q(H_0)$ be dense in $\mathcal{H}$ and let W be form-small with respect to H_0. Define $H = H_0 + U + W$ as sum of quadratic forms and denote by $\hat{H}_0$ the self-adjoint operator associated to the closed positive quadratic form obtained by closing the quadratic form of H_0 restricted to $Q(H_0) \cap Q(W)$. If $\hat{H}_0$ is indecomposable so is also H.* ♢

Remark that if U satisfies $(\phi_0, U\phi_0) < +\infty$, and $Q(H_0) \cap Q(U)$ is dense in $Q(H_0)$, then $\hat{H}_0 = H_0$ and H is indecomposable.

Proof of Theorem 9 It is easy to verify that $P(Y)$ commutes with H iff $g \in Q(H)$ implies

$$P(Y)g \in Q(H), \quad (f, HP(Y)g) - (P(Y)f, Hg) = 0 \ \ \forall f, \ g \in Q(H) \tag{28}$$

In particular if $H \geq 0$ one has

$$P(Y)\, H\psi = H\, P(Y)\psi, \quad \psi \in D(H) \Rightarrow P(Y)\, \sqrt{H}\phi = \sqrt{H}\, P(Y)\phi, \ \ \phi \in D(\sqrt{H}) \tag{29}$$

Assume that $P(Y)$ commutes with H. Since $P(Y)$ commutes with U and W, if $Q(H)$ is dense in $Q(H_0)$ it follows from (29) that

$$(f, H_0\, P(Y)g) = (P(Y)f, H_0)g) \quad \forall f,\ g \in Q(H_0) \tag{30}$$

and therefore either $\mu(Y) = 0$ or $\mu(X - Y) = 0$, since by assumption H_0 is indecomposable. If $Q(H)$ is not dense in $Q(H_0)$ notice that $(P(Y)g, H_0P(Y)g) = (g, H_0g)$ if $g \in Q(H)$. Therefore the map $g \to P(Y)g$ is continuous in the topology of $Q(H_0)$.

It follows that $g \in Q(\hat{H}_0) \Rightarrow P(Y)g \in Q(\hat{H}_0)$ and that $g \to P(Y)g$ is continuous in the topology of $\hat{H}_0$. Therefore (30) holds also for $\hat{H}_0$ and $P(Y)$ commutes $\hat{H}_0$.

♡

It is important to have criteria which guarantee that a given self-adjoint operator be the generator of a positivity preserving semigroup. Of particular interest are criteria that refer only to the quadratic form associated to the operator. The basic results are due to Beurling and Deny [2].

Theorem 10 (Beurling–Deny I) *Let $H \geq 0$ on $L^2(X, d\mu)$ and define $(\psi, H\psi) = +\infty$ if $\psi \notin D(H)$. The following statements are equivalent*
(a) e^{-tH} is positivity preserving for each $t > 0$
(b) $(|u|, H|u|) \leq (u, Hu) \quad \forall u \in L^2(X, d\mu)$
(c) e^{-tH} preserves reality for all $t > 0$ and

$$(u_+, Hu_+) \leq (u, Hu)\forall u \in L^2(X, d\mu)$$
$$(u_+(x) \equiv max\{u(x), 0\})(u_+, Hu_+) + (u_-, Hu_n) \leq (u, Hu) \quad u = u_+ - u_- \tag{31}$$

◇

Remark that the thesis of the theorem have a simpler form if expressed in terms of the corresponding quadratic forms. Denote by $\mathcal{E}_H$ the quadratic form associated to the operator H and with $Q(\mathcal{E}_H)$ its form domain.

In what follows we omit the suffix H. With these notations conditions (b)–(d) become

(b′)

$$u \in Q(\mathcal{E}) \Rightarrow |u| \in Q(\mathcal{E}) \qquad \mathcal{E}(|u|, |u|) \leq \mathcal{E}(u, u) \tag{32}$$

(c′)

$$u \in Q(\mathcal{E}) \Rightarrow u_+ \in Q(\mathcal{E}), \quad \mathcal{E}(u_+, u_+) \leq \mathcal{E}(u \cdot u) \tag{33}$$

(d′)

$$u \in Q(\mathcal{E}) \Rightarrow u_+,\ u_- \in Q(\mathcal{E}), \quad \mathcal{E}(u_+, u_+) + \mathcal{E}(u_-, u_-) \leq \mathcal{E}(u \cdot u) \tag{34}$$

◇

Proof of Theorem 10 In the applications we shall see that the interesting part of the theorem is $(a) \Leftrightarrow (b)$.

This the only part which we shall prove. The proof of the other implications is similar.

Proof (*a*) ⇒ (*b*)

One has

$$(u, Hu) = lim_{t\to\infty}\frac{1}{t}(u, (I - e^{-tH})u) \tag{35}$$

$$(u, e^{-tH}u) = |e^{-\frac{t}{2}H}u|^2 \leq |e^{-\frac{t}{2}H}u||^2 = (|u|, e^{-tH}|u|) \tag{36}$$

Therefore

$$(u, (I - e^{-tH})u) \geq (|u|, (I - e^{-tH})|u|) \tag{37}$$

Passing to the limit $t \to \infty$ (b) follow. One may notice that the result is obtained in the form 2′ because (37) holds for $u \in L^2(X, d\mu)$ and the limit in (31) exists for $u \in Q(\mathcal{E})$ and equals $\mathcal{E}(u, u)$.

(*b*) ⇒ (*a*)

Let $u \geq 0, \quad \lambda > 0$. Define

$$w \equiv (H + \lambda I)^{-1} \tag{38}$$

We want to prove $w \geq 0$. This shows the the resolvent is positivity preserving and then the semigroup has the same property. Set

$$\mathcal{E}(u, u) \equiv (\phi, H\phi) + \lambda(\phi, \phi) \tag{39}$$

Performing the calculations one obtains

$$\mathcal{E}(\phi + \psi, \phi + \psi) = \mathcal{E}(\phi, \phi) + \mathcal{E}(\psi, \psi) + 2\,Re((H + \lambda)\phi, \psi) \tag{40}$$

If $Re(V) > 0$

$$\mathcal{E}(w + v, w + v) = \mathcal{E}(w, w) + \mathcal{E}(v, v) + Re(u, v) \geq \mathcal{E}(w, w) + \mathcal{E}(v, v) \tag{41}$$

One may notice the analogy with the inequality which characterizes dissipative operators. If one has equality in (41) then $v = 0$ because $u \geq 0$. Set $v = w - w$. Then

$$\mathcal{E}(|w|, |w|) \geq \mathcal{E}(w, w) + \mathcal{E}(|w| - w, |w| - w) \tag{42}$$

and identity holds if $v = 0$. From (40) one derives $v = 0$ since by assumption $\mathcal{E}(|w|, |w|) \leq \mathcal{E}(w, w)$. ♡

6 Contractions

Theorem 11 *Let $H \geq 0$ be a self-adjoint operator on $L^2(X \cdot \mu)$ generator of a positivity preserving semigroup.*

Define $(f \wedge 1)(x) \equiv inf\{f(x), 1\}$. The following statements are equivalent to each other

(a) For all $t > 0$ the operator e^{-tH} is a contraction on L^p, $1 \leq p \leq \infty$
(b) For all $t > 0$ the operator e^{-tH} is a contraction on L^∞
(c) For all f one has $(f \wedge 1, H f \wedge 1) \leq (f, Hf)$.
(d) If F is such that $|F(x|) \leq |x|$ and $|F(x) - F(y)| \leq |x - y| \quad \forall x, y \in R$, then $(F(f), HF(f)) \leq (f, Hf) \quad \forall f \in L^2$. ◇

Remark that we have use the term *defines a contraction* because initially the operator e^{-tH} is defined on $L^2(X, \mu)$. One obtains the extension to L^p, $p \neq 2$ by first restricting the operator to $L^2 \cap L^p$ and extending the result to all L^p (e^{-tH} is by assumption bounded with bound one on $L^2 \cap L^p$ in the topology of linear operators L^p).

Also in this theorem the best formulation is by means of quadratic forms. For example, points (c) and (d) become

(c′)

$$f \in Q(\mathcal{E}) \Rightarrow f \wedge 1 \in Q(\mathcal{E}), \qquad \mathcal{E}(f \wedge 1), f \wedge 1) \leq \mathcal{E}(f \cdot f) \tag{43}$$

(d′)

$$|F(x)| \leq |x|, \quad |F(x) - F(y)| \leq |x - y| \Rightarrow f \in Q(\mathcal{E}) \rightarrow F(f) \in Q(\mathcal{E}) \tag{44}$$

and $\mathcal{E}(F(f), F(f)) \leq \mathcal{E}(f, f)$. Notice that F is a contraction with Lipshitz norm ≤ 1. Therefore (d′) is the requirement that $x \rightarrow F(f(x))$ leave invariant the form domain and operate as a contraction.

Proof of Theorem 11 The implication $(d) \rightarrow (c)$, $\quad (b) \rightarrow (a)$, $\quad (c) \rightarrow (b)$ are easy to prove. We now prove $(c) \rightarrow (b)$, $\quad (a) \rightarrow (d)$.

$(c) \rightarrow (d)$

Let $u \in L^2$, $\quad 0 \leq u(x) \leq 1 \quad \forall x$. Define for $v \in Q(\mathcal{E})$

$$\psi(v) = (v, Hv) + \|u - v\|^2 = (v, (H + I)v) + \|u\|^2 - 2Re\ (u, v) \tag{45}$$

and set $R_1 \equiv (H + I)^{-1}$. Then $\psi(R_1 u) = \|u\|^2 - (u, R_1 u)$ and

$$((R_1 u - v), (H + I)(R_1 u - v)H(R_1 u - v)) = (u, R_1 u) + (v, (H - I)v) - 2Re\ (u, v) \tag{46}$$

Therefore

$$\begin{aligned}\psi(v) &= \psi({}_1u) + ((R_1 u - v), (H + I)(R_1 u - v)H(R_1 u - v)) \\ &= (u, R_1 u) + (v, (H - I)v) - 2Re\ (u, v)\end{aligned} \tag{47}$$

It follows that $\psi(v)$ reaches the maximum value only in $v = R_1 u$. Since $u \leq 1$ one has

$$|u(x) - sup(v(x), 1)| \leq |(u(x) - v(x)| \tag{48}$$

Therefore $\psi((R_1 \wedge 1)) \leq \psi(R_1\ u)$. Since $R_1 u$ is a minimum point $(R_1 \wedge 1) = R_1$ and therefore $R_1 u \leq 1$, It follows that R_1 is a contraction in $L^\infty(X, d\mu)$. In the same way one proves that R_ϵ is a contraction in $L^\infty(X, d\mu)$ and hence $e^{-tH} \equiv lim_{n\to\infty}(I + \frac{tH}{n})^n$ is a contraction in $L^\infty(X, d\mu)$.
$(a) \to (d)$
It suffices to prove that under the hypotheses made of F

$$(F(f), (I - e^{-tH})F(f)) \leq (f, e^{-tH} f) \tag{49}$$

From (45) one obtains (d) dividing by t and passing to the limit $t \to 0$. Consider a partition α of X in measurable disjoint sets $S_1, \ldots S_{N(\alpha)}$. Let Π_α be the projection operator on the space of functions that are constant in each set. These functions are often called *simple*.
By density it suffices to prove that for any finite partition

$$(F(\Pi_\alpha f), (I - e^{-tH})F(\Pi_\alpha f)) \leq (\Pi_\alpha f, e^{-tH}\Pi_\alpha f) \tag{50}$$

If ξ_S is the indicator function of the set $S$and $b_{k,h} \equiv (\xi_{S_k}, (I - e^{-tH})\xi_{S_k})$ we must prove

$$\sum_{h,k} \bar{F}(\alpha_h)F(\alpha_k)b_{h,k} \leq \sum \bar{\alpha}_k\ \alpha_h\ b_{h,k} \tag{51}$$

under the assumption $|F(\alpha)| \leq \alpha, \quad |F(\alpha) - F(\beta)| \leq |\alpha - \beta|$. Set

$$\lambda_k \equiv (\xi_k, \xi_k), \quad b_{h,k} \equiv \lambda_k \delta_{h,k} - a_{h,k}, \quad a_{h,k} \equiv (\xi_{S_k}, e^{-tH}\xi_{S_k}) \tag{52}$$

One has $\sum_h a_{h,k} \leq \lambda_k$ and therefore

$$\sum \bar{z}_h\ z_k b_{h,k} = \sum_{h<k} a_{h,k}|z_h - z_k|^2 + \sum_k m_k|z_k|^2, \quad m_k \equiv \lambda_k - \sum_h a_{h,k} \geq 0 \tag{53}$$

Define $z_k \equiv F(\alpha_k)$; one has

$$\begin{aligned}\sum \bar{F}(\alpha_h)\ F(\alpha_k)b_{h,k} &= \sum_{h<k} a_{h,k}|F(\alpha_h) - F(\alpha_k)|^2 + \sum_k m_k|F(\alpha_k)|^2 \\ &\leq \sum_{h<k} a_{h,k}|\alpha_h - \alpha_k|^2 + \sum_k m_k|\alpha_k|^2 = \sum_{h,k} \bar{\alpha}_h\ \alpha_k b_{h,k}\end{aligned} \tag{54}$$

♡

7 Positive Distributions

A further characterization is based on the following Lemma [2].

Lemma 3 (Levy–Kintchine) *Let $F(x)$ be a complex-valued function on R^d, Re $F(x) \geq -c$. Define $e^{-F(i\nabla)} = \mathcal{F}e^{-F(x)}$ ($\mathcal{F}$ is Fourier tranform). The following statements are equivalent*
(a) The operator $e^{-F(i\nabla)}$ is positivity preserving.
(b) $\forall t \geq 0$, $e^{-tF(x)}$ is a positive distribution (in Bochner's sense).
(c)

$$\bar{F}(x) = F(-x), \quad \sum_1^m F(z_i - z_j)\bar{z}_i z_j \leq 0, \quad \forall x_i \in R^d, \quad z \in C^m \quad \sum_1^m z_i = 0. \quad (55)$$

◇

Proof $(b) \to (a)$
Set $G(x) \equiv e^{-tF(x)}$ and denote by $*$ convolution. One has

$$(f, G(-i\nabla)g) = (2\pi)^{-\frac{d}{2}}(\hat{G} * (\hat{g} * \hat{f}))(0) \quad (56)$$

Therefore if $\hat{G}$ is a positive measure then $(f, G(-i\nabla)g) \geq 0$.
$(a) \to (b)$
Assume $G(-i\nabla)$ preserves positivity. Set $g_y(x) \equiv f(x+y)$. Taking Fourier transforms

$$(2\pi)^{-\frac{d}{2}}(\hat{G} * (\hat{f} * \hat{f}))(y) = (2\pi)^{-\frac{d}{2}}(\hat{G} * (\hat{g}_y * \hat{f}))_0 = (f, G(-i\nabla)g) \geq 0 \quad (57)$$

Since Re $F(x) \geq -C$ one has $G(x) \leq e^{tC}$ $\forall x$. Therefore $G(x)$ is a tempered distribution and so is $\hat{G}$.

Defining $f(x) = j_\epsilon(x)$ where j_ϵ is an approximated δ and passing to the limit $\epsilon \to 0$ one has $\hat{g}_\epsilon(k)\hat{G}(k) \to \hat{G}(k)$ uniformly over compact sets. It follows that $\hat{G}(k)$ is a positive measure.
$(b) \leftrightarrow (c)$
Denote by A the matrix with elements $a_{i,j}$ and with $M(t)$ the matrix with elements $e^{ta_{i,j}}$. We must prove that $M(t)$ is positive definite iff is positive definite the restriction of A to the subspace $\sum_k \xi_k = 0 \equiv (\iota, \xi)$ (ι is the vector with all components equal to one).

The condition is *necessary*: from $M(0)_{i,j} = 1$ follows $(\xi, M(0)\xi) = 0$ if $(\xi, \hat{1}) = 0$. Since $(\xi, M(t)\xi) \geq 0$ $\forall t \geq 0$ one has

$$(\xi, A\xi) \equiv \frac{d}{dt}(\xi, M(t)\xi)_{t=0} > 0 \quad (58)$$

The condition is *sufficient*: denote by $I - P$ the orthogonal projection on $\hat{1}$. By assumption $PAP > C\,I$. One has

$$A = PAP + (I_P)A(I_P) + PA(I - P) + (I - P)AP \tag{59}$$

and $a_{i,j} = \tilde{a}_{i,j} + \bar{b}_i + b_j$ where $\tilde{A}$ is positive definite. Hence $M(t)_{i,j} = e^{t\bar{b}_j}\tilde{M}(t)_{i,j}e^{tb_i}$ i.e. the matrix M is obtained from the positive matrix $\tilde{M}$ through a linear transformation with positive coefficients and is therefore positive.

♡

A generalization of the theorem II of Beurling–Deny has been given by M. Fukushima. It provides a *one-to -one correspondence* between positivity preserving semigroups and Dirichlet forms having spacial properties.

Theorem 12 (Fukushima [2]) *In theorem II of Beurling–Deny, the semigroup* improves positivity *iff the corresponding Dirichlet form is* strictly contractive *i.e.*

$$|f| \geq c > 0, \quad E(|f|,|f|) = E(f,f) \Rightarrow f = \alpha|f| \tag{60}$$

◇

For a proof of this theorem and for a detailed description of the relation between quadratic forms and Markov processes on can see [2].

Notice that if T is a d-dimensional torus and H is the laplacian defined on T with periodic boundary conditions, if $f \in L^2(T)$ for any $t > 0$ one has $e^{tH}f \in C^\infty(T)$. In fact

$$\mathcal{F}(e^{tH}f) = \sum_{k=1}^{d} \sum_{n_k \in N} e^{-n_k^2} f_{n_1,\dots n_d} \tag{61}$$

and the series is uniformly convergent for every $t > 0$. The same holds true if X is a compact Riemann manifold and H is the Laplace–Beltrami operator.

In the case of non-compact manifolds and for a general probability space the improvement in regularity is of different nature and is a generalization of the hypercontractivity bound we have mentioned in this lecture.

Let μ a probability measure on X. The following inclusions hold

$$L^p(X,\mu) \subset L^q(X,\mu), \qquad 1 \leq p \leq q \leq \infty \tag{62}$$

and the inclusions are strict unless the measure μ is supported by a finite number of points. Define

$$\|e^{-tH}\|_{q\to p} \equiv sup\|e^{-tH}f\|_p, \quad f \in L^q \cap L^p \quad \|f\|_q \leq 1 \tag{63}$$

The relation (57) means $\|e^{-tH}\|_{q\to p} \leq 1 \quad q \geq p$. The regularization property we want to discuss considers the case $q < p$.

Definition 8 The semigroup e^{-tH} is said to be (q, p, t_0) *hyper-contractive*, with $q < p$, if there is $t_0 > 0$ such that

$$\|e^{-t_0 H}\|_{q \to p} \leq 1, \quad q < p \tag{64}$$

◇

Remark that if (60) holds for $t = t_0$ it also holds for $t > t_0$. The (q, p, t_0) hyper-contractivity property holds for singular perturbations of the Laplace–Beltrami (which have no $L^2 \to L^\infty$ regularization property).

It also holds and also in some cases of infinite-dimensional spaces, e.g. R^∞ if one makes use of Gauss measure in some models of Quantized Field Theory and of the Dobrushin–Ruelle measure (generalization of Gibbs measure) and in models of Statistical Mechanics for infinite particles systems. If zero is a simple eigenvalue of $H \geq 0$, inequality (60) implies that it is isolated.

References

1. L. Gross, J. Funct. Anal. **10**, 52–109 (1972)
2. M. Fukushima, *Dirichlet Forms and Markov Processes* (North-Holland, Amsterdam, 1980)

Lecture 15: Hypercontractivity. Logarithmic Sobolev Inequalities. Harmonic Group

We ended the previous Lecture with an analysis of conditions under which the semigroup e^{-tH} has suitable regularizing properties.

In this lecture we exploit these results. For example if T is a d-dimensional torus and H is the laplacian defined on T with periodic boundary conditions, if $f \in L^2(T)$ for any $t > 0$ one has $e^{tH}f \in C^\infty(T)$ as one proves noticing that upon taking Fourier transform on has

$$\mathcal{F}(e^{tH}f) = \sum_{k=1}^{d} \sum_{n_k \in N} e^{-n_k^2} f_{n_1,\ldots,n_d} \tag{1}$$

and the series is uniformly convergent for every $t > 0$. The same hold true if the manifold is smooth manifold and H is minus the Laplace–Beltrami operator.

In the case of non-compact manifolds and for a general probability space the improvement in regularity is of different nature and is a generalization of the hypercontractivity property.

For a probability measure on a Banach space X one has $L^p(X, \mu) \subset L^q(X, \mu)$, $1 \leq p \leq q \leq \infty$ and the inclusions are strict unless the measure μ is supported by a finite number of points. Define

$$\|e^{-tH}\|_{q\to p} \equiv sup\|e^{-tH}f\|_p, \quad f \in L^q \cap L^p \quad \|f\|_q \leq 1 \tag{2}$$

Therefore $\|e^{-tH}\|_{q\to p} \leq 1 \quad q \geq p$.

Definition 1 The semigroup e^{-tH} is said to be (q, p, t_0)-*hyper-contractive*, with $q < p$, if there is $t_0 > 0$ such that

$$\|e^{-t_0 H}\|_{q\to p} \leq 1, \quad q < p \tag{3}$$

◇

G. Dell'Antonio, *Lectures on the Mathematics of Quantum Mechanics II: Selected Topics*, Atlantis Studies in Mathematical Physics: Theory and Applications 2, DOI 10.2991/978-94-6239-115-4_15

Remark that if (3) holds for $t = t_0$ it also holds for $t > t_0$. The (q, p, t_0) hyper-contractivity property holds for *singular perturbations* of the Laplace–Beltrami operator which have no $L^2 \to L^\infty$ regularization property.

It also holds and also in some cases of infinite-dimensional spaces, e.g. R^∞ if one makes use of Gauss measure.

This property is used in some models of Quantized Field Theory and in models of Statistical Mechanics for infinite particles systems (Dobrushin–Ruelle measure, a generalization of Gibbs measure).

We shall show that if zero is a simple eigenvalue of $H \geq 0$, inequality (3) implies that it is isolated. It is therefore interesting to give a characterization of the generators of the semigroups are hyper-contractive.

1 Logarithmic Sobolev Inequalities

Definition 2 Let (X, μ) be a probability space, and let $\mathcal{E}(f)$ be a non-negative *closed* quadratic form densely defined on $L^2(X, d\mu)$. We will say that $\mathcal{E}$ *determines* (or *satisfies*) a logarithmic Sobolev inequality (in short $L.S.$) if there exists a positive constant K such that

$$K|\int_X |(f(x)|^2 log \frac{|f(x)|}{\|f\|_2} d\mu(x)| \leq \mathcal{E}(f), \quad \forall\, f \in Q(\mathcal{E}) \cap L^2, \ f \neq 0 \tag{4}$$

◇

The greatest constant K for which the inequality is satisfied will be called *logarithmic Sobolev constant* relative to the triple $(\mathcal{E}, \mu, X)$. We remark that by construction both terms in (3) are homogeneous of order two for the map $f \to \lambda f \quad \lambda \in R^+$. Therefore (3) can be written

$$K|\int_X |(f(x)|^2 log|f(x)| d\mu(x)| \leq \mathcal{E}(f), \quad \|f\|_2 = 1 \tag{5}$$

We will show that (3) provides a *necessary and sufficient condition* that the Friedrichs extension associated to the quadratic form $\mathcal{E}$ be the generator of a hyper-contractive semigroup. Before proving this, let us compare in the case $X = R^d$, $d < \infty$ and μ = Lebesgue measure, inequality (3) with the classic Sobolev inequalities i.e.

$$\|f\|_q \leq C_{p,d} \|\nabla f\|_p, \quad \frac{1}{q} = \frac{1}{p} - \frac{1}{d}, \quad 1 \leq p \leq q \leq +\infty \tag{6}$$

where $C_{p,d}$ are suitable positive constants. The inequalities (6) are established first for $f \in C_0^\infty(R^d)$, and remain valid by density and continuity for functions such that the right-hand side is defined. We shall denote these inequalities with $S_{d,p}$ (S for Sobolev).

Comparing (3) with (5) one sees that $S_{d,p}$ contains more information than $S.L.$ on the possible local singularity of f.

However these information becomes less relevant when d increases and lose interest in the limit $d \to \infty$. In this limit the $L.S.$ inequality give useful information

For the behavior of the functions at infinity (if X is not compact), notice that inequalities $S_{d,p}$ are valid only for those functions that are contained in the closure of the C_0^∞ in the norm $\|\nabla f\|_p$.

This set *does not* contain all function which have finite L_p norm. In the case X is not compact the comparison should be rather with the *coercive Sobolev inequalities*

$$\|f\|_q \le c_p \|\nabla f\|_p + b_p \|f\|_p \qquad \frac{1}{q} = \frac{1}{p} - \frac{1}{d}, \quad 1 \le p \le \infty \tag{7}$$

for suitable constants $c_p,\ b_p\ > 0$. In (7) the symbol $\|f\|_p$ means

$$\|f\|_p^p = \int_X |f(x)|^p d\mu(x) \tag{8}$$

where μ is a measure continuous with respect to Lebesgue measure.

For completeness we remark that in R^d (or on a non-compact manifold of dimension d) inequalities (3) with $b_p > 0$ imply $L.S.$

To see this, e.g. in case $p = 2$, choose f positive and set $d\nu = f^2 d\mu$. Jensen's inequality gives

$$\frac{2}{q-2}\int log f^{q-2} d\nu \le \frac{2}{q-2} log \int f^{q-2} d\nu == \frac{q}{q-2} log \|f\|_q^2 \le \frac{q}{q-2}(\|f\|_q^2 - 1) \tag{9}$$

(in the last inequality we have used $\alpha \ge 1 \to log\ \alpha \le \alpha - 1$.

These inequalities imply that if f is positive there are constants $a > 0,\ b > 1$ such that

$$\int f^2 log \frac{f^2}{\|f\|_2^2} d\mu \le c\frac{q}{q-2}\int |\nabla f|^2 d\mu + (b-1)\frac{q}{q-2}\int |f|^2 d\mu \tag{10}$$

If X is not compact, it is not possible to derive $L.S.$ from the Sobolev inequalities because $L.S.$ require more stringent conditions to the behavior of the function at large distances.

However $L.S.$ *can be derived* from $S_{d,2}$ if one requires that the function satisfies the following *Poincaré inequality*.

$$\alpha_d \|f - E(f)\|_2^2 \le \int |\nabla f|^2 d\mu(x) \equiv \mathcal{E}(f,f) \tag{11}$$

where we have denoted with $\mathcal{E}$ the energy form, α_d is a suitable constant and

$$E(f) \equiv \int f(x)d\mu(x), \quad f \in C^{\infty}(R^d) \tag{12}$$

Notice that Schwartz inequality implies that $\mathcal{E}(f)$ is well defined since $f \in C^{\infty}(R^d) \cap L^2(R^d, d\mu)$.

Roughly speaking, if a function satisfies Poincaré inequality, then its norm $\|f\|_2$ is controlled by it mean absolute value and the L^2 norm of its gradient.

From the Poincaré inequality one derives that, if the mean of f is zero, then $\alpha_d \|f\|_2^2 \leq \mathcal{E}(f,f)$ and therefore the logarithmic Sobolev inequality is implied, for $d < \infty$, by the Sobolev inequality $S_{d,2}$.

One should note, however, that α_d in (11) is such that $lim_{d \to \infty} \alpha_d = 0$. If $E(f) \neq 0$ set

$$\tilde{f} \equiv f - E(f) \tag{13}$$

If μ is a finite measure one has $\tilde{f} = \pi(f)$, where π is the orthogonal projection in $L^2(X, d\mu)$ on the constant function. Explicit computation shows

$$\int |f(x)|^2 log\left(\frac{|f(x)|^2}{\|f\|}\right) d\mu(x) \leq \int |\tilde{f}(x)|^2 log\left(\frac{|\tilde{f}(x)|}{\|\tilde{f}\|}^2\right) d\mu(x) + 2 \int |\tilde{f}(x)|^2 d\mu(x) \tag{14}$$

and therefore there exists a constant K_d for which

$$K_d \int f^2(x) log\left(\frac{f(x)}{\|f\|}\right) d\mu(x) \leq \mathcal{E}(f,f) \quad K_d^{-1} = C_d + \frac{b_d + 2}{\alpha_d} \tag{15}$$

Suppose now that on $L^2(X, d\mu)$ acts a semigroup T_t, has the contractive and Markov properties and that its generator is the Friedrichs extension associated to the positive quadratic form $\mathcal{E}$

$$lim_{t \to 0} t^{-1}(T_t f - f) = \mathcal{E}(f,f) = -(f, Lf) \tag{16}$$

The function ι identically equal to one is a simple eigenvector L and the corresponding eigenvalue is zero. If it is isolated

$$Sp\, L \subset \{0\} \cup [\alpha, \infty, \quad \alpha > 0 \tag{17}$$

and from spectral theory

$$\alpha^2 \|f - E(f)\|^2 \leq \mathcal{E}(f,f) \tag{18}$$

We shall see that the $L.S.$ inequality (15) implies (18) (with $2K \leq \alpha$).

2 Relation with the Entropy

The presence of a logarithm in $L.S.$ suggests a relation between $L.S.$ and the entropy.

Recall that the *relative (von Neumann) entropy* of a probability measures μ on X relative to another measure ν is denoted $H(\mu|\nu)$ and is defined as follows
(1) if μ is not absolutely continuous with respect to ν, $H(\mu|\nu) = \infty$
(2) If μ is absolutely continuous with respect to ν with Radon–Nikodym derivative f_μ then

$$H(\mu|\nu) \equiv \int_X f_\mu(x) log f_\mu(x) d\nu(x) \tag{19}$$

It is easy to verify the reflexive property

$$H(\mu|\nu) = H(\nu|\mu) \tag{20}$$

and that $H(\mu|\nu) = 0$ iff $\mu = \nu$. For the Radon–Nikodym derivative of f_μ of μ with respect to ν one has $\int f(x)_\mu d\nu(x) = 1$.

Making use of Schwarz inequality and of the inequalities

$$3(y-1)^2 \leq (4+2y)(y\ logy - y + 1) \qquad ylogy - y + 1 \geq 0 \ \ \forall y > 0 \tag{21}$$

one derives

$$\begin{aligned} 3\|\mu - \nu\|_{var} \equiv 3\|f_\mu(x) - 1\|_{L^1(\nu)} &\leq \|(4 - 2f_\mu)^{\frac{1}{2}} (f_\mu log f_\mu - f_\mu + 1)\|^{\frac{1}{2}}_{L^1(\nu)} \\ &\leq \|4 + 2f\|_{L^1(\nu} \|f_\mu\ log\, f_\mu - f_\mu + 1\|_{L^1(\nu)} \equiv 6H(\mu|\nu) \end{aligned} \tag{22}$$

In the last identity we have used

$$\|f_\mu log f_\mu - f_\mu + 1\|_{L^1(\nu)} \equiv \int [f_\mu(x) log f_\mu(x) - f_\mu(x) + 1] d\nu(x) = \int f_\mu(x) log f_\mu(x) d\nu(x) \tag{23}$$

We study now the relation between the logarithmic Sobolev inequality and the spectral properties of the Laplace–Beltrami operator on a compact Riemann surface. We shall then generalize to semigroups on probability spaces.

Let X be a compact Riemann surface. Denote by μ the Riemann-Lebesgue measure which satisfies $P_t^* \mu \equiv \mu.P_t = \mu$.

We have denoted P_t the semigroup generated by the Laplace–Beltrami operator $\mathcal{L}$ defined by $(u, \mathcal{L}u) = -\int |\nabla u|^2 d\mu,\ u \in D(\mathcal{L})$. Denote with f_t the Radon–Nikodym derivative of $P_t^* \nu$ with respect to μ.

$$f_t \equiv \frac{d(\nu P_t)}{d\mu} \tag{24}$$

A straightforward calculation gives

$$\frac{d}{dt}H(P_t^*\nu|\mu) = \int_X (\mathcal{L}f_t) \tag{25}$$

(we have integrated by parts and used the relation $(P_t^*\nu)(g) = \nu(P_t g)$. Equations (22) and (25) imply

$$\frac{d}{dt}H(P_t^*\nu, \mu) \leq -4\mathcal{E}(f_t^{\frac{1}{2}}, f_t^{\frac{1}{2}}) \tag{26}$$

In these notation the *L.S.* inequality reads $K\ H(\nu|\mu) \leq \mathcal{E}(f_t^{\frac{1}{2}}, f_t^{\frac{1}{2}})$. Equation (26) implies

$$\frac{d}{dt}H(P_t^*\nu|\mu) \leq -4K\ H(P_t^*\nu|\mu) \tag{27}$$

and therefore

$$H(P_t^*\nu|\mu) \leq e^{-4K}H(\nu|\mu) \tag{28}$$

From (27) one derives

$$\|P_t^*\nu - \mu\|_{var} \leq \sqrt{2H(\mu|\nu)}e^{-2Kt}, \quad t \geq 0 \tag{29}$$

which can be rewritten as

$$\|P_t f - \iota\|_{L^1}(\mu) \leq \sqrt{2H(\mu|\nu)}e^{-2Kt}, \quad \forall f \in L_\mu^1, \quad \|f\|_{L_\mu^1} = 1 \tag{30}$$

We conclude that the semigroup generated by the Laplace–Beltrami operator converges strongly in $L^1(\mu)$ with exponential speed to the projection onto the ground state.

If the converence takes place also in the L_μ^2 topology, the spectrum of the operator $\mathcal{L}$ is contained in $\{0\} \cup [2K, +\infty)$ and zero is a simple eigenvalue.

Inequalities of the type (26) can hold in more general contexts, and is useful in the study the case $X = R^\infty$ with a suitable measure.

It is sufficient that one can define a quadratic form

$$\mathcal{E}(u, u) = \sum_{n=1}^{\infty} |\frac{\partial u}{\partial x_n}|^2 \mu(dx) \tag{31}$$

and that integration by parts (to define (26)) be legitimate.

The bilinear form $E(u, v)$ in (31) can be defined for functions on R^∞ which depend only on a finite number of coordinates (cylindrical functions) and are in the domain of the partial derivative with respect to these coordinates.

Denote by D_0 the collection of such functions. It can be shown, under suitable conditions on $\mu(dx)$, that the quadratic form defined by (31) on D_0 is closable.

3 Estimates of Quadratic Forms

We have seen that the constant K in the logarithmic Sobolev inequality gives an estimate from below of the gap between the lowest eigenvalue and the rest of the spectrum. For this reason the following problem is relevant:

Let μ a probability measure on a Riemann on a d-dimensional manifold X. Consider the quadratic form

$$\mathcal{E}(\phi, \phi) \equiv \int_X |\nabla\phi|^2 d\mu(x) \tag{32}$$

defined on $C_0^\infty(X)$ and closable. Assume that $S.L.$ is satisfied with constant K

$$K \int |\phi(x)|^2 log \frac{\phi(x)}{\|\phi\|_{L^2_\mu}}^2 d\mu(x) \leq \mathcal{E}(\phi, \phi) \tag{33}$$

for any real-valued function $\phi \in D(\mathcal{E}) \cap L^2_\mu$. For any given function $U \in C^\infty(X)$ integrable with respect to ν define a new probability measure ν_U on X by

$$\nu_U(dx) \equiv Z^{-1} e^{-U(x)} \mu(dx) \tag{34}$$

(Z is a normalization factor). Consider now the quadratic form $\mathcal{E}_U$

$$\mathcal{E}_U \equiv \int |\nabla\phi|^2 d\nu_U(x) = Z^{-1} \int |\nabla\phi|^2 e^{-U(x)} d\mu(x) \tag{35}$$

Lemma 1 ([1, 2]) *If $U \in C_0^\infty(X)$ the quadratic form $\mathcal{E}_U(\phi, \phi) \equiv \int_X |\nabla\phi(x)|^2 d\nu_U(x)$ satisfies a logarithmic Sobolev inequality. Moreover $K_U \geq K e^{-osc(U)}$ where the oscillation of U (denoted by $osc(U)$) is defined as $osc(U) \equiv max_{x\in X} U(x) - min_{x \in X} U(x)$.* ◇

Proof For any probability measure ν on X, for any real valued function $\phi \in L^2_\nu$ and for any $t \in R^+$ the following holds

$$0 \leq \phi^2(x) log \frac{\phi^2(x)}{\|\phi\|^2_{L^2(\nu)}} \leq \phi(x)^2 log(\phi(x)^2) - \phi^2(x) log\, t^2 - \phi^2(x) + t^2 \tag{36}$$

(the term to the left is a convex function of t that reaches its minimum at $t = \|\phi\|_{L^2(\nu)} \equiv \|\phi\|_\nu$). Integrating with respect to $\nu(x) \equiv \phi(x)^2 d\mu(x)$, choosing $t = \|\phi\|_\mu$ and keeping onto account that $\mathcal{E}$ satisfies $L.S.$ one has

$$\int \phi^2(x)\left(log\frac{\phi(x)}{\|\phi\|_\nu}\right)^2 d\nu$$
$$\leq \int \phi(x)^2 log\phi(x)^2 d\nu - \int \phi(x)^2 log\|\phi\|_\nu - \int \phi(x)^2 d\nu + \int \|\phi\|^2 d\nu$$
$$\leq \frac{e^{-minU(x)}}{Z}\int \phi(x)^2\frac{log(\phi(x)}{\|\phi\|^2} \leq \frac{e^{-minU(x)}}{Z}\int |\nabla\phi|^2 d\mu \leq \frac{e^{-oscU}}{K}\int |\nabla\phi|^2 d\nu \tag{37}$$

♡

If a quadratic form Q is defined on $\mathcal{H} \equiv \otimes\mathcal{H}_n$ by $Q = \sum_{n=1}^N Q_n$, and each Q_n satisfies $L.S.$ with constant K_n then Q satisfies $L.S.$ with constant not smaller that the minimum of the K_n. We shall use later this property to prove that the Gauss–Dirichlet form, defined by

$$\mathcal{E}(f,f) \equiv \int_{R^d} |\nabla f|^2 d\mu_G(x) \tag{38}$$

satisfies $L.S.$ Here $d\mu_G$ is a Gauss probability measure.

We shall use the fact that the quadratic form $Q(f)$ defined on $X = \{1\} \cup \{-1\}$ by $Q(f) = \frac{1}{4}(f(1) - f(-1))^2$ satisfies $L.S.$ With an analogous procedure one can prove that $L.S.$ inequalities hold for the are verified for the Gauss–Dirichlet form in R^∞.

This property is at the basis of the analysis by E. Nelson of the properties of the scalar free quantum field and of some interacting ones.

4 Spectral Properties

We discuss now some spectral properties that are derived form the fact that the generator of the semigroup satisfies the $L.S.$ inequalities.

Theorem 1 ([2]) (Federbush, Gross, Faris) *If the quadratic form $\mathcal{E}$ satisfies L.S. with constant K and $V(x)$ is a real valued function on X and satisfies $\|e^{-V}\|_2 < \infty$, then the following holds*

$$\frac{1}{K}\mathcal{E}(f) + (f, Vf) \geq -log\|e^{-V}\| \, \|f\|_2^2 \quad \forall f \in Q(\mathcal{E}) \tag{39}$$

Conversely, if $\|e^{-V}\|_2 < \infty$ implies that (39) holds for every $f \in L^2(X) \cap Q(\mathcal{E})$, then $\mathcal{E}$ satisfies L.S. with constant K. ◇

Proof For the first part of the theorem we consider in detail only the case $\|e^{-V}\| < \infty$ and V bounded from above. The general case follows by interpolation and continuity.

The integral $\int_X V(x)|f(x)|^2 d\mu(x)$ is well defined. Using the inequality $st \leq s\, log\, s - s + e^t \quad s \geq 0, \quad t \in R$ and setting $s = t^2$ one has

$$-(Vf,f) \le \frac{1}{2}\int_X [|f(x)|^2 log|f(x)|^2 - |f(x)|^2] d\mu(x) + \frac{1}{2}\int_X e^{-2V(x)} d\mu(x)$$
$$\le \frac{1}{K}\mathcal{E}(f) + \|f\|^2 log\|f\| - \frac{1}{2}\|f\|^2 + \frac{1}{2}\|e^{-V}\| \quad (40)$$

Therefore

$$\frac{1}{K}\mathcal{E}(f) + (Vf,f) \ge -\|f\|^2 log\|f\| + \frac{1}{2}(\|f\|^2 - \|e^{-V}\|^2) \quad (41)$$

Since the $L.S.$ inequalities are homogeneous (invariant under $f \to \lambda f$) it suffices to verify them for $\|f\| = \|e^{-V}\|$. But in this case coincides with $L.S.$

We prove now the second part. Consider a generic function $f \in Q(\mathcal{E}) \cap L^2(X, d\mu)$ and set $V(x) \equiv -log|f(x)|$. Then $\|e^{-V}\| = \|f\| < \infty$. By assumption (39) holds. Therefore

$$\frac{1}{K}\mathcal{E}(f) - \int_X |f(x)|^2 log|f(x) d\mu(x) \ge -\|f\|^2 log\|f| \quad (42)$$

Hence $L.S.$ holds for f and f is arbitrary in $Q(\mathcal{E}) \cap L^2(X, d\mu)$. Thus $\mathcal{E}$ satisfies the $L.S.$ inequality. ♡

The next theorem states, under some supplementary assumptions, that if Q satisfies $L.S.$ the lower boundary of the spectrum of the Friedrichs extension is an isolated simple eigenvalue.

Theorem 2 ([2]) (Rothaus, Simon) *Let $\mu(X) = 1$ and let $\mathcal{E}(f)$, f real, satisfy L.S. with constant K and moreover*
(i) $\mathcal{E}(\iota) = 0$
(ii) $L^\infty(X) \cap Q(\mathcal{E})$ is a core for $\mathcal{E}$

Then for any real g

$$g \perp \iota \to \mathcal{E}(g) \ge K\|g\|_2^2 \quad (43)$$

(we have denoted by ι the function identically equal to one). This implies that there is a gap in the spectrum and gives a lower bound to it. ◇

Proof Denote by $\mathcal{E}(f, g)$ the bilinear form obtained from $\mathcal{E}(f)$ by polarization. From $\mathcal{E}(f, \iota) \le \mathcal{E}(f)\mathcal{E}(\iota)$ it follows $\mathcal{E}(f, \iota) = 0 \ \ \forall f \in Q(\mathcal{E})$ and

$$\mathcal{E}(\iota + sg) = s^2\mathcal{E}(g), \quad \|\iota + sg\| = 1 + s^2\|g\|^2 \quad \forall g \in Q(\mathcal{E}) \quad (44)$$

Let $g \in L^\infty$ and of mean zero. For s sufficiently small we can develop $log(1 + sg(x))$ in powers of s; inserting in $L.S.$ one obtains

$$\int_X (1 + 2sg(x) + s^2g^2(x))\left(sg(x) - \frac{s^2g^2(x)}{2}\right) d\mu(x) \quad (45)$$

$$\leq \frac{1}{K}s^2(g) + \frac{1}{2}(1 + s^2\|g\|^2 + O(s^3)) \tag{46}$$

By assumption $\int_x g(x)dx = 0$ and then

$$Ks^2|g|^2 \leq s^2\mathcal{E}(g) + O(s^3) \tag{47}$$

Dividing by s^2 and passing to the limit $s \to 0$ we obtain (42) for all $g \in L^\infty$, $(g, \iota) = 0$.

Let now $f \in Q(\mathcal{E})$, $(f, \iota) = 0$. From assumption (ii) there exists a sequence $\{f_n\}$, $f_n \in Q(\mathcal{E}) \cap L^\infty$ such that

$$lim_{n\to\infty}\|f_n - f\| = 0 \qquad lim_{n\to\infty}\mathcal{E}(f_n - f) = 0 \tag{48}$$

Since $lim_{n\to\infty}(f_n - f, \iota) = 0$ we can substitute $f_n(x)$ with $f_n(x) - (\iota, f_n)f_n(x)$ and assume $(f_n, \iota) = 0 \quad \forall n$. Taking the limit we obtain (42) for all functions $f \in Q(\mathcal{E})$, $(\iota, f) = 0$. ♡

5 Logarithmic Sobolev Inequalities and Hypercontractivity

We study next the relation between the logarithmic Sobolev inequality and the hypercontractivity of the semigroup generated by the Friedrichs operator associated to a positive quadratic form.

We shall use the following notation: for $p > 1$ $f_p \equiv signf \ |f|^{p-1}$ (with the convention that $sign0 = 0$).

Definition 3 (*Principal symbol*) Let Ω, μ be a probability space and let $p \in (1, \infty)$. An operator H on $L^p(\mu)$ is *a Sobolev generator of index p* if it is the generator of a continuous contraction semigroup in $L^p(\mu)$ and there exist constants $K > 1$ and $\gamma \in R$ such that

$$\int |f|^p log|f|d\mu - \|f\|_p^p log\|f\|_p \leq K \ Re((H+\gamma)f, f_p) \quad f \in D(H) \tag{49}$$

The constant K is called *principal symbol* of H and γ is its *local norm* ◇

Notice that If $p = 2$ and $f \geq 0$ the inequality (47) is the logarithmic Sobolev inequality

Definition 4 (*Sobolev generator*) Let $0 \leq a < b \leq \infty$. The operator is *a Sobolev generator in the interval* (a, b) if the exist in this interval functions $K(s)$, $\gamma(s)$ and a family of strongly continuous semigroups e^{-tH_s} on L^s such that

$$e^{-tH_s}|_{L^r} = e^{-tH_r} \qquad a < s < r < b \tag{50}$$

and the generator of the semigroup e^{-tH_s} has principal symbol $K(s)$ and local norm $\gamma(s)$. ◇

Using Jensen's inequality and (46) one can prove

$$\|(H+\gamma+\lambda)f\|_p \geq \lambda \|f\|_p \tag{51}$$

and therefore, according to the Hille–Yosida theorem $\|e^{-t(H+\gamma)}\|_{p,p} \leq 1$. In particular if $\gamma(p)=0$ the semigroup e^{-tH} contracts in L^p. ◇

Theorem 3 ([2]) *If H is a Sobolev generator in* (a, b) *then the semigroup generated by H is hyper-contractive.* ◇

Proof We shall give the proof only in the case

$$(Hf, f) \equiv \int_{R^d} |\nabla f|^2 d\mu(x) \tag{52}$$

where μ is absolutely continuous with respect to Lebesgue measure with a Radon–Nikodym derivative of class C^∞.

The theorem holds in greater generality under the condition that H be self-adjoint on $L^2(\Omega, d\mu)$ where Ω, μ is a probability space and that e^{-tH} be positivity preserving and act as a contraction in L^∞ (for a proof, see e.g. [1]).

We limit ourselves to the case $f \in C^\infty$ and positive. Derivation of composite functions gives

$$|\nabla(f(x))|^{\frac{p}{2}} = \frac{p}{2}^2 (f(x)^{\frac{p}{2}-1})^2 |\nabla f(x)|^2 \tag{53}$$

and also

$$\nabla f(x) \cdot \nabla f^{p-1}(x) = (p-1) f^{p-2}(x) |\nabla f|^2 \tag{54}$$

Therefore

$$\frac{p^2}{4(p-1)^2} |(\nabla f, \nabla f^{p-1})| = |\nabla f^{\frac{p}{2}}|^2 \tag{55}$$

and, if H satisfies

$$\int f^2(x) log f(x) d\mu(x) \leq K(f, Hf) + \|f\|^2 log \|f\| \tag{56}$$

then, substituting f with j $f^{\frac{p}{2}}$ one obtains

$$\int f^p(x) log f(x) d\mu(x) \leq \frac{Kp}{4(p-1)}(f, Hf) + \|f\|_p^p log \|f\|_p \tag{57}$$

The proof in the case f is positive and in the domain of H is obtained by approximation. ♡

An important relation between Sobolev generators and hyper-contractivity is given by the following theorem that we quote here without proof.

Theorem 4 ([2]) *Let H be a Sobolev generator in (a, b) with principal coefficient $K(s)$ and local norm $\gamma(s)$. If $q \in (a, b)$ denote by $p(t, q)$ the solution of*

$$K(p)\frac{dp}{dt} = p, \qquad p(0) = q \quad t \geq 0 \tag{58}$$

and set

$$M(t \cdot q) \equiv \int_0^t \gamma(p(s, q))ds1 \tag{59}$$

($\gamma(t, q)$ e $M(t, q)$ are defined if $p(t \cdot q) < b$). One has then

$$\|e^{-tH}\|_{q,p(t,q)} \leq e^{M(t,q)} \tag{60}$$

◇

We remark that if the local norm is zero, the semigroup generated by H is a contraction from L_q to$L_{p(t,q)}$.

6 An Example: Gauss–Dirichlet Operator

To exemplify Theorem 4 we shall now prove the following hyper-contractive result for the Gauss–Dirichlet form in R^d, due to E. Nelson.

The result, with a similar proof, holds in R^∞ and can be used to study the free relativistic field. It leads to rigorous results for polynomial interactions in the theory of relativistic quantized field in two space-time dimensions.

Let ν be Gauss measure on R^d with mean 0 and covariance 1. Denote by N the Gauss–Dirichlet operator associated to the quadratic form

$$(Nf, f)_{L^2(\nu)} \equiv \int_{R^d} (\nabla f, \nabla f)d\nu(x) \tag{61}$$

Integrating by parts

$$(Nf, f)_{L^2(\nu)} \equiv \sum_{j=1}^{d}\left[-\frac{\partial^2 f}{\partial^2 x_j} + x_j\frac{\partial f}{\partial x_j}\right] \quad f \in D(N) \tag{62}$$

Theorem 5 (Nelson [3]) *If* $1 \le q, p < \infty \quad e^{-2t} \le \frac{q-1}{p-1}$ *then*

$$\|e^{-tN}\|_{q \to p} = 1 \tag{63}$$

◇

Proof ([1]) From (60) and (61), substituting $f \ge 0$ with $f^{\frac{p}{2}}$ one obtains

$$\int_{R^d} f(x)^p log f(x) d\nu(x) \le \frac{p}{2(p-1)}(Nf, f^{p-1}) + \|f\|_p^p log \|f\|_p \tag{64}$$

We have made use of the fact that L satisfies a logarithmic Sobolev inequality with local norm zero and coefficient one.

We will prove this fact later in this lecture. Therefore for the local norm of the function f is zero and its principal coefficient is $K(p) = \frac{p}{2(p-1)}$.

The semigroup e^{-tN} is positivity preserving and contracts in L^p for all $p \in [1, \infty)$. We can apply therefore Theorem 5.

In the present case the solution of (55) is

$$p(t, q) = 1 + (q-1)e^{2t}, \qquad q \ge 2, \quad t \ge 0 \tag{65}$$

Moreover $\gamma = 0$, $e^{-tN}\iota = \iota \;\; \forall t, \iota \in L^p$ for all p, and $\|e^{-tN}\|_{q \to p(q,t)} \le 1 \qquad q \ge 2.$

♡

Using the duality between L^p and L^q, $\; q \equiv \frac{p}{p-1}$ it is possible to prove the thesis of Theorem 6 hold for any $1 < q < p < \infty$.

Nelson theorem proves hyper-contractivity of the heat semigroup in R^d. It is an *optimal result* as seen in the following lemma.

Lemma 2 ([3]) *Let N be the hamiltonian of the harmonic oscillator in* $d = 1$. *If* $p > 1 + e^{2t}(q-1)$ *the operator* e^{-tN} *is unbounded from* L^q *to* L^p, $t \ge 0$. ◇

Proof The Kernel of the semigroup e^{-tN} is

$$(e^{-tN}f)(x) = \int_R f(e^{-t}x + \sqrt{1-e^{-2t}}y)e^{-\frac{y^2}{2}} dy \tag{66}$$

Consider the function $f_\lambda(x) \equiv e^{\lambda x} \;\; \lambda \in R$. It belongs to the domain of N and

$$(e^{-tN}f_\lambda)(x) = e^{\frac{\lambda^2}{2}(1-e^{-2t})} f_\lambda e^{-\lambda t} \tag{67}$$

A straightforward computation gives

$$\|(e^{-tN}f_\lambda)\|_P = e^{\frac{\lambda^2}{2}[e^{-2t}(p-1)+1-q]} \|f\|_q \tag{68}$$

This quantity is not bounded above as function of $\lambda \in R$ if $p - 1 > e^{2t}$ $(q-1)$.

♡

Theorem 6 ([2]) *Let T_t be an hyper-contracting semigroup on $L^2(X,\mu)$ such that $T_t : L^\infty \to L^\infty$. Then for all $1 < q < p < \infty$ there exists a positive a $C_{q,p}$ and time $t_{q,p} > 0$ such that*

$$\|T_t u\|_p \le C_{q,p}\|u\|_q \qquad t \ge t_{q,p}, \quad \forall u \in L^q(X,\mu) \tag{69}$$

◇

Proof Since $T_{t_0} \; : \; L^2(X,\mu) \to L^{p_0}(X,\mu)$ and $T_{t_0} \; : \; L^\infty \to L^\infty$ the interpolation theorem of Riesz–Thorin provides a constant C such that for every $r \ge 2$ $\|T_{t_0}u\|_r \le C\|u\|_{\frac{p_0}{2}r}$ holds.

Consider two cases:
(a) If $q \ge 2$ choose n large enough in order t satisfy $2(\frac{p_0}{2})^n > p$. Then $\|T_{nt_0}u\|_{2\frac{(p_0}{2)}^2} \le C^n\|u\|_2$ Since T_t is a contraction in every L^p, $1 \le p < \infty$

$$\|T_{nt_0}u\|_{2\frac{(t_0}{2)}^2} \le C^n\|u\|_2 \le C^n\|u\|_q \tag{70}$$

(b) if $q < 2$ choose n large enough so that

$$2\left(\frac{p_0}{2}\right)^n > p > q > C\alpha \qquad \alpha^{-1} + \left(2\left(\frac{p_0}{2}^n\right)\right)^{-1} = 1 \tag{71}$$

Since T_{nt_0} is a bounded map from L^2 to $L^{2(\frac{p_0}{2})^n}$, by duality $T^*_{nt_0}$ is a bounded map from L^C to L^2.

We have assumed that T coincides with its adjoint and therefore T_{nt_0} is bounded from $L^{C\alpha}$ to $L^{2(\frac{p_0}{2})^n}$.

The thesis of the theorem follows then from the remark that , by construction, $C < q < p < 2(\frac{p_0}{2})^n$.

♡

7 Other Examples

Example 1 The harmonic oscillator is hyper-contractive
We shall now give an alternative proof that the harmonic oscillator semigroup is hyper-contractive. Recall that in $L^2(R, \frac{1}{2\pi}e^{-x^2}dx)$, the operator of the harmonic oscillator is

$$H_0 \equiv -\frac{1}{2}\frac{d^2}{dx^2} + x\frac{d}{dx} \tag{72}$$

The operator H_0 is essentially self-adjoint on the finite linear combinations of Hermite polynomials P_n and $H_0P_n = nP_n$. Using this information one can see that $T_t \equiv e^{-tH_0}$ acts as contraction semigroup on L^p for all $p \ge 1$. For each T the semigroup e^{-tH_0} preserves positivity.

It is a contraction from L^∞ to L^∞ and also on L^1 since $e^{-tH_0}\iota = \iota$. By interpolation it contracts on every L^p, $1 \le p \le \infty$). We show that there exist $t_0 > 0$ such that

$$\|e^{-tH_0}u\|_4 \le C\|u\|_2 \quad u \in L^2(R, \mu_G), \quad t \ge t_0 \tag{73}$$

Setting $x = \frac{1}{\sqrt{2}}(a + a^*)$, $[a, a^*] = 1$ one has

$$P_n(x) = \frac{1}{\sqrt{n!}}(a^*)^n P_0 = \frac{1}{\sqrt{n!}} 2^{\frac{n}{2}} : (a + a^*)^n : P_0 \tag{74}$$

where : .. : is Wick ordering of a polynomial in a, a^* (obtained placing the a^s to the right of the a^*)

$$\|x^n P_n(x)\|_{L^2(R, d\mu_G} \le 2^n \left(\frac{(2n)!}{(n!)^2}\right)^{\frac{1}{2}} \le 4^n \tag{75}$$

It is easy to verify that $\|x^n\|_{L^4(R,d\mu_G)} = \|x^n P_n\|^{\frac{1}{2}}_{L^2(R,dx)}$. Setting

$$\phi = \sum_n a_n x^n \in L^2(R, d\mu_G) \cap \mathcal{S}(R) \tag{76}$$

one has

$$\begin{aligned}\|e^{-tH_0} \sum a_n x^n\|_{L^2} &\le \sum |a_n| e^{-tn} \|p_n(x)\|_{L^4(R,d\mu_G)} \\ &\le \left(\sum |a_n|^2\right)^{\frac{1}{2}} \left(\sum e^{-2tn} 4^n\right) \frac{1}{2} \le C\|\phi\|_{L^2(R,\mu_G)}\end{aligned} \tag{77}$$

for $t > \frac{1}{2} log\ 4$ ♡

Example 2 Let

$$X \equiv \{1, -1\} \qquad \mu(\{1\} = \mu\{-1\} = \frac{1}{2} \tag{78}$$

If $f : X \to R$ define $\nabla f = \frac{1}{2}[f(1) - f(-1)]$. Define the quadratic form

$$Q(f) \equiv \int_X |\nabla f|^2(x) d\mu(x) = \frac{1}{4}(f(1) - f(-1))^2 \tag{79}$$

Then Q satisfies a logarithmic Sobolev inequality with constant one. ◇

Proof Since $Q(|f|) \le Q(f)$ it suffices to consider the case $f > 0$. Any function on X has the form $f(x) = a + bx$ and the condition $f \ge 0$ gives $a > 0$, $|b| < 1$.

Due to homogeneity it suffices to take $a = 1$ and by symmetry it suffices to consider $0 \le b \le 1$.

Set $f_s(x) = 1 + sx, \quad 0 \le s \le 1$. One has $\|f_s\|^2 = 1 + s^2$. Define the function (entropy)

$$H(s) = \int f_s log f_s d\mu - \int \|f_s\|^2 log \|f_s\| d\mu \tag{80}$$

Explicit calculation shows

$$H(s) = \frac{1}{2}(1+s)^2 log(1+s) + (1-s)^2 log(1-s)] - \frac{1}{2}(1+s^2) log(1+s^2) \tag{81}$$

From the definition of Q one has $Q(f_s) = s^2$. Therefore to verify the $L.S.$ inequality it suffices to prove $H(s) \le s^2, \quad 0 \le s \le 1$.

Since $H(0) = 0$ it suffices to prove $H'(s) \le 2s$ and since $H'(0) = 0$ it suffices to prove that $H''(s) \le 2$. One easily computes

$$H''(s) = 2 + log \frac{1-s^2}{1+s^2} - \frac{2s^2}{1+s^2} \tag{82}$$

and the $L.S.$ is satisfied since for $0 \le s \le 1$ the second and third terms are non positive.

♡

Example 3 ([3]) The Gauss–Dirichlet quadratic form is

$$\mathcal{E}(f, f) \equiv \int_{R^d} |\nabla f|^2 d\mu_G(x) \tag{83}$$

The gradient is meant in distributional sense and μ_G is Gauss measure with mean zero an covariance one R^d.

We must prove that if $f \in Q(\mathcal{E}) \cap L^2_\mu$ then

$$\int_{R^d} |f(x)^2| log |f(x)| d\mu(x) - \|f\|^2 log \|f\| \le \int_{R^d} |\nabla f(x)||^2 d\mu(x) \tag{84}$$

Because of the additivity theorem it suffices to give the proof for $d = 1$.

Identify, as measure space, R with Gauss measure with the direct product of denumerable copies of the measure space used in Example 2, and use the additivity property.

We have previously employed this procedure to give a representation of Brownian motion as measure on the space of continuous trajectories. Set

$$\Omega_K \equiv \Pi_{j=1}^K X_j, \quad \mu_K \equiv \Pi_1^K \mu_j \tag{85}$$

where X_j, μ_j are identical copies of X, μ. The additivity theorem gives

$$\int_{\Omega_K} f(x)^2 log|f(x)|d\mu_K(x) \leq \mathcal{E}_K(f) + \|f\|_{\mu_K} log\|f\|_{\mu_K} \tag{86}$$

where

$$\mathcal{E}_K(f) = \sum_j \int_{\Omega_K} (\delta_j f)^2 d\mu_K(\delta_j f)(x)$$

$$= \frac{1}{2}[f(x_1, x-2, x_{j-1}, 1, x_{j+1} \dots x_K) - f(x_1, x-2, x_{j-1}, -1, x_{j+1} \dots x_K) \tag{87}$$

Set $y = \frac{1}{\sqrt{K}}(x_1 + \cdots + x_K)$ and evaluate (85) on a function $\phi(y) \in C_0^\infty$.

The Central Limit Theorem, applied to the sum of gaussian random variables identically distributed with mean zero and variance one, states that the left hand side of inequality (85) converges when $K \to \infty$ to

$$\int_R |\phi(t)|^2 log|\phi(t)|d\nu(t) \tag{88}$$

For the same reason the right hand side converges to

$$\|\phi\|^2 log\|\phi\|, \qquad \|\phi\|^2 \equiv \frac{1}{2\pi}\int |\phi(t)|^2 e^{-\frac{t^2}{2}} dt \tag{89}$$

It remains to be proved that $\mathcal{E}_K(f)$ verifies

$$lim_{K\to\infty}\mathcal{E}_K(f) = \int |\phi'(t)|^2 d\mu(t) \tag{90}$$

Since $\phi \in C_0^\infty$ Dini's theorem gives the existence of a bounded function $g(t, x, h)$ on $R \times \{1, -1\} \times (0, 2)$ such that

$$\frac{1}{2}[\phi(t - hx + h) - \phi(t - hx - h)] - \phi'(t)h = h^2 g(t, x, h) \tag{91}$$

One has

$$(\delta_j(f)(x) = \frac{1}{2}[\phi(y - hx_j + h) - \phi(y - hx_j - h) = \Phi'(y)h + h^2 g((y, x_j, h) \tag{92}$$

and then $\sum_{j=1}^K |(\delta_j f)(x)|^2 = |\phi'(y)|^2 + \psi_K(x, h)$ where ψ_K is the sum of K terms each of which is of order h^3 o h^4. Taking $h = \frac{1}{\sqrt{K}}$ one derives

$$\mathcal{E}_K(f) = \int_X |\phi'(y)|d\mu(y) + \int_X \psi(x, h)d\mu(x) \tag{93}$$

with $\psi(x, h) = O(h)$ uniformly in x. Using once more the central limit theorem

$$lim_{k\to\infty}\mathcal{E}_K(f) = \int_R |\phi'(t)|^2 d\nu(t) \tag{94}$$

This proves (88) when $f \in C_0^\infty$.

To extend the proof to $Q(\mathcal{E}) \cap L^2(R, d\nu)$ one makes use of a limiting procedure. If $f \in L^2(R, d\nu)$ and its distributional derivative satisfies $f' \in L^2(R, d\nu)$, there exists a sequence of C_0^∞ functions f_n which converges to f in the $\|f\|_{L^2(\nu)} + \|f'\|_{L^2(\nu)}$ norm.

The function $t^2 logt$ is bounded below for $t \geq 0$ and therefore one can apply Fatou's lemma (passing if needed to a subsequence that converges almost everywhere).

The logarithmic Sobolev inequality is thereby proved for any $f \in Q(\mathcal{E}) \cap L^2(R, d\nu)$.

♡

References

1. D. Stroock, *An Introduction to the Theory of Large Deviations* (Springer, New York, 1984)
2. L. Gross, J. Funct. Anal. **10**, 52–109 (1972)
3. E. Nelson, J. Funct. Anal. **12**, 211–227 (1973)

Lecture 16: Measure (Gage) Spaces. Clifford Algebra, C.A.R. Relations. Fermi Field

In Book I of these Lecture Notes we have studied the Weyl algebra and its infinitesimal version, i.e. canonical commutation relations.

In that context we have considered the Fock representation, which can be extended to the case of an infinite (denumerable) number of degrees of freedom to construct the free Bose field.

In the preceding lectures we have also mentioned that this field can be constructed by probabilistic techniques through the use of gaussian measures and conditional probabilities.

In this lecture we seek an analogous procedure for algebra of canonical anti-commutation relations

$$a_h a_k^* + a_k^* a_h = \delta_{k,h} \qquad a_k a_h + a_h a_k = 0$$

but this time we have to resort to *non-commutative integration*.

We start with a general outlook on *non-commutative integration on gage spaces* [1–3, 6].

1 Gage Spaces

Definition 1 (*gage spaces*) A *gage space* (regular measure space) is a triple $\{\mathcal{H}, \mathcal{A}, m\}$ where $\mathcal{H}$ is a separable Hilbert space, $\mathcal{A}$ is a concrete von Neumann algebra of operators on $\mathcal{H}$ with identity e and m is a non-negative function on the projection operators P in $\mathcal{A}$ with the following properties
(1) m is completely additive
(2) $m(U^*PU) = m(P)$ for every unitary operator $U \in \mathcal{B}(\mathcal{H})$
(3) $m(e) < +\infty$ (e is the unit element of the algebra)
(4) m is *regular* i.e. $P \neq 0 \to m(P) > 0$ ◇

G. Dell'Antonio, *Lectures on the Mathematics of Quantum Mechanics II: Selected Topics*, Atlantis Studies in Mathematical Physics: Theory and Applications 2, DOI 10.2991/978-94-6239-115-4_16

Under these conditions there exist *a unique function* (the Dixmier trace) that extends $m(P)$ to $\mathcal{A}$ [6]. We shall denote it with the symbol Tr.

The trace has the following properties:
If $A > 0$ then $TrA > 0$.
If the Hilbert space $\mathcal{H}$ has dimension N and P projects onto a Mdimensional space, then $TrP = \frac{M}{N}$.

One has $Tre = 1 \; \forall N$ and this property holds also in the infinite-dimensional case.

The trace Tr is *normal* (completely additive). One has $Tr(AB) = Tr(BA)$ and if $A > 0$ then $Tr(A) > 0$.

If A is self-adjoint with spectral projections E_λ, using Riesz theorem and the Gelfand construction one has

$$TrA = \int \lambda dm(E_\lambda)$$

We define for $a \in \mathcal{A}$

$$||a|| = (a^* a)^{\frac{1}{2}}, \quad \|a\|_p = [Tr(a^* a)^{\frac{p}{2}}]^{\frac{1}{p}} \quad 1 \leq p \leq +\infty \tag{1}$$

With these definition $\|a\|_\infty$ coincides with $\|a\|$ i.e. with the operator norm of a.

We define $L^p(\mathcal{A})$ to be the completion of $\mathcal{A}$ in the L^p norm. Notice that L^p for $1 \leq p \leq +\infty$ can be regarded as a space of *unbounded operators* on $\mathcal{H}$.

With these definitions Hölder inequalities hold.

The space of operators which are measurable is closed for *strong sum* (closure of the sum), strong product and conjugation.

If the algebra $\mathcal{A}$ is commutative, by the Gelfand isomorphism one recovers the usual structure of integration theory (the projection operators are the indicator functions of the measurable sets).

Since

$$\|ax\|_p\| \leq \|a\|_\infty \|x\|_p \qquad \|xa\|_p\| \leq \|a\|_\infty \|x\|_p \tag{2}$$

one can define in every L^p *left and right multiplication* by an element of $a \in \mathcal{A}$. We shall denote them by the symbols L_a, R_a.

Definition 2 (*Pierce subspace*) We define *Pierce subspace* $\mathcal{P}_e$ associated to the projection e the range of $P_e \equiv L_e \, R_e$ i.e. the closure of $e\mathcal{A}$. ◇

Pierce subspaces *are closed* in all L^p and also in L^∞.

Notice that we are defining a *non-commutative integration* from the point of view of functional integration, i.e. defining non commutative L^p spaces. This is possible because we have a functional (the trace) that has the same property of an integral.

In the commutative case (Lebesgue) integration over a space X can be defined through the introduction of measurable sets and their indicator functions. In the non-commutative gage theory that we are considering the Pierce subspaces play the role of measurable sets.

Notice that if $e_1\ e_2 = 0$ then $\mathcal{P}_{e_1} \perp \mathcal{P}_{e_2}$ (if the product of two measurable characteristic functions is zero then they have with disjoint support).

More generally an integration theory (in the sense of defining L^p spaces) may be defined if the algebra admits a cyclic and separating vector as we have seen in Vol 1 of these Lecture Notes.

Indeed in this case the Tomita–Takesaki theory establishes a duality between the algebra $\mathcal{A}$ and the algebra of functions over the $\mathcal{A}$ and therefore allows for the definition of non commutative integration based on measurable sets.

If a tracial state exists, as in a gage theory, the probability space has measure one and one can define a non-commutative *measure theory* (which is the basis for non-commutative *Lebesgue integration theory*).

If a tracial state does not exists a *Lebesgue-like*, non-commutative algebraic integration theory is still possible (algebraic in the sense that L^p spaces and Radon–Nikodym derivatives can be defined) but the construction of a *non-commutative measure theory* requires a different approach.

For an introduction to non-commutative *measure theory* which leads to a non-commutative *Lebesgue integration theory* one can consult [3–5].

In non-commutative gage theory the algebra $\mathcal{A}$ is a subalgebra of $\mathcal{B}(\mathcal{H})$. If the projector P projects on $\phi \in \mathcal{H}$, then the Pierce subspace associated to P is the set generated by the action of $\mathcal{A}$ on ϕ.

By definition the *support* of $a \in \mathcal{A}$ is the union Range a $\cup$ Range a^*. If $z \in \mathcal{A}$ is real then there a unique pair $x, y \in \mathcal{A}^+$ such that $z = x - y$.

The *positive cone* is the set of positive elements $x \in \mathcal{A}^+$ such that there does not exist a projection e for which $exe = 0$.

We say that $B \in \mathcal{L}(\mathcal{A})$ (the set of linear functionals on $\mathcal{A}$) is *positivity preserving* if $a \geq 0$ implies $Ba \geq 0$. It follows for the definitions that the following lemma holds.

Lemma 1 *Let $\{\mathcal{H}, \mathcal{A}, m\}$ be a regular gage space. Set*

$$(\alpha, \beta) \equiv Tr(\alpha\beta) \tag{3}$$

If $\alpha \geq 0$ and $\beta \geq 0$ then $(\alpha, \beta) \geq 0$ and if $(\alpha, \beta) = 0$ then α and β have orthogonal supports. ◇

We use this lemma to prove [3]

Theorem 1 *Let $\{\mathcal{H}, \mathcal{A}, m\}$ be a regular gage space. Let $a \in L^2(\mathcal{A})$ be positivity preserving. If a does not leave invariant any Pierce subspace, and if $\lambda = \|a\|$ is an eigenvalue, then this eigenvalue* has multiplicity one *and the associated eigenvector is strictly positive.* ◇

Proof Let z be an eigenvector associated to λ. Let $z = x - y\ x, y \geq 0$. Then one has

$$\lambda(|z|, |z|) = (Az, z) = (Ax, x) + (Ay, y) - (Ax, y) - (Ay, x)$$
$$\leq (A|z|, |z|) \leq \|A\|(|z|, |z|) \qquad (4)$$

and the equality sign holds only if $|z|$ is an eigenvalue. Therefore $ax = \lambda x\ ay = \lambda y$. If π is the projector onto the null space of x consider $P_\pi \equiv L_\pi R_\pi$ and let $b \geq 0$. Then

$$(x, aP_\pi b) = (ax, P_\pi b) = \lambda(x, P_\pi b) = \lambda(x, P_\pi b) = \lambda(P_\pi x, b) = 0 \qquad (5)$$

and the Pierce subspace of π is invariant under $\mathcal{A}$.

It follows that $\pi = 0$ and therefore the range of x is the entire space $\mathcal{H}$. ♡

Definition 3 (*ergodic*) A map T of the algebra $\mathcal{A}$ is *ergodic* if for any $x, y \in L^2(\mathcal{A})$, $x, y \geq 0$ there exist $n \in Z$ such that $(T^n x, y) > 0$. We say that the algebra $\mathcal{A}$ is *indecomposable* if it leaves invariant no Pierce subspace. ◇

Proposition 1 *If $\mathcal{A}$ preserves positivity and is bounded over $L^2(\mathcal{A})$ then it is ergodic if and only if it is indecomposable.* ◇

Proof

$\Rightarrow$

If $\pi \in \mathcal{A}, \pi \neq 0$ and $aP_\pi = P_\pi a,\ a \in \mathcal{A}$ for any element $x, y \geq 0$, $P_\pi x = x, P_\pi y = y$ one has

$$(a^n x, y) = (a^n P_\pi x, y) = (P_\pi a^n x, y) = (a^n x, P_\pi y) = 0 \quad \forall n \qquad (6)$$

$\Leftarrow$

Let T be not ergodic. Choose $x, y \geq 0$ and $(T^n x, y) = 0\ \forall n \in N$. Denote by $\mathcal{N}(B)$ the null space of B. Then the projection π onto $\mathcal{N}(A^n)x$ belongs to $\mathcal{A}$ and is not the null element.

If $c \in L^2(\mathcal{A}), 0 \leq c \leq P_e c$ one has $(a^n c, c) = 0\ \forall n$. Since $a^n x \geq 0$, and $c \geq 0$ it follows that the range of $\mathcal{A}c$ is contained in $\mathcal{N}(A^n x)$ for all n. It follows $P_l(\mathcal{A}c) = \mathcal{A}c$ and the range of P_l is left invariant by $\mathcal{A}$. ♡

2 Interpolation Theorem

In the present non-commutative setting one has the following non-commutative equivalent of Stein's interpolation theorem.

Proposition 2 ([1–3, 6]) *Let $\{\mathcal{H}, \mathcal{A}, m)\}$ and $\{\mathcal{K}, \mathcal{B}, n\}$ be two finite regular gage spaces.*

For every $z \in C$, $0 \leq Rez \leq 1$ let T_z be a norm continuous map from $\mathcal{A}$ to $L^1(\mathcal{B})$.

Assume that for all $a \in \mathcal{A}, b \in \mathcal{B}$ the function $\Psi(z) \equiv Tr_{\mathcal{B}}(T_z(a)b)$ be bounded and continuous for $0 \leq Rez \leq 1$ and analytic in $0 < Imz < 1$.

Choose $1 \leq p_0, p_1 q_0, q_1$ and define, for $0 \leq s \leq 1$

$$\frac{1}{p} \equiv (1-s)\frac{1}{p_0} + \frac{s}{p_1} \quad \frac{1}{q} \equiv (1-s)\frac{1}{q_0} + \frac{s}{q_1} \tag{7}$$

Assume moreover that there exist a_0, a_1 such that

$$\|T_{iy}A\|_{q_0} \leq a_0 \|A\|_{p_0} \quad \|T_{1+iy}(A)\|_{p_1} \leq a_1 \|A\|_{p_1} \forall y \in R \forall A \in \mathcal{A} \tag{8}$$

Then for all $a \in \mathcal{A}$ one has

$$\|T_s(A)\|_q \leq a(s)\|A\|_p \quad \forall A \in \mathcal{A} \quad a(s) = a_0^{1-s} a_1^s \tag{9}$$

◇

We do not here give the proof of Proposition 2.

It can be reduced to the commutative case by using the polar decomposition of the elements in $\mathcal{A}$ and the spectral decomposition of the positive elements of $\mathcal{A}$ as operators on $\mathcal{H}$.

3 Perturbation Theory for Gauge Spaces

We now give some basic elements of perturbation theory in gage spaces.

In the non-commutative setting it is natural, in the description of perturbation of a free hamiltonian, to substitute the *potential* with the sum of right and left multiplication by a real (=selfadjoint) element of the algebra. This operation preserves reality.

Let H_0 be a positive operator on $\mathcal{H}$ with 0 as simple eigenvalue with eigenvector I. Choose $\alpha \in L^2(\mathcal{A})$ and define

$$H_\alpha = H_0 + L_\alpha + R_\alpha \tag{10}$$

With this definition the operator H_α is symmetric.

Assume that H_α is self-adjoint on $D(H_0) \cap D(L_\alpha) \cap D(R_\alpha)$, and that $H_\alpha \geq -C$. Assume moreover that $\mathcal{A} \cap D(H_\alpha)$ coincides with $\mathcal{A} \cap D(H_0)$. This is certainly true if $\alpha \in \mathcal{A}$.

In this case one has $H_\alpha b = H_0 b + \{a, b\}$. Then one has

Proposition 3 ([1, 2]) *There is no Pierce subspace that is left invariant by the operator H_α* ◇

We end this brief outline of non-commutative integration by a condition on the existence and uniqueness of the ground state. Also in this case the proofs follows the same lines as in the commutative case.

Proposition 4 ([1, 3]) *Let $\{\mathcal{H}, \mathcal{A}, m\}$ be a finite regular gage space and let $H_0 \geq 0$ on $L^2(\mathcal{A})$.*

Assume
(1) e^{-tH_0} is a contraction in L^p for all $T > 0$ and there exists $T_0 > 0$ such that e^{-TH_0} is a contraction from $L^2(\mathcal{A})$ to $L^p(\mathcal{A})$.
(2) e^{-tH_0} is positivity preserving.
(3) $H_0\phi = 0 \to \phi = e \in \mathcal{A}$.
(4) v is a self-adjoint element in $L^2(\mathcal{A})$, $v \in L^p(\mathcal{A})$ for some $p > 2$ and $e^{-v} \in L^p$ for every $p < +\infty$.

Set $V = L_v + R_v$. Then
(a) $H_0 + V$ is essentially self-adjoint on $D(H_0)$ and its closure is bounded below.
(b) Define $E_0 = \inf \Sigma(H)$. Then E_0 is a simple eigenvalue and the corresponding null space is trivial. ◇

Proof Since $e^{-tV}u = e^{-\alpha v}ue^{-\alpha v}$ one has of v is bounded

$$e^{-t(H_0+V)} = s - lim_{n\to\infty}\left[e^{-t\frac{V}{n}}e^{-t\frac{H_0}{n}}\right]^n \tag{11}$$

Therefore $e^{-t(H_0+V)}$ is positivity preserving (if V is unbounded, $V = \int \lambda dE(\lambda)$ one considers the sequence $V_n = \int_{-n}^{n} \lambda dE(\lambda)$).

Notice then that if a sequence $\psi_n \in L^2(\mathcal{A})$ is such that $\|\psi_n\| \leq c\ \forall n$ then $\|\psi\|_p \leq c$ for all $p > 2$ (use $Tr(\psi_n, \phi) \leq c\|\phi\|$ if $\frac{1}{p} + \frac{1}{p'} = 1$).

Point b is proved along the lines of the commutative case [3]. ♡

4 Non-commutative Integration Theory for Fermions

We now apply the theory of integration in gage spaces to formulate an integration theory for particles which satisfy the Fermi-Dirac statistics and are therefore quantized according the canonical anti-commutation relations (C.A.R.).

We recall that the algebra C.A.R. of canonical *anti-commutation relations* for a system of $N \leq +\infty$ degrees of freedom is the C^* algebra $\mathcal{A}_N$ generated by elements that satisfy the relations

$$a_i a_k^* + a_k^* a_i = \delta_{i,k} \quad a_i a_k + a_k a_i = 0 \quad i, k = 1, \ldots, N \tag{12}$$

As a consequence of these relations $a_i^* a_i + a_i a_i^* = 1$ and since both terms are positive it follows that on any realization as operators on a Hilbert space the operators a_i have norm bounded by one.

The Fock representation of the algebra C.A.R is obtained by requiring that in the Hilbert space there exists a vector Ω for which $a_k \Omega = 0\ \forall k$.

From the defining relations it follows that for each value of the index k the pair a_k, a_k^* can be realized faithfully and irreducibly by two dimensional complex-valued matrices and if N is finite the entire algebra can be realized faithfully and irreducibly in the Hilbert space C^{2N}.

If $N < \infty$ all irreducible faithful representations of $\mathcal{A}_N$ are equivalent and in any such representation there is a vector Ω_N (called *vacuum*) such that $a_k \Omega_N = 0$, $k = 1 \dots N$.

A basis in this representation is made of the vectors

$$|i_1, \quad i_K >\equiv a_{i_1}^* \cdots a_{i_K}^* \Omega \tag{13}$$

where $0 \le K \le N$ and the indices are all distinct.

Correspondingly the representation is called *Fock representation* and each element of the basis is labelled by a sequences N of numbers n_i which are zero and one according to whether the index appears in (14).

The operators a_k^* are called *creation operators* (since they change a zero in a one) and a_k are called destruction operators.

Notice that according to (13) one has $a_k^* |i_1, \quad i_K >= 0$ if $k \in \{i_1, \quad i_K\}$ (the *occupation number* for each index is at most one).

For this reason the algebra *CAR* is suitable for the description of particles which satisfy the Fermi-Dirac statistics (the Pauli exclusion principle holds).

5 Clifford Algebra

We give now a connection of the algebra *CAR* with the *Clifford algebra.*

Definition 4 (*orthogonal space*) Given a topological vector space M we define *orthogonal space* the space

$$(M \oplus M^*, S) \tag{14}$$

where M^* is the topological dual of M and S is the quadratic (symplectic) form

$$S(x \oplus \lambda, x' \oplus \lambda') = \lambda'(x) - \lambda(x') \tag{15}$$

Definition 5 (*Clifford structure*) Let $L = M \oplus M^*$, and let $\{L, S\}$ be an orthogonal space. A *Clifford structure* on $\{L, S\}$ is a pair $(\mathcal{K}, \phi)$ where $\mathcal{K}$ is a complex Hilbert space and ϕ is a linear continuous map from L to $\mathcal{B}(\mathcal{K})$ such that

$$\phi(z)\phi(z') + \phi(z')\phi(z) = S(z, z')I \tag{16}$$

◇

Definition 6 (*Clifford system*) Let $\mathcal{H}$ be a complex Hilbert space, and let $\mathcal{H}^*$ its presentation as a pair of real Hilbert spaces. Let $S(z, z') \equiv Re(z, z')$.

Then $(\mathcal{H}^*, S)$ is a *Clifford system* on $\mathcal{H}^*$. If there is a self-adjoint operator on $\mathcal{H}$ such that $S(z, z') = Re(z, Az')$ we will say that the pair $\{\mathcal{H}^*, S\}$ is a *Clifford system with covariance A*. ◇

The relation of the *CAR* with a Clifford algebra is as follows:

Definition 7 (*Clifford algebra*) Let $\mathcal{H}$ be a complex Hilbert space and set $\mathcal{H}^* = \mathcal{H}_r \oplus \mathcal{H}_{im}$.

Then the Clifford algebra is *the only associative algebra* on the field of real numbers generated by $\mathcal{H}^*$ and by a unit e and defined by the following relations

$$xy + yx = Re(x, y)\ e \tag{17}$$

◇

Notice that if $\mathcal{H}$ is finite dimensional for the Clifford algebra there exist *a unique functional E* such that

$$E(ab) = E(ba)\ \forall a, b \in \mathcal{B}(\mathcal{H}) \qquad E(e) = 1 \tag{18}$$

and a unique adjoint map such that $x^* = x\ \forall x \in \mathcal{H}^a st.$

Definition 8 (*Clifford field* [1]) Let $\mathcal{H}$ be the closure of Cl with respect to the scalar product $< a, b >= E(b^*a)$. Let $a \in Cl$ and denote by L_a (left multiplication by the element a) the map $b \to ab$, $b \in \mathcal{A}$.

Similarly denote by R_a (right multiplication by the element a) the map $b \to ba$. It is easy to verify that the following holds true for $z \in \mathcal{H}^*$

$$< a, L_z b >= E(b^* z a) \tag{19}$$

This identifies L_z with an hermitian operator densely defined in $\mathcal{H}$.

It extends uniquely to a self-adjoint operator which is bounded since $L_z^2 = \frac{1}{2}\|z\|^2 I$. We shall call L_z *Clifford field* and denote it with the symbol $\psi(z)$.

If V is an orthogonal map on $\mathcal{H}$ (it preserves S) and $\psi(x)$ is a Clifford fields, also $\psi_V(x) \equiv \psi(Vx)$ is a Clifford field.

Moreover if ψ and ϕ are anti-commuting Clifford fields,

$$\psi(x)\phi(y) = -\phi(y)\psi(x) \tag{20}$$

and a, b are real numbers with $|a|^2 + |b|^2 = 1$ also

$$\psi'(x) \equiv a\psi(x) + b\phi(x) \tag{21}$$

is a Clifford field.

Denote by T the automorphism $z \to -z$ in Cl. Then T anticommutes with L_z and with R_z and $T^2 = I$. It follows that

$$z \to iL_zT, \qquad z \to iR_zT \tag{22}$$

define Clifford system and L_zT and R_wT anticommute for every $z, w \in \mathcal{H}$. It follows that for every $a, bc \in R^+$ the map

$$z \to aL_z + ibR_zT \tag{23}$$

defines a Clifford system with variance c such that $|a|^2 + |b|^2 = c^2$.

Remark that $\mathcal{H}^* = \mathcal{H}_r \oplus \mathcal{H}_l$ is regarded as a real Hilbert space and S is a symplectic form, while Cl is the algebra over the complex field generated by $\mathcal{H}^*$.

For the Clifford field on $\mathcal{H}^*$ one can define [1] creation and annihilation operator by

$$c(z) = \frac{1}{\sqrt{2}}[\phi(z) - i\phi(-z)] \qquad c(z^*) = \frac{1}{\sqrt{2}}[\phi(z) + i\phi(-z)] \tag{24}$$

These operators are bounded and satisfy the canonical anticommutation relations

$$c(z)c(w)^* + c(w)^*c(x) = C(z, w) \quad c(z)c(w) + c(w)c(z) = 0, \quad c(iz) = ic(z) \tag{25}$$

Conversely, every system of operators on a complex Hilbert space $\mathcal{H}$ which satisfies (25) define a Clifford field on $\mathcal{H} = \mathcal{H}_R \oplus \mathcal{H}_R$.

In the case of a finite-dimensional Hilbert space all the irreducible representations of (26) are equivalent; this is not the case if the Hilbert space is infinite-dimensional.

The conditions for equivalence are the same as in the case of the Canonical Commutation relations as discussed in Vol I.

6 Free Fermi Field

Definition 9 (*Free Fermi field I* [1, 4]) The *free Fermi field* on the *complex* Hilbert space $\mathcal{H}$ is a Clifford system together with
(1) A map which satisfies (26) with C(z, w) = (z, w)
(2) A continuous representation Γ of the unitary group on $\mathcal{H}$ on the unitary group of $\mathcal{K}$ which satisfies

$$\Gamma(u)c(z)\Gamma^{-1}(u) = c(uz) \quad \forall z \in \mathcal{H} \quad \forall u \in U \tag{26}$$

(3) An element $\nu \in \mathcal{H}$ which is cyclic for the algebra generated by the $c(z)$ and such that $\Gamma(u)\nu = \nu \; \forall u \in U$. ◇

Let A be a non-negative element of $\mathcal{B}(\mathcal{H})$ and denote by $\partial\Gamma(A)$ the generator of the unitary group $\Gamma(e^{itA})$.

Then $\partial\Gamma(A)$ is positive. One has [1].

Theorem 2 (Segal) *The free Fermi field as defined above is unique up to unitary transformations.* ◇

We shall later see a different but equivalent definition. We do not give here the proof of Theorem 2. It follows the same lines as the analogous theorem in the bosonic case proved in Volume I of these Lecture Notes.

The explicit construction of the Fock representation can be done as in the bosonic case (with the simplifying feature that all operator considered are bounded).

7 Construction of a Non-commutative Integration

We are interested here in the construction of a *non-commutative integration* on function of the Clifford algebra (as one constructs a *commutative gaussian integration* theory in the bosonic case).

Recall that if $\mathcal{H}_r$ is a *real* Hilbert space of dimension $2n$ we have defined *Clifford algebra* Cl_n on $\mathcal{H}_r$ with variance C the C^*-algebra $\mathcal{A}$ over the complex field which is the norm closure of the algebra generated by the unit element e and by elements in $\mathcal{B}(\mathcal{H}_r)$ which satisfy the relation

$$xy + yx = C(x, y)e \quad \forall x, y \in \mathcal{H}_r \tag{27}$$

If $n > m$ there is a natural injection of $\mathcal{B}(R^{2m})$ in $\mathcal{B}(R^{2n})$ given by $C \to C \otimes I_{2n-2m}$. Therefore Cl_m is naturally immersed as a subalgebra of Cl_n for $n > m$ by the map $b \to b \otimes I_{2(n-m)}$.

Each of the algebras Cl_n is a C^* algebra with the natural norm. The immersion preserves the norm and satisfies obvious compatibility and immersion relations if one considers a sequence $n_1 < n_2 < \cdots$.

In the infinite dimensional case one can therefore consider therefore the Clifford algebras Cl_n as subalgebras of a normed algebra Cl. We denote by $\mathcal{A}$ the norm closure of Cl. It is isomorphic to the algebra of *canonical anticommutation relations*.

Theorem 3 ([1, 2, 5]) *There exists on $\mathcal{A}$ a* unique *functional E with the properties*

$$E(e) = 1 \quad E(ab) = E(ba) \qquad \forall a, b \in \mathcal{A} \tag{28}$$

(e is the unit of $\mathcal{A}$). ◇

The functional E has the properties of a trace. With this functional we construct an integration theory. We shall denote by η the canonical injection of $\mathcal{H}_r$ in the complex Hilbert space $\mathcal{H}$.

The functional E is constructed the following way. If the dimension $2n$ is finite, the algebra $\mathcal{A}$ is made of all real matrices of rank $2n$ and E is the usual trace *normalized to one on the identity.*

If $n = \infty$ the algebra $\mathcal{A}$ is generated (as norm closure) by the algebras which are constructed over a finite-dimensional space.

Continuity and uniqueness follow from the fact that the finite-dimensional algebras are unique and uniformly continuous.

To prove existence notice that every element of Cl is based on R^{2m} for some finite n and there is a natural immersion of $\mathcal{B}(R^{2m})$ in $\mathcal{B}(R^{2n})$ $n > m$ given by $D \to D \otimes I_{2n-2m}$.

This immersion does not alter the value of the functional E (recall that it is normalized to one on the unit element). Therefore E is defined on a dense set, is continuous (and bounded) and extends to $\mathcal{A}$.

Remark that steps we have followed to define the functional E are the same as those followed to define a probability measure on the infinite product of measure spaces on each of which is defined a probability measure satisfying suitable compatibility conditions.

Therefore the construction of the functional E parallels in the non-commutative case the construction of a theory of integration in the commutative setting.

The functional E has been constructed over the C^*-algebra $\mathcal{A}$. The GNS construction based on the functional E provides a representation $\pi_0(\mathcal{A})$ of $\mathcal{A}$ as an algebra of bounded operators on a Hilbert space $\mathcal{H}_0$. The representation can be extended to the weak closure of $\pi_0(\mathcal{A})$.

Notice that this representation *is different from the Fock representation*. In the infinite-dimensional case they *are inequivalent.*

It indeed easy to verify from the construction that on the projection operators in $\pi_0(\mathcal{A}_F)$ the functional E takes values which cover the interval (0, 1].

It is important to notice that if P is a projection operator in $\mathcal{A}$ it projects *on a infinite dimensional subspace.*

We conclude that in the infinite-dimensional case the representation π_0 of the C.A.R. is a von Neumann algebra of type II in von Neumann classification. In this representation there is a vector Ω

$$E(a_1^* a_2^* * \dots a_n^*) = (\Omega, \pi(a_1^*)\pi(a_2^*) \dots \pi(a_n^*)\Omega) \tag{29}$$

where $\pi(a)^*$ is either the creation or the destruction operator.

8 Dual System

Definition 10 If (K, Φ) is a real Clifford system the *dual system* is defined as

$$\{K, P(x), Q(x)\} \quad P(x) = \phi(x), \quad Q(x) = \phi(ix), x \in \mathcal{H}_r \tag{30}$$

◇

Notice that this definition depends on the choice of the conjugation in $\mathcal{H} = \mathcal{H}_r \oplus \mathcal{H}_i$. Conversely of $\{K, P(x)\ Q(x)\}$ is a dual system, the real system is given by $\phi(z) = P(x) + Q(y)$ if $z = x + iy$.

The complex representation can be regarded as the analog of the Segal–Bargmann representation for bosons.

Since there is no complex quadratic form which is invariant under the unitary group, in the Clifford algebras $\mathcal{A}$ only real space are considered and the complex representations depend on the choice of conjugation.

Define $\phi \to \bar{\phi}$ the conjugation in $Cl(H^*)$, $H^* = H_r \oplus H_l$. It is the unique operation that extends $\eta(x) + i\eta(y) \to \eta(x) - i\eta(y)$.

The connection of the algebra $\mathcal{A}$ with the *fermionic free field* is as follows:

Theorem 4 *Let $\mathcal{H}$ be a Hilbert space, and let $\mathcal{K}'$ be the space of function s which are* anti-holomorphic *in $L^2(Cl(\mathcal{H}))$. For $x \in \mathcal{H}$ define the operator $\phi(x)$ as*

$$\phi(x) = \frac{1}{\sqrt{2}}[L_x + iR_{ix}] \tag{31}$$

For every unitary on $\mathcal{H}$ let $\Gamma_0(U)$ the second quantization of U. Let ι be the function identically equal to one in $L^2(Cl(H), E)$. The space $\mathcal{K}'$ is left invariant under the action of $\phi(x)$ and of $\Gamma_0(U)$. Denote by $\phi(x)'$, $\Gamma_0(U)'$ the restriction of these operators to $\mathcal{K}'$.

Then the algebra generated by the operators $\phi(x)'$ is isomorphic the algebra $\mathcal{A}$. ◇

9 Alternative Definition of Fermi Field

Definition 11 ([2, 4]) The *free Fermi field* on $\mathcal{H}$ is the quadruple $\mathcal{K}', \phi', \Gamma_0', \iota$. ◇

Theorem 5 *The free Fermi field is self-adjoint and satisfies the Clifford relations.* ◇

Proof If $z \in \mathcal{H}^*$, $z = \eta(x) - i\eta(ix)$ the following relations hold true

$$\phi(z) = \frac{1}{\sqrt{2}}[L_z - iR_z] \quad \phi(\bar{z}) = \frac{1}{\sqrt{2}}[L_{\bar{z}} + iR_{\bar{z}}] \tag{32}$$

Moreover if $U(t) = e^{iht}$ one has

$$\eta(U(t)x - i\eta(iU(t))x = e^{iht}(\eta(x) - i\eta(ix)) \tag{33}$$

♡

Recalling the definition of gage space (Definition 1) we see that the free Fermi field is an example of *non-commutative integration theory*.

In the case of the free Fermi field one can define a gage as follows. Consider the Hilbert space

$$\Lambda(\mathcal{H}) \equiv \sum_{n=1}^{\infty} \Lambda^n(\mathcal{H}) \tag{34}$$

where $\Lambda^n(\mathcal{H})$ is the Hilbert space of the antisymmetric tensors of rank n on the complex Hilbert space $\mathcal{H}$.

Let J be a conjugation in $\mathcal{H}$. Define for each $x \in \mathcal{H}$

$$B_x = C_x + A_{Jx} \qquad A_x = C_{x^*} \tag{35}$$

where

$$C_x u = (n+1)^{\frac{1}{2}} x \wedge u \tag{36}$$

n is the rank of the tensor u and $A_x = C_x^*$. Let $\mathcal{M}$ be the smallest von Neumann algebra that contains all B_x, $x \in \mathcal{H}$. These data *define a gage space.*

Theorem 6 ([2, 4]) $\{\mathcal{H}, m, \mathcal{M}\}$ *above define a gage if one takes* $m(u) = (u\Omega, \Omega)$ *where* Ω *is the* vacuum state *i.e. the unit of* $\wedge^0(\mathcal{H})$. *Moreover* $u \to u\Omega$ *extends to a unitary operator from* $L^2(Cl)$ *onto* $\Lambda(\mathcal{H})$. ◇

Proof Let $\mathcal{C}_1$ the algebra generated (algebraically) by the B_x. One has $B_x^* = B_{Jx}$ and therefore $\mathcal{C}_1$ is self-adjoint. Let $\mathcal{M}$ be its weak closure.

The function Tr defined by $Tr(u) = (u\Omega, \Omega)$ is positive and $Tr(I) = 1$. Repeated use of $A_{Jx}\Omega = 0$ and $A_y C_x + C_x A_y = (x, y)I$ leads to

$$(B_x B_{y_1} \dots B_{y_n}\Omega, \Omega) = \sum_{j=1}^{n} (-i)^{j-1} (x, y_j)(B_{y-1} \dots \hat{B}_{y_j} \dots B_{y_n}\Omega, \Omega) \tag{37}$$

where the hat signifies that the symbol must be omitted. In the same way one has

$$B_{y_1} \dots B_{y_n} C_x \Omega = \sum_{j=1}^{n} (-1)^{n-j} B_{y_1} \dots \hat{B}_{x_j} \dots B_{y_n}\Omega \pm C_x B_{y_1} \dots B_{y_n}\Omega \tag{38}$$

It follows

$$(B_{x_1} \dots B_{x_n}\Omega, \Omega) = \sum_{j=1}^{n} (-1)^{n-j} (x_j, x)(B_{x_1} \dots \hat{B}_{x_j} \cdot B_{x_n}\Omega, \Omega) \tag{39}$$

Define $B_{\mathbf{x}}^n = (B_{x_1} \dots B_{x_n})$. If n is even, one has

$$(B_{\mathbf{x}}^n B_y \Omega, B_y \Omega) = B_y B_{\mathbf{x}}^n B_y, \Omega) \tag{40}$$

If n is odd, $B^n_{\mathbf{x}}\Omega$ is a tensor of odd rank, therefore $(B^n_{\mathbf{x}}B_y\Omega, B_y\Omega) = 0$. It follows that $(BC\Omega, \Omega) = (CB\Omega, \Omega)$ for every $B, C \in \mathcal{A}$.

Therefore the function

$$TrB = (\Omega, B\Omega) \tag{41}$$

is a *central trace*. The map $A \to A\Omega$ is faithful since

$$\begin{aligned} \|AB\Omega\|^2 &= (B^*A^*\Omega, AB\Omega) = Tr(B^*A^*AB) \\ &= Tr(BB^*A^*A) = Tr(A^*AB^*B) = (BB^*A^*A\Omega) \end{aligned} \tag{42}$$

and therefore

$$A\Omega = 0 \to AB\Omega = 0 \quad \forall B \tag{43}$$

Moreover the function $(\Omega, K\Omega)$, $K \in \mathcal{A}$ is clearly σ-additive and for every unitary U one has $Tr(U^*KU) = TrK$. Therefore

$$\{\mathcal{H},\ \mathcal{A},\ m\} \qquad m(A) \equiv (\omega, A\Omega) \tag{44}$$

is a regular finite gage ♡

10 Integration on a Regular Gage Space

We shall give here some results of integration theory on a regular gage space. Later we shall give an outline of the integration of a fermionic field in presence of an external field.

We begin by giving a definition that is equivalent to the support of a function in the case of a measure space. Recall the definition

Definition (*Pierce subspace*) Let $\{\mathcal{H}, \mathcal{A}, m\}$ be a regular finite gage space, and e a projection operator in $\mathcal{A}$.
Define $P_e = L_e R_e$. The range of P_e is called *Pierce subspace* of e. ◇

Definition 12 (*positivity preservation*) A bounded operator A on $L^2(\mathcal{A})$ is *positivity preserving* if $\phi \geq 0 \to A\phi \geq 0$. The *support* of a densely defined operator B is the convex closure of the union of the range of B and the range of B^*. ◇

Lemma 2 *Let* $\{\mathcal{H}, \mathcal{A}, m\}$ *be a regular finite gage space, if* $a \geq 0, b \geq 0$ *then* $Tr(ab) \geq 0$, *if* $tr(ab) = 0$ *the elements a and b have disjoint support.* ◇

Proof The first statement is obviously true. For the second, notice that $Tr(a^{\frac{1}{2}}ba^{\frac{1}{2}}) = o$ implies $a^{\frac{1}{2}}ba^{\frac{1}{2}} = 0$. Setting $b = c^2$ with c self-adjoint and measurable one has $\|ca^{\frac{1}{2}}x\| = 0$ for every x in the support of $a^{\frac{1}{2}}ba^{\frac{1}{2}} = 0$. Therefore $ca^{\frac{1}{2}} = 0$ on a dense set, and then $ca^{\frac{1}{2}} = 0$ and $ba = 0$. ♡

Theorem 7 (Gross [3]) *Let $\{\mathcal{H}, \mathcal{A}, m\}$ be a regular finite gage space. Let A on $L^2(\mathcal{A})$ positivity preserving. Suppose that $\|A\|$ is an eigenvalue of A and that A does not invariant any proper Pierce subspace. Then $\|A\|$ has multiplicity one.* ◇

Proof By assumption A maps self-adjoint operators to self-adjoint operators and has a self-adjoint eigenvector to the eigenvalue $\|A\|$.

It is easy to see that the positive and negative part of this eigenfunction separately belong to the eigenspace to the eigenvalue $\|A\|$.

Let now $x \geq 0$ belong to the eigenspace to the eigenvalue $\|A\|$ and let e be the projection to the null space of x. Set $P_e \equiv L_e R_e$ and let $b \in L^2(\mathcal{A})$. Then

$$(x, AP_e b) = (Ax, P_e b) = \|A\|(x, P_e b) = \|A\| Tr(P_e x, b) = 0 \qquad (45)$$

But $AP_e b \geq 0$ and therefore the support of AP_e is contained in the range of P_e. Therefore the Pierce subspace of e is invariant under the action of $\mathcal{A}$.

The eigenspace associated to $\|A\|$ is therefore spanned by its self-adjoint elements and these can be chosen to be positive. It follows that the eigenspace has dimension one. ♡

Definition 13 (*strongly finite*) A regular gage $\{\mathcal{H}, \mathcal{A}m\}$ is *strongly finite* if $\mathcal{A}$ contains a family $\mathcal{A}_\alpha$ of finite-dimensional subalgebras, directed by inclusion, and such that $\cup_\alpha \mathcal{A}_\alpha$ is dense in $L^2(\mathcal{A})$. ◇

Theorem 8 (Gross [3]) *Let $\{\mathcal{H}, \mathcal{A}m\}$ be a regular strongly finite gage. Let A be a bounded operator positivity preserving. If the exist a number $p > 2$ such that*

$$\|A\phi\|_p \leq M\|\phi\|_2 \qquad M > 0 \;\; \forall \phi \in L^2(\mathcal{A}) \qquad (46)$$

then $\|A\|$ is an eigen value of A. ◇

Remark that the hypothesis $p > 2$ is an hypothesis of *hyper-contractivity*. This theorem has a counterpart in the integration theory on the Bosonic Fock space based on gaussian integration.

Hypercontractivity is at the root of the construction given by Nelson [4] of the free Bose field as a measure in the space of distributions.

Proof Let P_α be conditional expectation with respect to $\mathcal{A}_\alpha$. By definition it is the only element of $\mathcal{A}_\alpha$ such that

$$Tr(P_\alpha x, y) = Tr(x, y) \quad \forall y \in \mathcal{A}_\alpha \qquad (47)$$

This defines P_α for every $x \in L^{\mathcal{A}}$; when restricted to $L^1(\mathcal{A})$ it is the orthogonal projection on $\mathcal{A}_\alpha$.

It is now easy to prove that P_α preserves positivity. Moreover

$$\|P_\alpha x\|_p = sup\{Tr(P_\alpha x)y) \ \ y \in \mathcal{A}_\alpha, \ \ \|y\|_q \le 1\} = sup\{Tr(xy); \ y \in \mathcal{A}_\alpha, \ \ \|y\|_q \le 1\}$$
$$\le \{Tr(xy); \ y \in \mathcal{A}, \ \ \|y\|_q \le 1\} = \|x\|_p \quad \frac{1}{p} + \frac{1}{q} = 1 \tag{48}$$

It follows that the restriction of P_α to $L^p(\mathcal{A})$ has norm one.

Since $\cup_\alpha \mathcal{A}_\alpha$ is dense in $L^2(\mathcal{A})$ the net P_α converges strongly to the identity map. If $A \in \mathcal{A}$ define $A_\alpha = P_\alpha A P$.

The operator A_α preserves positivity, leaves $\mathcal{A}_\alpha$ invariant, and therefore by the Perron–Frobenius theorem has an eigenvector $\Phi_\alpha \in \mathcal{A}_\alpha$ to the eigenvalue λ_α.

From the fact that P_α increases to the identity it follows $\lambda_\alpha \le \|A\|$ and $lim_\alpha \lambda_\alpha = \|A\|$. On the other hand, by density, for each $\Phi \in L^2(\mathcal{A})$ there exist an index β such that that for every $\psi \in L^2(\mathcal{A})$

$$|(\psi, P_\beta \psi - \psi)| < \epsilon \rightarrow (A\phi, P_\beta \psi - \psi) < \epsilon \tag{49}$$

It follows that weakly

$$P_\beta \psi \rightarrow \psi \quad A\psi = \|A\| \psi \tag{50}$$

We must now show that ψ is not the zero element of $L^2(\mathcal{A})$. For this we use the hyper-contractivity assumption For any choice of $a; b$ with $\frac{1}{a} + \frac{1}{b} = 1$ we have by interpolation [5, 6].

$$\|f\|_2 \le \|f\|_1^q \|f\|_p^p \qquad a = \frac{p-2}{2(p-q)} \qquad b = \frac{p}{2(p-1)} \tag{51}$$

Since P_α has norm one in $L^p(\mathcal{A})$ one has

$$\|A\| \|\psi_\alpha\|_p = \|A_\alpha \Psi_\alpha\|_p \le M \|\psi_\alpha\|_2 = M \tag{52}$$

It follows

$$1 = \|\psi_\alpha\|_2 \le \|\psi_\alpha\|_1 \frac{M}{\|A\|}^b \tag{53}$$

and therefore

$$\|\psi_\alpha\|_1 \ge \left(\frac{\|A\|}{M}\right)^{\frac{p}{p-2}} \tag{54}$$

Since $\psi_\alpha \ge 0$ for all α one has

$$(\psi, I) = \lim_\alpha (\psi_\alpha, 1) = lim \|\psi_\alpha\| = \left(\frac{\|A\|}{M}\right)^{\frac{p}{p-2}} \tag{55}$$

Therefore $\psi \neq 0$. ♡

In the proof of the previous theorem we have used the non-commutative version of Stein's lemma [5, 6].

For a comparison, notice that in the Bose case the fields $\phi(x)$ and $\pi(x)$ are real valued distributions, and therefore

$$\phi(f) = \int f(x)\phi(x)dx, \quad \pi(g) = \int g(x)\pi(x)dx \tag{56}$$

are symmetric operators that are self-adjoint in the Fock representation.

Therefore for them integration theory hods in the classical sense if one makes use of suitable gaussian measures.

11 Construction of Fock Space

As an application of the theory of gage spaces we formulate now a theorem that is useful in the construction of the representation for a free Fermi field. We begin with a construction of *Fock space*. Let A be a self-adjoint operator on the complex Hilbert space $\mathcal{H}$. Denote by $\Gamma(e^{itA})$ the strongly continuous group of unitary operators defined by

$$\Gamma(e^{itA}) = \oplus_n e^{itA} \otimes e^{itA} \ldots \otimes e^{itA} \tag{57}$$

where the nth term act on antisymmetric tensors of rank n and by convention the first term is the identity. Also here the map Λ is called *second quantization*. We have discussed it in Volume I in the case of the Bose Field. Denote by $d\Gamma(A)$ he infinitesimal generator of $\Gamma(e^{itA})$ so that formally

$$\Gamma(e^{itA}) = e^{itd\Gamma A} \tag{58}$$

Denote by $\Lambda(\mathcal{H})$ the direct sum of antisymmetric tensors over $\mathcal{H}$.

Lemma 3 ([3, 4]) *Let D be the extension of the map $u \to u_\nu$ of an unitary operator from $L^2(Cl)$ to $\Lambda(\mathcal{H})$. Define*

$$\beta = \Gamma(-1) \qquad a = B_x \tag{59}$$

Then

$$DL_xD^{-1} = C_x + A_{Jx} \quad DR_xD^{-1} = (C_x - A_{Jx})\beta \tag{60}$$

The operator β is one on the even forms and minus one on the even forms (this reflects the anti-commutation properties of the Dirac-Fermi field). ◇

Proof The first relation follows from

$$DL_xD^{-1}Du = DL_xD^{-1}u_\nu = DL_xu - Dau = B_xu \tag{61}$$

For the second relation notice that for any $y \in \mathcal{H}$ one has

$$[C_x - A_{Jx}, B_y] = 0 \tag{62}$$

It follows that setting $E = C_x - A_{Jx}\beta$ one has

$$Eu\Omega = uE\Omega = uC_x\Omega = u(C_x + Ja_x)\Omega = ua\Omega + R_x\Omega = R_x\Omega \tag{63}$$

Therefore

$$(ED - DR_x)\Omega = 0 \tag{64}$$

and by (69) the same relation holds in $L^2(Cl)$. ♡

Lemma 4 *Let $x, y \in \mathcal{H}$. Define*

$$\sigma \equiv \frac{1}{2}B_xB_y - \frac{1}{2}(x, Jy)I \tag{65}$$

Then $\sigma \in Cl$, $Tr\sigma = 0$ and

$$D(L_a + R_a)D^{-1} = C_xC_y + A_{Jx}A_{Jy} \qquad D\sigma = \frac{1}{2}C_xC_y\Omega \tag{66}$$

◇

Proof From the Clifford relations it follows

$$B_xB_y \mp B_yB_x = 2(x, y)I \tag{67}$$

Defining

$$R_\sigma = (x, y)I - R_uR_v \tag{68}$$

from the preceding lemma

$$DR_\sigma D^{-1} = (x, y)I - \frac{1}{2}(C_y - C_{Jx})\beta(C_y - A_{Jx})\beta - \frac{1}{2}(x, y)I \tag{69}$$

Using $\beta^2 = I$ and $\{C_y - A_{Jx}, \beta\} = 0$ and the preceding lemma one has

$$DR_\sigma D^{-1} = \frac{1}{2}(C_c + A_{Jx})(C_y + A_{Jx}) - \frac{1}{2}(x, y)I \tag{70}$$

To conclude the proof of Lemma 4 note that $Tr(B_xB_y) = (x, y)$. Acting on Ω with A_{Jx} and C_y and using $D\Omega = I$ we have

$$2D\sigma = C_xC_y\Omega \tag{71}$$

♡

We can formulate the following theorem [3, 4].

Theorem 9 [3, 4] *Let $\mathcal{H}$ be a complex Hilbert space, J a conjugation. Let A be a self-adjoint operator in $\mathcal{H}$, $A \geq mI$, $m > 0$. Set*

$$H = D^{-1}d\Gamma(A)D \tag{72}$$

If A commutes with J then
(1) e^{-tH} is a contraction in $L^p(Cl)$ for every $t \geq 0$ and a contraction on $L^p(Cl) \cup L^2(\mathcal{H})$ for every $p \in [1, +\infty]$
(2) If $t \geq \frac{1}{2}log3$ then e^{-tH} is a contraction from $L^2(Cl)$ to $L^4(Cl)$
(3) For every $t \geq 0$ the semigroup e^{-tH} is positivity preserving. ◇

To simplify the presentation, we will prove first this theorem assuming the validity of Lemmas 5 and 6 below. We shall then prove these lemmas.

Lemma 5 ([3]) *Let $U = D^{-}1d\Gamma(I)D$. If $t \geq \frac{1}{2}log3$ then e^{-tH} is a contraction from $L^2(Cl)$ in $L^4(Cl)$.* ◇

Lemma 6 ([4]) *Let*

$$A \geq 0 \quad [A, J] = 0, \quad H = D^{-1}d\Gamma(A)D \tag{73}$$

Then for every $t \geq 0$ the operator e^{-tH} is positivity preserving. Moreover it is a contraction in $L^\infty(Cl)$ and in $L^1(Cl)$. ◇

Proof of Theorem 9 **assuming the validity of Lemmas** 5 **and** 6.
(1) If $H \geq 0$ and if a sequence of operators $A_n \geq 0$ is such that

$$e^{-tA_n} \to e^{-tH} \tag{74}$$

then

$$e^{-td\Gamma(A_n)} \to e^{-td\Gamma(H)} \tag{75}$$

This follows because the sequence is uniformly bounded.
(2) If A has finite range and commutes with J, then J leaves invariant the range R_A. In fact, define

$$\Lambda(\mathcal{K}) = Cl(\mathcal{K}) \tag{76}$$

with $C_l(\mathcal{K})$ based on $R(A)$. By Lemma 6 one has

$$u \geq 0 \rightarrow e^{-tH_A} u \geq 0, \quad H_A \equiv D^{-1} d\Gamma(A) D \tag{77}$$

and moreover by Lemma 4 $\|e^{-tH_A} u\| \leq \|u\|$.

The union of of subspaces that are invariant under J and which contain R_A is dense in $L^2(Cl)$ and also dense in $L^1(Cl)$ due to Lemma 6 (contraction implies convergence of the iterations).

Therefore for any $u \in L^2(Cl)$ there exists a sequence $u_n \in L^2(C_n)$ which converges to u in the L^2 norm and then

$$(e^{-tH} u, \phi) \leq u\|_1 \|\phi\|_\infty \quad \forall\, \phi \in Cl \tag{78}$$

and moreover

$$(e^{-tH} u, \phi) \geq \phi > o \quad \rightarrow e^{-tH} u \geq 0 \tag{79}$$

If $A \geq 0$ and bounded and not of finite range, one can repeat this procedure with A_n of finite range. If $A > 0$ self-adjoint unbounded with spectral projections E_λ, take

$$A_n = \int_0^n \lambda d\lambda \qquad [E(\lambda), J] = 0 \tag{80}$$

and consider

$$A_n \rightarrow A; \quad e^{-tA_n} \rightarrow e^{-tA} \tag{81}$$

It follows that $e^{-t(D\Gamma A)D^{-1}}$ preserves positivity and is a contraction in $L^1(Cl)$. By duality it is a contraction in $L^\infty(Cl)$ and by the Riesz–Thorin theorem it is a contraction from $L^2(Cl)$ to $L^4(Cl)$ if $mt > \frac{log3}{2}$.

Now set $N = d\Gamma(I)$ (in Fock space this is the *number operator*). Acting on any finite-dimensional subspace $\mathcal{K}$ the operator $e^{-tD^{-1}ND}$ leaves $C^1(\mathcal{K})$ invariant and is a contraction form $L^1(Cl)$ to $L^4(Cl)$.

Since the finite-dimensional space $\mathcal{K}$ is arbitrary

$$|(e^{-tD^{-1}ND} u, \phi)| \leq |u|_2 |\phi|_{\frac{4}{3}} \tag{82}$$

and this inequality extends by continuity to all $L^2(Cl)$. If $A \geq mI$, one has $d\Gamma(A) \geq mN$ and therefore $e^{-td\Gamma(A)} \leq e^{-tN}$. It follows that $E \equiv e^{mtN} e^{-td\Gamma(A)}$ has norm not greater than one and

$$e^{-tH_0} \equiv e^{-mtD^{-1}ND} D^{-1} E D \tag{83}$$

is a contraction from $L^2(Cl)$ to $L^4(Cl)$ if $mt \geq \frac{log3}{2}$. ♡

We now prove Lemmas 5 and 6,

Proof of Lemma 5 It is sufficient to prove the lemma in the case A has discrete spectrum. In this case by factorization it sufficient to give the proof in the one-dimensional case.

Then every element of $\mathcal{H}_2 \equiv \{x \in \mathcal{H}, Jx = x\}$ can be written as

$$w = u + av \qquad a = B_{x_1} \qquad x_1 \in \mathcal{H}_r \tag{84}$$

and one has $e^{-tH}u = u, e^{-tH}v = v$. Recall that $a^1 a = I, a^* = aa = Bx_1$ and set $z = r + sa$. Then

$$z^* z = r^* r + e^{-2t} s^* s - e^{-t}(s^* ar + r^* as) \tag{85}$$

We have $|z|^4 = (z^* z)^2$ and $\|z\|_4^4 = Tr|z|^4$. Making use of the cyclic property of the trace and of the expression of $z^* z$ one verifies

$$\|z\|^4 = Tr(r^* r + e^{-2t} s^* s)^2 + e^{-2t}(s^* ar) + r^* as)^2 \tag{86}$$

and therefore

$$\|z\|^4 \leq \|u\|_2^4 + e^{-4t}\|v\|_2^4 + 6e^{-2t}\|u\|_2^2\|v\|^2 \tag{87}$$

If $T \geq \frac{log 3}{2}$ one has $6e^{-2t} \leq 2$ and therefore

$$\|z\|_2^4 \leq (\|u\|_2^2 + \|v\|_2^2) \tag{88}$$

Since $\|w\|_2^2 = Tr((u + av^*(u + av^* = Tr(u^* u + v^* v)$ the case $N = 1$ implies the generic case. ♡

Proof of Lemma 6 Let $\mathcal{K}$ be finite-dimensional and let

$$[A \cdot J] = 0, \quad A \geq 0 \quad H = D^{-1} d\Gamma(A) D \tag{89}$$

Then e^{-tH} is positivity preserving and is a contraction on $L^p(Cl)$ for $p = 1$ and $p = \infty$ If A is a one-dimensional projection Lemma 5 gives

$$e^{-tH}(w^* w) = e^{-t} w^* w + (1 - e^{-t})(u^* U + v^* v) \geq o \tag{90}$$

If A is not a one-dimensional projection, let $A = \sum \lambda_i P_i$ where P_i are one-dimensional projections. Then

$$e^{-tH} = \Pi_k e^{-t\lambda_k H_k} \qquad H_k = D^{-1} d\Gamma(P_k) D \tag{91}$$

and each factor is positivity preserving. To prove the contraction property, begin again with the case in which A is a rank-one projector. Then one has

$$U^{-1}(u + av)U = u - av \tag{92}$$

where if $A = P_i$ then U is the unitary operator which corresponds to the operation $x_i \to . - x_i, x_j \to x_j$ for $j \neq i$. Notice that

$$e^{-H} w = u + e^{-t} av = \frac{1 + e^{-t}}{2}(u + av) + \frac{1 - e^{-t}}{2}(u - av) \tag{93}$$

This implies $\|e^{-tH} w\|_\infty \leq \|w\|_\infty$. If $A = \sum_i \lambda_i P_i$ one proceeds similarly. It follows also that e^{-tH} is a contraction in L^1 and since L^1 and L^∞ are dual for the coupling $< u, v >= Tr(v^+ u)$ and e^{-tH} is auto-adjoint for this coupling since $(e^{-tH} v)^* = e^{-tH} v^*$.

Notice finally that if a map is a contraction both in L^1 and in L^∞ then it is a contraction in L^p for $1 \leq p \leq +\infty$. ♡

References

1. I. Segal, Ann. Math. **57**, 401–457 (1953). **58**, 595–596 (1953)
2. S. Kunze, Trans. Am. Math. Soc. **89**, 519–540 (1958)
3. L. Gross, J. Funct. Anal. **10**, 52–109 (1972)
4. E. Nelson, J. Funct. Anal. **15**, 103–116 (1974)
5. U. Haagerup, Coll. Int. CNRS **N 274**, 175–184 (1978)
6. H. Kosaki, J. Funct. Anal. **56**, 29–78 (1984)

Index

G. Dell'Antonio, *Lectures on the Mathematics of Quantum Mechanics II: Selected Topics*, Atlantis Studies in Mathematical Physics: Theory and Applications 2, DOI 10.2991/978-94-6239-115-4

J

K

L

M

FSC
www.fsc.org
MIX
Papier aus verantwortungsvollen Quellen
Paper from responsible sources
FSC® C105338

If you have any concerns about our products,
you can contact us on
ProductSafety@springernature.com

In case Publisher is established outside the EU,
the EU authorized representative is:
Springer Nature Customer Service Center GmbH
Europaplatz 3, 69115 Heidelberg, Germany

Printed by Libri Plureos GmbH
in Hamburg, Germany